5th International Conference on Geological, Geographical, Aerospaces and Earth Sciences (5th AeroEarth 2017)

IOP Conference Series: Earth and Environmental Science
Volume 88

Kuta, Indonesia
20 - 21 May 2017

Editors:

Ford Lumban Gaol

ISBN: 978-1-5108-5110-8
ISSN: 1755-1307

Printed from e-media with permission by:

Curran Associates, Inc.
57 Morehouse Lane
Red Hook, NY 12571

CURRAN ASSOCIATES INC.
proceedings
.com

Some format issues inherent in the e-media version may also appear in this print version.

Printed by Curran Associates, Inc. (2017)

For permission requests, please contact the Institute of Physics
at the address below.

Institute of Physics
Dirac House, Temple Back
Bristol BS1 6BE UK

Phone: 44 1 17 929 7481
Fax: 44 1 17 920 0979

techtracking@iop.org

Additional copies of this publication are available from:

Curran Associates, Inc.
57 Morehouse Lane
Red Hook, NY 12571 USA
Phone: 845-758-0400
Fax: 845-758-2633
Email: curran@proceedings.com
Web: www.proceedings.com

Table of contents

Volume 88

5th International Conferences on Geological,Geographical, Aerospaces and Earth Sciences 2017 (5th AeroEarth 2017)

20–21 May 2017, Kuta, Bali, Indonesia

Accepted papers received: 4 October 2017
Published online: 30 October 2017

Preface

Papers

IOP Conf. Series: Earth and Environmental Science **88** (2017) 011001 doi:10.1088/1755-1315/88/1/011001

Preface

The 5th International Conferences on Geological,Geographical, Aerospaces and Earth Sciences 2017 (The 5th AeroEarth 2017) was held at Kuta Paradiso Hotel - Kuta, Bali, Indonesia during 20 - 21 May 2017.

The 4th International Conferences on Geological, Geographical, Aerospaces and Earth Sciences 2016 (the 4th AeroEarth 2016), was held at Four Points by Sheraton Bali, Kuta, Bali, Indonesia during 04 – 05 June 2016.

The 1st AeoroEarth 2013, 2nd AeroEarth 2014, 3rd AeroEarth 2015 and 4th AeroEarth 2016 already held succefully in Jakarta on 2013, Kuta Bali on 2014, Jakarta on 2015 and Kuta Bali on 2016 respectively.

The publication of The 1st AeoroEarth 2013, 2nd AeroEarth 2014, 3rd AeroEarth 2015, and the 4th AeroEarth 2016 already published by EES IOP in http://iopscience.iop.org/1755-1315/19/1, http://iopscience.iop.org/1755-1315/23/1, http://iopscience.iop.org/1755-1315/30/1, http://iopscience.iop.org/issue/1755-1315/38/1 respectively.

The theme of AeroEarth 2017 is "The Frontiers Research in Earth Sciences"

Earth provides resources and the exact conditions to make life possible. However, with the advent of technology and industrialization, the Earth's resources are being pushed to the brink of depletion. Non-sustainable industrial practices are not only endangering the supply of the Earth's natural resources, but are also putting burden on life itself by bringing about pollution and climate change.

A major role of earth science scholars is to examine the delicate balance between the Earth's resources and the growing demands of industrialization. Through research and development, earth scientists have the power to preserve the planet's different resource domains by providing expert opinion and information about the forces which make life possible on Earth.

We would like to express our sincere gratitude to all in the Technical Program Committee who have reviewed the papers and developed a very interesting Conference Program as well as the invited and plenary speakers.

This year, we received 139 papers and after rigorous review, 31 papers were accepted.

The participants come from 6 countries. There are 6 (Six) Parallel Sessions and 2 (Two) Keynote Speakers.

It is an honour to present this volume *of IOP Conference Series: Earth and Environmental Science (EES)* and we deeply thank the authors for their enthusiastic and high-grade contribution.

Finally, we would like to thank the conference chairmen, the members of the steering committee, the organizing committee, the organizing secretariat and the financial support from the conference sponsors that allowed the success of AeroEarth 2017.

The Editor of the AeroEarth 2017 Proceedings.

Prof. Dr. Ford Lumban Gaol
Bina Nusantar University

IOP Conf. Series: Earth and Environmental Science **88** (2017) 011001 doi:10.1088/1755-1315/88/1/011001

IOP Conf. Series: Earth and Environmental Science **88** (2017) 011001 doi:10.1088/1755-1315/88/1/011001

ISTIE
The 4th International Conference on

IOP Conf. Series: Earth and Environmental Science **88** (2017) 011002 doi:10.1088/1755-1315/88/1/011002

Keynote Speakers

1. Professor Dr Ah-Hwee Tan

Professor of Computer Science | Associate Chair (Research) | School of Computer Science and Engineering | Nanyang Technological University - Singapore

Title: Biologically-inspired Machine Learning Theory for Knowledge Discovery

Abstract:

Machine learning and knowledge discovery are two critical intertwined functions in next generation intelligent information systems. This talk will present a family of self-organizing neural networks, collectively known as fusion Adaptive Resonance Theory (fusion ART), for building intelligent knowledge-based systems with real-time learning capabilities. By extending the Adaptive Resonance Theory (ART) models, consisting of a single input pattern field, into a multi-channel architecture, fusion ART unifies a number of important neural network designs developed over the past decades. Based on a universal set of neural encoding and adaptation principles, fusion AT supports a myriad of machine learning paradigms, notably unsupervised learning, supervised learning, and reinforcement learning. In addition, domain knowledge in the form of symbolic rules can be inserted into fusion ART and subsequently refined as part of the network's dynamics, which maximizes exploitation of the existing knowledge while retaining the plasticity of exploring new solutions. Case studies will be presented, illustrating how such self-adaptive intelligent systems may be used as autonomous Non-Player Characters (NPC) in first-person shooting games, Computer Generated Forces (CGF) in air combat simulation, and human-like avatars in 3D virtual environment.

Biography:

Dr Ah-Hwee Tan received Ph.D. in Cognitive and Neural Systems from Boston University, Master of Science and Bachelor of Science (First Class Honors) in Computer and Information Science from the National University of Singapore. He is currently a Professor of Computer Science and the Associate Chair (Research) at the School of Computer Science and Engineering (SCE), Nanyang Technological University. Prior to joining NTU, he was a Research Manager at the A*STAR Institute for Infocomm Research (I2R), heading the Text Mining and Intelligent Agents research programmes. His current research interests include cognitive and neural systems, brain-inspired intelligent agents, machine learning, and text mining. Dr. Tan has published ten edited books/proceeding volumes and over 200 technical papers in major international journals and conferences of his fields. He holds two US patents, five Singapore patents, and has spearheaded several A*STAR projects in commercializing a suite of knowledge management and text mining software. He serves as Associate Editor/Editorial Board Member of several journals, including IEEE Access, IEEE Transactions on Neural Networks and Learning, and IEEE Transactions on SMC Systems. He is a Senior Member of IEEE, a Member of Web Intelligence (WI) Technical Committee and Web Intelligence (WI) Conference Steering Committee, and Vice Chair of IEEE CIS Task Force on Towards Human-Like Intelligence.

IOP Conf. Series: Earth and Environmental Science **88** (2017) 011002 doi:10.1088/1755-1315/88/1/011002

2. Professor Dr. Ida Ayu Dwi Giriantari

Title: The role of Geothermal Energy in Indonesia

Abstract: Geothermal energy is heat energy generated and stored in the Earth. Thermal energy is the energy that determines the temperature of matter. The geothermal energy of the Earth's crust originates from the original formation of the planet and from radioactive decay of materials (in currently uncertain but possibly roughly equal proportions). The geothermal gradient, which is the difference in temperature between the core of the planet and its surface, drives a continuous conduction of thermal energy in the form of heat from the core to the surface.

In the 20th century, demand for electricity led to the consideration of geothermal power as a generating source. Indonesia Power Electricity tested the first geothermal power generator on 4 July 1978, at the same dry steam field where geothermal acid extraction began. It successfully lit four light bulbs. Later, in 1990, the Indonesia's first commercial geothermal power plant was built.

Short Bio:

Ida Ayu Dwi Giriantari is Professor in Electrical Engineering Udayana University. She hold PhD from The University of New South Wales in 2003. Her research areas are in Power Plant, Renewable Energy and Power Transformers.

She is Head of Magister Program of Electrical Engineering, Udayana University.

Peer review statement

All papers published in this volume of *IOP Conference Series: Earth and Environmental Science* have been peer reviewed through processes administered by the proceedings Editors. Reviews were conducted by expert referees to the professional and scientific standards expected of a proceedings journal published by IOP Publishing.

Characterizing Geological Facies using Seismic Waveform Classification in Sarawak Basin

Afiqah Zahraa Ahmad Zailani, Deva Prasad Ghosh

Universiti Teknologi PETRONAS
32610 Seri Iskandar, Perak Darul Ridzuan,
Malaysia

afiqahzahraa@gmail.com

Abstract. Numerous effort have been made to build relationship between geology and geophysics using different techniques throughout the years. The integration of these two most important data in oil and gas industry can be used to reduce uncertainty in exploration and production especially for reservoir productivity enhancement and stratigraphic identification. This paper is focusing on seismic waveform classification to different classes using neural network and to link them according to the geological facies which are established using the knowledge on lithology and log motif of well data. Seismic inversion is used as the input for the neural network to act as the direct lithology indicator reducing dependency on well calibration. The interpretation of seismic facies classification map provides a better understanding towards the lithology distribution, depositional environment and help to identify significant reservoir rock

1. Introduction

Reservoir characterization is a very risky task and multiple methods and techniques have been developed to reduce the uncertainty during the stage of exploration and production. There are few factors that contribute to the uncertainty including the limitation of seismic resolving power, insufficient well logs, the approximation of physical model and natural variability in rock [9][5].Hence, a method using statistical approach such as automated facies recognition of neural network had been performed to reduce the uncertainty. It is also known as artificial intelligence and automatic clustering and has become popular on 90s due to its successful application in other field of study such as biology and economy.

Hence, more study were conducted to see the ability of artificial neural network to recognize the populations of similar multi-attribute response [16] and to select the right attributes from the list of multi-attribute in the input process [19].The use of multi-attributes in the seismic facies recognition is very important as it enables the interpreter to analyse different aspects of seismic response (energy, frequency, phase, geometry and texture, etc) simultaneously, generating a map of facies or correlating seismic response to engineering/production data [15][12][1][2][4][6]. The other important use of neural network in reservoir characterization is to classify the seismic facies and pattern recognition in which various algorithms have been developed, proposed and applied to improve the network [3][18]. Since seismic data contain huge amounts of data samples and they are highly continuous, greatly redundant and significantly noisy [8], neural network especially self-organizing maps (SOM) able to delineate the structure of the data and to reduce the dimensionality. Seismic facies classification algorithms attempt to mimic human pattern recognition and can be applied to a more arbitrary collection and greater number of attributes [19].Neural network also has been used to extract structural lineaments [10], lithological classification [17] and replacing and to improve seismic inversion techniques [14][13]. The approach has been proven to have a better result than any other technique especially when it is used in highly heterogeneous formation [7].

Neural network is the best automatic and supervised method used to identify and to classify the seismic facies to multiple classes due to its ability to suppress the noise, handle the complex and large database and to eliminate the need of complex statistical technique. Hence, the objective of this paper is to characterize and to predict the lithofacies distribution of the study area by interpreting the seismic facies classes and relating them to geological facies. As now the stratigraphic trap is becoming widely popular, identification of lithofacies distribution helps to reduce the risk for the future exploration. The lithofacies distribution can be predicted by using seismic waveform classification of neural network which plays important role in reducing the subtleness of the changes of seismic amplitude and also phase. The seismic used for the classification is the P-impedance seismic as it acts as the direct lithofacies indicator with less dependency on the wells. The study is important to discover the ability of neural network to propagate the clustering in the seismic especially away from the wells The interpreted lithofacies map can also help to reduce uncertainty in characterizing the reservoir by highlighting the prospect (gas sand) and non-prospect area (brine sand).

2. Geological Settings

K field is located in Borneo Basin of Malaysia. The main reservoir is F3 sand with the thickness about 23m and depositional environment ranging from holomarine inner to outer neritic or possible ponded turbidites. The coarsening upward trend of the sand probably characterized by the transition of environment from holomarine outer-middle-inner which formed during the highstand system tract [8]. Being the post carbonate siliciclastic and due to interaction between eustacy and progradation behaviour of the sand against the carbonate build-ups, it is believed the hydrocarbon migration of F3 to be originated from underlying carbonate itself as it is found to be underfilled. Due to stacking pattern of forced regression, the vertical migration from carbonate to upper silt and sand is found possible proven by interpretation of multiple seismic section regionally.

3. Data Scope

Seismic dataset was acquired on 2004. The full-fold 3D area comprised an area of 1871.5 km^2 forming a rectangle 50 km in length by 37.8 km in breadth. The quality of 3D seismic data is good. Most of major markers above 2.5 seconds (Precarbonate) have good lateral continuity and clear discontinuity at fault locations. Structural and stratigraphic features can be clearly seen from the 3D data. The dominant frequency at major reservoir zone from full stacking volume is around 35Hz to 40Hz, but the seismic wavelet shows mixed phase characteristic. Three wells are provided in the field; K-1, K-2 and K-3. All wells are used for the seismic-to-well correlation. Basic wireline logs such as gamma ray, neutron, density, resistivity, S-slowness, P-slowness and saturation are provided for further use. However, in K field, the porosity log are not used as the reading shows inconsistency with the lithological data. The type of rock mineral might contribute to this issue, but no further research is made.

4. Research Method

There are two types of seismic waveform classification in the study; voxel based and trace based. The time interval used for waveform analyses is different according to the method. Voxel based consists of three data axis; inline, crosslines and time/depth meanwhile trace based consists of only inline and xline. Voxel based method performed the waveform classification in voxel or 3D seismic pixel manner which is more details compare to the trace based method. Trace based takes certain interval of seismic volume and utilize the average shape or morphology of seismic waveform for classification purpose. Both methods can utilize multiple attributes volumes as the main input for classification. However, in voxel based, the final output will consist of seismic facies volume whereas a map in trace based method. Principal component analysis (PCA) is conducted to reduce the redundant data and noise in multi-attributes volume or even a single seismic volume before being used as input for seismic waveform classification. In this study, same neural network principle is used for the classification in both of the methods with the same type of input which is P-impedance. However, in voxel based, the wells are calibrated and each of the classification class is being defined based on the lithofacies response in the wells. The well calibration in trace based method is not possible due to clustering of the wells in which the reservoir interval remains consistent with only slight variety in terms of lithology, porosity and thickness in all three wells. Hence, it is difficult to represent the changes of reservoir behaviour laterally and its effect to the seismic waveform morphology. However, the differences can be noticed vertically as the shale transitioning to sand with various thickness and porosity and these differences can only be captured by using the voxel based method.

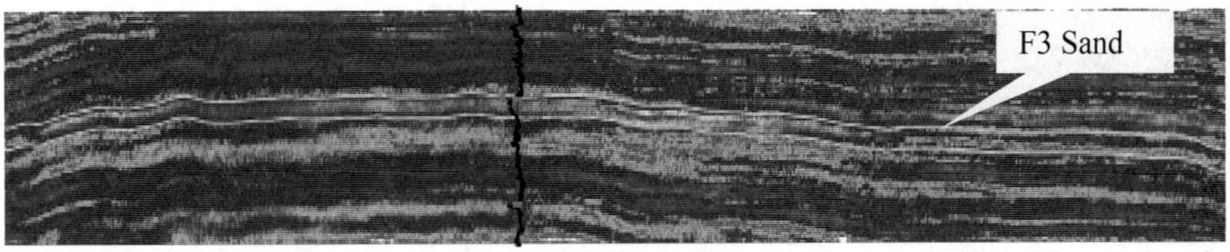

Figure 1 : The interpreted reservoir horizon of F3 with the thickness of about 23m. Below is F3A which appears to be thinner than F3

Most of the seismic waveform classification study use the reflectivity seismic data as the main input beside other multi attribute such as amplitude. But due to the lack of well representation for seismic waveform, the focus is now shifted in using only seismic as the direct rock or lithology indicator by reducing the use of well calibration. The rock parameter identified to be the best in discriminating the lithology is the P-impedance. Hence, pre-stack inversion or also known as simultaneous inversion is conducted and being used as the input for seismic waveform classification.

5. Application and Results

5.1. Trace based

The time interval of -4ms+19ms is used with the P-impedance seismic cube as the input data. The training of neural network is conducted using 100 iterations and seven class number. Each of the class if represented by the trace model which is also known as the cluster centre or the best representation of the class. The distance curve shows constant sloping line which indicate the cumulative difference between the neighbour classes are equal. If too much class used, the slope will be in straight line meaning no much differences between the classes but if the number of class is too small, the slope will be drastically steep indicating the differences is too big and very geologically unlikely. The seismic waveform classification shows the final result map (Figure 2).

Based on the inversion result, the lowest P-impedance is shown by the gas sand follow by shale and sand with the highest quality or coarsest grain which is represented by the highest impedance. As seismic inversion can be used as the direct rock indicator, the class 1 or the lowest impedance based on the seismic morphology, can be defined as the gas sand, followed by class 2 and class 3 which possibly correlated with the shale lithology. The sudden sharp notch at the distance curve in class 4 might indicate the changes of lithology from shale to sand. As the P-impedance trace model is becoming higher in impedance, the lithology as well changes to coarser, cleaner type of sandstone. This coarser and cleaner type of sandstone can also be known as good brine sand and if the poor quality sandstone, lower in impedance is known as poor brine sand. Multiple seismic attributes such as

RMS amplitude, sweetness and RMS frequency are also generated to provide more evidence for the gas sand (Figure 4).

a)
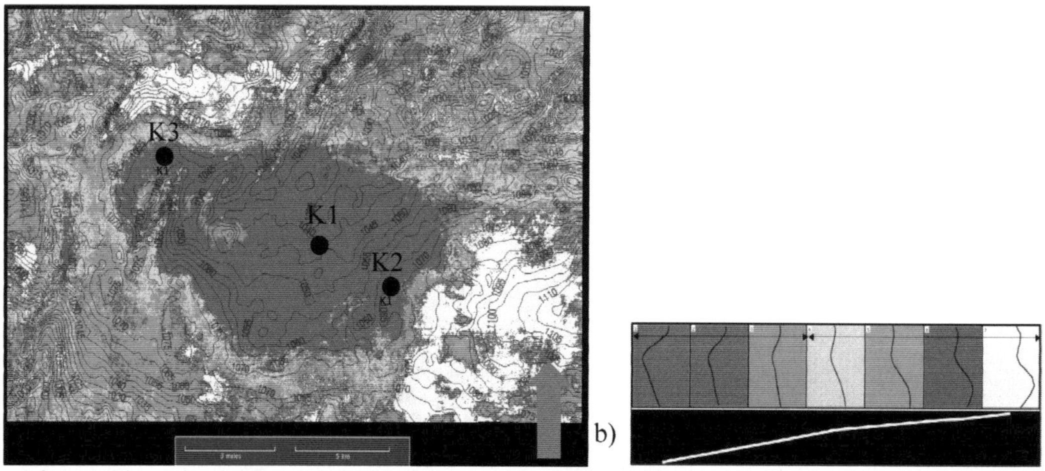

Figure 2 : a) The seismic facies classification map produced from trace based method. Each of the colour corresponds to different classes b) The above is the trace model and its corresponding colour code. The class is labelled first from left to seventh to the most right. Distant curve below shows cumulative difference between the neighbour classes.

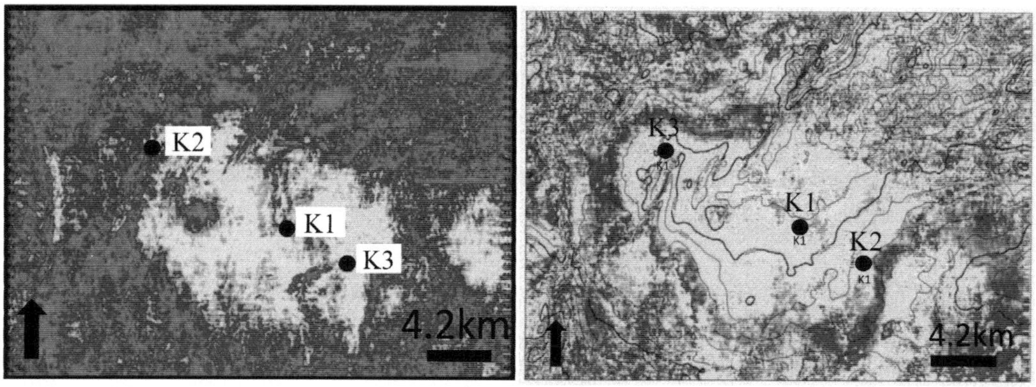

Figure 3 : a) The sweetness attribute b) RMS frequency attributes

5.1.1. Principal Component Analysis (PCA). Principal component analysis (PCA) is used to reduce the noise and the redundant data. It finds the best representation of data in multiple dimension according to the input data. It is an unsupervised, represented by two principal components called eigenvector/eigenvalue in which the first component denoted the direction of maximum data distribution or variances meanwhile the second component is smaller than the first and perpendicular to it. Since in this study, pre-stack inversion is used, PCA is utilized to determine the best cloud or cluster data and any data outside this cloud is best to be removed as it does not contribute much to distribution of data. These outside data can also be called noise. PCA can only be used in trace based method as the input data is count as traces and not treated as the volume as in the voxel based method. In voxel method, one seismic volume is enough for one main data principal representation. Five components are used as the input for the trace method (Figure 4). Any data with low contribution to the cloud data with eigen values are usually less than one, are eliminated from being analysed in neural network. These outside data might be noise and if included, can give a lot of effect during the classification process.

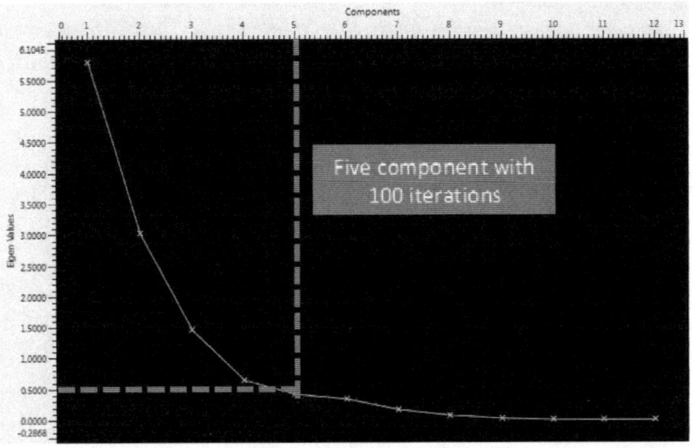

Figure 4 : Principal component analysis of P-impedance cube. The components consist of seismic traces and eigen values indicate the contribution of the components to data cloud

5.2. *Voxel Based*

Voxel-based method is a microslicing method in which the the seismic is being sliced into smaller interval based on the minimum/maximum, inflection points, zero sine or zero crossing. The seismic waveform classification is then conducted in smaller interval, hence the provided result is more detail in capturing the lithology changes vertically and laterally. As waveform classification in traces method only deal with inline and xline information, voxel based method is in 3D with the time as third axis or domain. The inflection points are used for the microslicing and -49ms+77ms of F3 horizon as the input for the neural network. Seven number of classes is selected with no PCA analyses. The results are shown below.

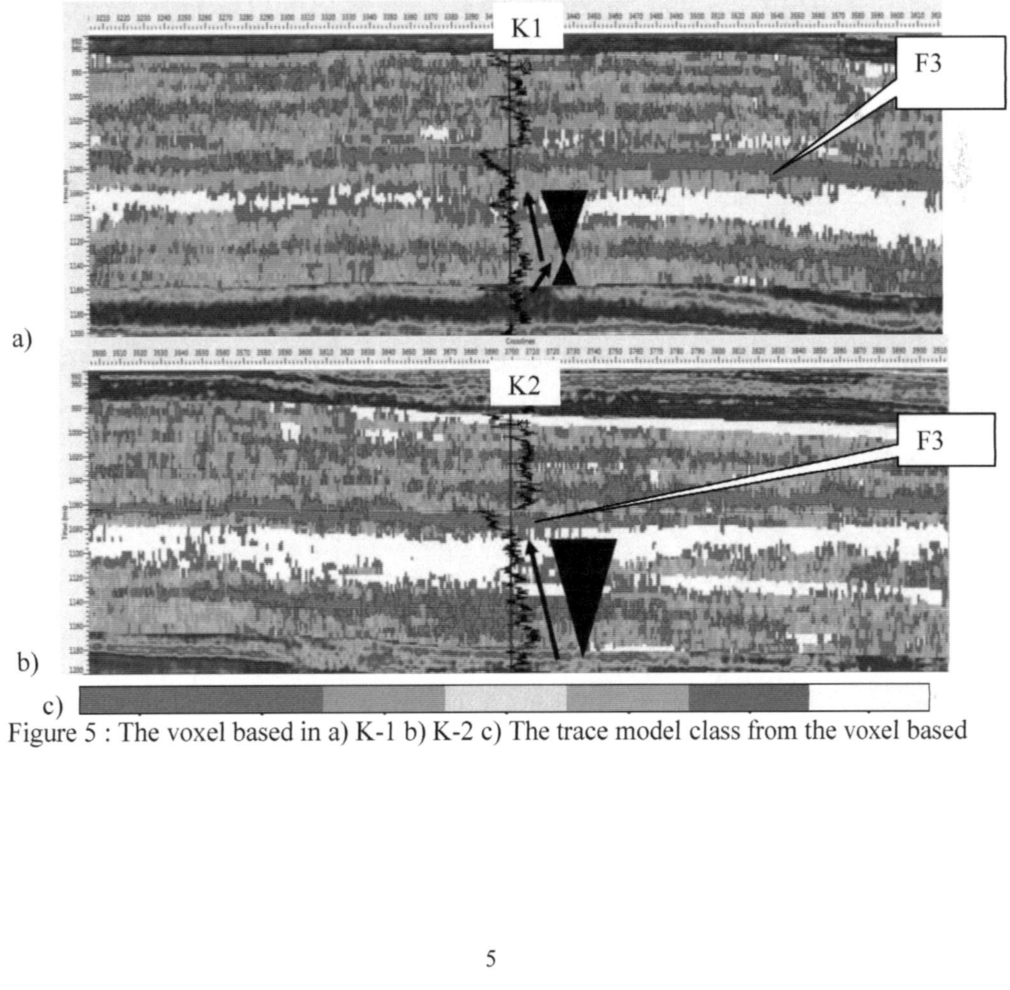

Figure 5 : The voxel based in a) K-1 b) K-2 c) The trace model class from the voxel based

The coarsening upward trend of gamma ray is marked by the transition of colour from red to yellow indicating the increase of impedance across the first class to the seventh class (Figure 5). The red colour class or the first class can be correlated to the gas sand in the well indicating the lowest impedance class. The red coloured class can also be seen in the upper interval of F3.signify these sands are also a gas sand. Some of the F3 gas sand is also being classified as light and dark green colour or the second and third trace class respectively indicating the increasing in impedance as the facies change. Below the gas sand, is the poorer quality of sand compared to gas sand, but still classified as colour yellow or good brine sand due to its high impedance.

In order to observe the lateral changes of these lithology class, time slice is generated by +4ms in difference (Figure 6). A1 shows the gas sand distribution which is started to be seen at 4ms below the F3 horizon and becoming less obvious as it reaches the bottom of the reservoir. A2 and A3 is mainly yellow colour coded class, corresponds to good brine sand can be seen as early as +2ms and become bigger throughout the reservoir. A4 sand comes later indicating the above part of A4 is mostly shale and silty and just like A2 and A3, the area increases towards the bottom of reservoir. A5 is green colour coded class and the lithology is consistent along the interval. By observing the time of the lithology formation and the classification of these lithology, it is then possible to understand the lithology distribution in the study area. A2 sand for example has better quality sand compare to A4 since A4 sand come later at 4ms below the F3 horizon meanwhile A2 already formed at 2ms. A4 is interpret to be a better brine sand and A2 is poor brine sand. Based on time of the lithology formation, A3 is also observed to be formed at the same time as A2. They are possibly in the same quality of brine sand. Since A5 is classified as in green coded class, it is inferred as shale lithology. A6 can also be inferred as shale due to characterization of this class by red and green colour code.

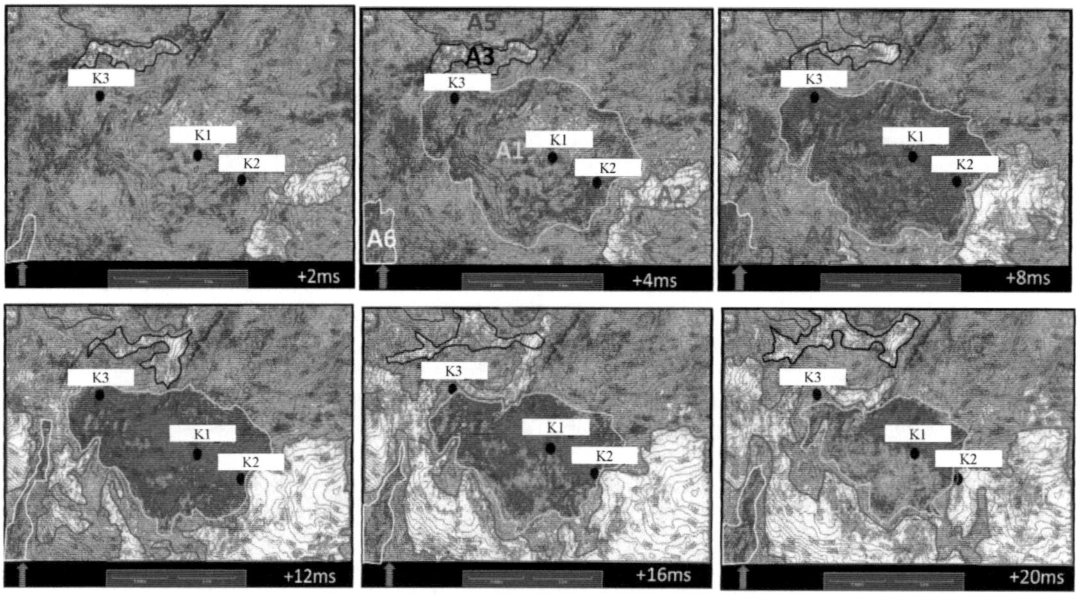

Figure 6 : The horizon slices for every +4ms and +2ms of voxel based classification volume. Each of the lithology area are marked with code A1, A2, A3, A4, A5 and A6

6. Discussion

The response of trace model in trace based method can be verified with additional information from the voxel based method. The development of the lithofacies along the reservoir interval can be observed using the voxel based method which use the well calibration to increase the accuracy of lithology interpretation. The transitioning of class from first class (red) to seventh (yellow) class indicate the coarsening upward sequence of gamma ray when being calibrated to the well information indicating the increase in the value of P-impedance. The integration of well and seismic shows that the highest quality of sand (based on the gamma ray value) is marked by the highest value of P-impedance and decreasing as it becomes shalier. The lowest P-impedance meanwhile marked by gas sand. Trace method however is very depending on the relationship of P-impedance to the lithology from the feasibility study and inversion data without integration of the lithofacies from well. However, by combining these two methods, the uncertainty in the trace based method is complemented by taking the advantage of using well in the voxel based method.

Using the trace based map, the seventh class is interpreted as good brine sand as P-impedance is high. This result is also agreeable with the result from the voxel method. Area A2 and A3 both shows the good brine sand. The brownish-orange colour class of fifth and sixth class are also conformable with the voxel result of A4 which is poor brine sand as the upper interval is mostly dominate by silt and shale. The shale area of A5 and A6 also shows the same responds of shale in the trace based method. Both of the methods are in conformable of each other which draw the conclusion on the final interpreted lithology map of K field (Figure 7).

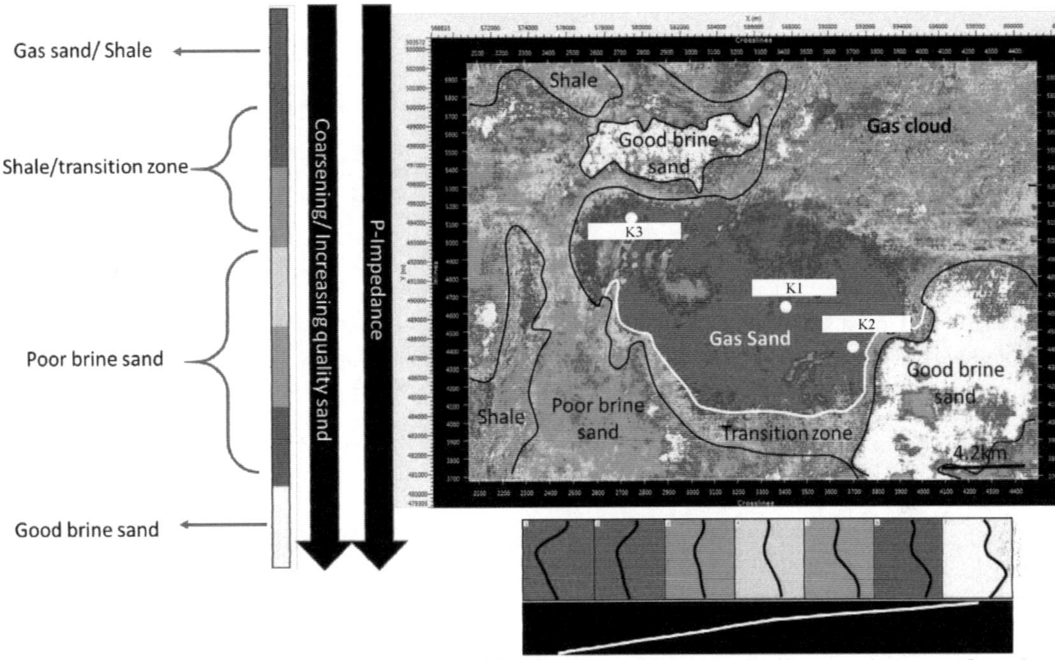

Figure 7 : The final interpreted seismic classification map. The seismic classes now are referred according to lithofacies types

7. Conclusion

The integration of wells and seismic increase the accuracy of the neural network classification result as each of the waveform class is well defined by different lithofacies, characterized by the gamma ray value of the well. However, with less dependency on the wells, the trace method is still proven to produce the best lithofacies classification as it appears conformable with the voxel based method which highly dependent on the well calibration. However by combining the result from voxel based and trace based, the final interpreted seismic facies map is highly convincing. It is now possible to delineate the area of brine sand and gas sand with the lithology various from poor to good.

Resolution of seismic is still the main issue in the study. The resolution of the seismic in the study area is around 10m. The reservoir under the resolution value cannot be detected which make it impossible to perform the waveform classification. The extent of transition zone in the interpreted map can be mapped if seismic able to resolve the thin layer or the real termination point of the reservoir. By knowing the zone, the final volumetric assessment can either be increased or decreased. However, neural network already able to solve multiple issue on noise, redundancy of data and dimensionality problem especially when multi attributes are used.

8. Acknowledgement

Special gratitude to PETRONAS for allowing the author to utilize the data and for providing financial support throughout the research. The author would like to extend the appreciation and thanks to project advisor Professor Dr. Deva Ghosh, who have been a tremendous mentor. The author also would like to thank the colleagues from Centre of Seismic Imaging of Universiti Teknologi PETRONAS whose provided insight and expertise that greatly assisted the research.

References

[1] Aminzadeh, F., & De Groot, P. (2004). Soft computing for qualitative and quantitative seismic object and reservoir property prediction. Part 1: Neural network applications. *first break*, 22(3).
[2] Aminzadeh, F., & De Groot, P. (2006). Neural networks and other soft computing techniques with applications in the oil industry. Eage Publications.
[3] Basman II, Y. V. (2015). Seismic waveform classification: renewing the interest in Barrolka field,
SW Queensland, Cooper Basin. *ASEG Extended Abstracts*, 2015(1), 1-3.
[4] Ghosh, D. (2010). Seismic attributes (Part 2) : Methodology E&P Application. *Seismic attributes*
adding a new dimension to prospect evaluation and geomorphology in the Malay and adjacent
Basins. Universiti Teknologi PETRONAS, pp 3-29
[5] Ghosh, D., Halim, M. F. A., Brewer, M., Viratno, B., & Darman, N. (2010). Geophysical issues and challenges in Malay and adjacent basins from an E & P perspective. *The Leading Edge*, 29(4), 436-449.
[6] Ghosh, D., Sajid, M., Ibrahim, N. A., & Viratno, B. (2014). Seismic attributes add a new dimension to prospect evaluation and geomorphology offshore Malaysia. *The Leading Edge*, 33(5), 536-545.
[7] Gonçalves, C. A., Harvey, P., & Lovell, M. A. (1995, January). Application of a multilayer neural
network and statistical techniques in formation characterisation. In *SPWLA 36th Annual Logging Symposium*. Society of Petrophysicists and Well-Log Analysts
[8] Coléou, T., Poupon, M., & Azbel, K. (2003). Unsupervised seismic facies classification: A review
and comparison of techniques and implementation. *The Leading Edge*, 22(10), 942-953.
[9] Grana, D., E. Paparozzi, S. Mancini, and C. Tarchiani (2013) Seismic driven probabilistic classification of reservoir facies for static reservoir modelling: *A case history in the Barents Sea: Geophysical Prospecting*, 61,613–629
[10] Kluesner, J. W., & Brothers, D. S. (2016). Seismic attribute detection of faults and fluid pathways
within an active strike-slip shear zone: New insights from high-resolution 3D P-Cable™ seismic data along the Hosgri Fault, offshore California. *Interpretation*, 4(1), SB131-SB148.
[11] Mazlan Madon et, al. (1999). The Petroleum Geology and Resources of Malaysia. *Petroliam Nasional Berhad (PETRONAS)*. ISBN 9839738100, 9789839738100
[12] Meldahl, P., Heggland, R., Bril, B., & de Groot, P. (2001). Identifying faults and gas chimneys using multiattributes and neural networks. *The Leading Edge*, 20(5), 474-482.
[13] Mohamed, I. A., El-Mowafy, H. Z., & Fathy, M. (2015). Prediction of elastic properties using seismic prestack inversion and neural network analysis. *Interpretation*, 3(2), T57-T68.
[14] Roethe, G., & Tarantola, A. (1991, January). Use of neural networks for inversion of seismic

data.
In *1991 SEG Annual Meeting*. Society of Exploration Geophysicists.

[15] Russell, B., Hampson, D., Schuelke, J., & Quirein, J. (1997). Multiattribute seismic analysis. *The Leading Edge*, 16(10), 1439-1444

[16] Strecker, U., & Uden, R. (2002). Data mining of 3D poststack seismic attribute volumes using Kohonen self-organizing maps. *The Leading Edge*, 21(10), 1032-1037

[17] Yi Guo, Hansen, R. O. and Harthill, N. (1992). Feature Recognitionfrom Potential Fields Using Neural Networks. *Artificial Intelligence I: Neural Networks in Geophysic.*

[18] Zhao, T., V. Jayaram, A. Roy, and K. J. Marfurt, 2015, A comparison of classification techniques for seismic facies recognition*: Interpretation*, 3, SAE29–SAE58

[19] Zhao, T., Zhang, J., Li, F., & Marfurt, K. J. (2016). Characterizing a turbidite system in Canterbury Basin, New Zealand, using seismic attributes and distance-preserving selforganizing maps. *Interpretation*, 4(1), SB79-SB89

IOP Conf. Series: Earth and Environmental Science **88** (2017) 012002 doi:10.1088/1755-1315/88/1/012002

Geological features based on Spectral Decomposition: Techniques and examples from the Malay Basin

A Shamsuddin[1], D Ghosh[2] and S Shahar[3]
PETRONAS University of Technology (Universiti Teknologi Petronas – UTP), 31750, Tronoh, Perak, Malaysia [1,2]
PETRONAS Carigali Sdn. Bhd., Kuala Lumpur City Center, Malaysia [3]
E-mail: anassash@gmail.com, dr deva@petronas.com.my, suhaile@petronas.com.my

Abstract. Seismic resolution is of interest in exploration and production of hydrocarbon area as it may hinder geological understanding. A frequency study-spectral decomposition was implemented in order to enhance geological understanding in the study area as it is related to the seismic resolution. Spectral Decomposition based on Fourier Transform analysis was conducted in the study area which is located in the Malay Basin. Four geological features were determined in the analysis 1 and three geological features were determined during analysis 2. The thickness and appearance of the geological features such as sharp edges, discontinuity were detected based on this analysis. The abandoned channel can be identified in the southeast part of the study area and was detected based on the modulation of the spectral decomposition analysis.

1. Introduction

Geophysical challenges in the Malay and adjacent basins have been addressed by Ghosh et. al (2010) including the appearance of seismic imaging below gas clouds, carbonates, basement and the geological features which are beyond seismic resolution. This paper is related to the seismic attributes - frequency in order to improve geological understanding in the South part of the Malay Basin as it requires higher seismic resolution and interpretation or can be known as spectral decomposition.

Spectral Attributes	
E&P Application	***Algorithms***
Seismic resolution	Wavelet Transform, Cepstrum, Instantaneous Frequency & Instantaneous Phase
Thin bed tuning	Spectral Decomposition
Absorption, Saturation	Qp/Qs

Table 1: Spectral attributes (Ghosh et. al, 2015)

Annur (2016) refers the classification of seismic attributes based on Ghosh et. al (2015) as he divided seismic attributes into three principal categories which are structure/geomorphological attributes, spectral attributes and fluid/lithology attributes. Main analysis of seismic attributes which are applied in this study is characterized in the class two (2) which is spectral attributes (Table 1).

2. Spectral Decomposition

Spectral decomposition is a tool for interpretive use, which is below resolution such as detecting stratigraphic features - sand thickness estimation and enhanced channel features especially abandon channel. The fourier transform is a one of spectral decomposition types which is calculated by using short window and a step size between the analysed frequencies. The other types of spectral decomposition are continuous wavelet transform, matching pursuit and exponential pursuit (Rocky et. al 2015).

2.1 Red-Green-Blue (RGB), Cyan-Magenta-Yellow (CMY), Hue-Saturation-Value (HSV) and Hue-Saturation-Lightness (HSL)

Corendering more attributes were a significant study that had been highlighted by Marfurt (2015) as it may enhance geologic insight. Red-green-blue (RGB), cyan-magenta-yellow (CMY), hue-saturation-value (HSV) and hue-saturation-lightness (HSL) are synonymous in corendering the attributes colour model. The red-green-blue (RGB) colour model is known to be used as a mixing light, for example, television screen or computer monitor, the cyan-magenta yellow- black (CMYK) colour model is important in printing and painting (Marfurt, 2015). The hue-lightness-saturation (HLS) model is more mathematical offers a means of modulation of one attribute of another, and hue-saturation-value as an alternative representation of the HLS model (Marfurt, 2015).

3. Frequency Modulation – Fourier Transform

Frequency modulation for this study is detected in the range of the target area. Three frequencies had been chosen in this study through the analysis of seismic are displayed with associate red in a lower frequency, green with a middle frequency, and blue with a higher frequency. Figure 1a-1c show three frequencies timeslices through the 10Hz - plotted against red, 45Hz - plotted against green, and 60Hz -plotted against blue for spectral magnitude components. For simplicity, the background was set to black. The simple addition of these color bars obtains an image such as that shown in Figure 1d. The other frequency modulation can be referred at Figure 2a-2c three frequencies timeslices through the 30Hz - plotted against red, 38Hz - plotted against green, and 40Hz - plotted against blue for spectral magnitude components. The addition of these color bars can be obtained an image such as that shown in Figure 2d.

3.1 Fourier Transform

The Fourier Transform analysis requires a short time-gate window and also a step-size between the analyzed frequencies which can be interpreted as the frequency resolution. Time gate in between -20ms and 20ms with 5 steps were applied in this study.

The output frequencies were divided into low, middle and high frequencies before modulation process. Analysis 1 involves 10Hz, 45 Hz and 60 Hz, while analysis 2 involves 30Hz, 38 Hz and 40 Hz. Analysis 1 can be defined in all three frequencies (low, middle and high) and analysis 2 only can be defined in two frequencies (low and middle) as the division of the frequencies are 10Hz-20Hz for low frequency, 20-40Hz for middle frequency and 40-60 Hz for high frequency.

4. Application: Malay Basin

A spectral decomposition study was conducted in the Malay Basin, which is located in the Southern part of the Gulf of Thailand, between Vietnam and Peninsular Malaysia with 80000km^2 coverage areas (Petronas, 1999) as there is appearance geological features that can be identified based on seismic analysis. It has a complex structural origin because of the combination of strike-slip and extensional regimes (Mansor 2014).

4.1 Geological identification

Based on the analysis - spectral decomposition (fourier transform), four geological features had been obtained .The first analysis is based on 10Hz-45Hz-60Hz and 30Hz-38Hz-40Hz for a second analysis. Figure 1 till figure 2 show the results in orange circles for geological features that can be analysed in this study. Based on the first analysis, feature 1 – northwest, feature 2 – southwest, feature 3 - center as well as feature 4 – center part of the study area can be defined in 10Hz-45Hz-60Hz and based on the second analysis feature 4 is absent with the appearance only for feature 1 till feature 3 (Table 2).

Analysis	Analysis 1 10Hz-45Hz-60Hz	Analysis 2 30Hz-38Hz-40Hz
Feature 1	10Hz (Red)	30Hz-38Hz
Feature 2	10Hz-45Hz	30Hz-38Hz
Feature 3	10Hz-45Hz	30Hz-38Hz
Feature 4	60Hz	None

Table 2: Results from Spectral Decomposition – Fourier Transform. Analysis 1 with 10Hz-45Hz-60Hz and analysis 2 with 30Hz-38Hz-40Hz.

4.2 Geological characterization

Analysis of geological features in this study is based on the thickness and also the appearance of the geological features.

i) Thickness of geological features

Based on the frequency, variation of thickness can be defined, especially at 10Hz-45Hz-60Hz (Figure 3d) and 30Hz-38Hz-40Hz (Figure 4d). The geological features are expected to be thicker at 10Hz-red in analysis 1 at 10Hz-45Hz-60Hz analysis compared to analysis 2 at 30Hz-38Hz-40Hz. The thickness is getting moderately thick at 30Hz-38Hz-40Hz in analysis 2 and 45Hz-60Hz in analysis 1.

Analysis	Analysis 1 10Hz-45Hz-60Hz	Analysis 2 30Hz-38Hz-40Hz
Thickness 1	10 Hz (Red): High	30Hz-38Hz-40Hz (Red, Green, Blue): Low
Thickness 2	45Hz-60Hz(Green, Blue): Moderate	30Hz-38Hz-40Hz (Red, Green, Blue): Moderate

Table 3: Results for thickness in analysis 1 with 10Hz-45Hz-60Hz and analysis 2 with 30Hz-38Hz-40Hz.

ii) Architecture of geological features (Sharp edge and discontinuity)

a. Sharp edge: Sharp edge of the geological features can be determined at 10Hz for analysis 1: 10Hz-45Hz-60Hz) – white arrow (3a) and this feature is expected to be classified as a structural feature – fault.

b. Discontinuity: The appearance of discontinuity at 30Hz (4a) and 38Hz (4b) at analysis 2: 30Hz-38Hz-40Hz can be determined.

IOP Conf. Series: Earth and Environmental Science **88** (2017) 012002 doi:10.1088/1755-1315/88/1/012002

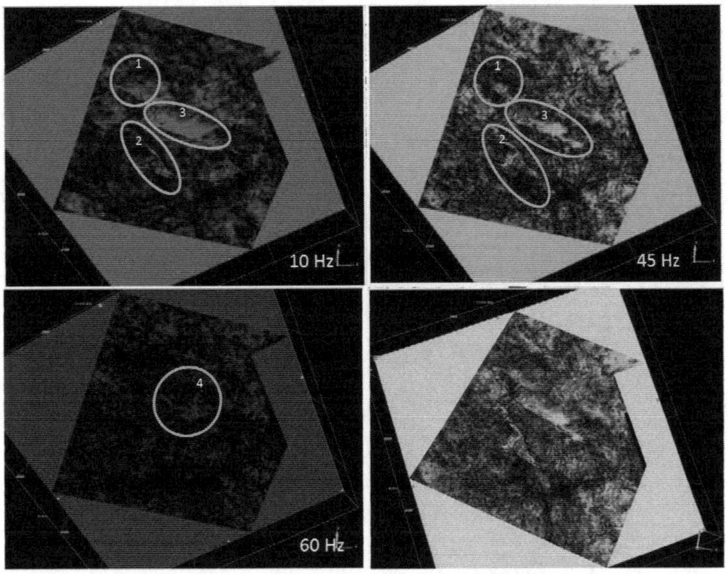

Figure 1: Timeslice through spectral magnitude components at (1a) 10 Hz plotted against red, (1b) 45 Hz plotted against green, and (1c) 60 Hz plotted against blue. (1d).The desired image obtained by color addition of the RGB components shown in panels (1a-1c) using external software.

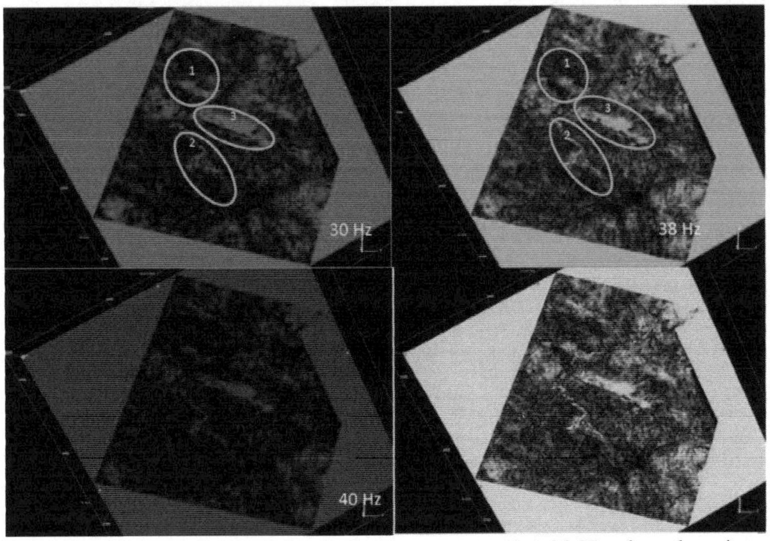

Figure 2: Timeslice through spectral magnitude components at (2a) 30 Hz plotted against red, (2b) 38 Hz plotted against green, and (2c) 40 Hz plotted against blue. (2d).The desired image obtained by color addition of the RGB components shown in panels (2a-2c) using external software

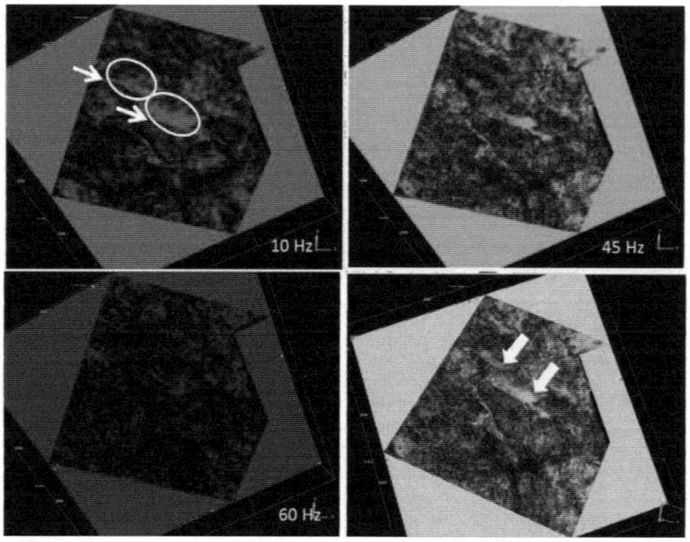

Figure 3: Time slice through spectral magnitude components 10Hz-45Hz-60Hz with a sharp edge – white arrow at 10Hz (3a) and variations of thickness can be analysed – yellow arrow in (3d).

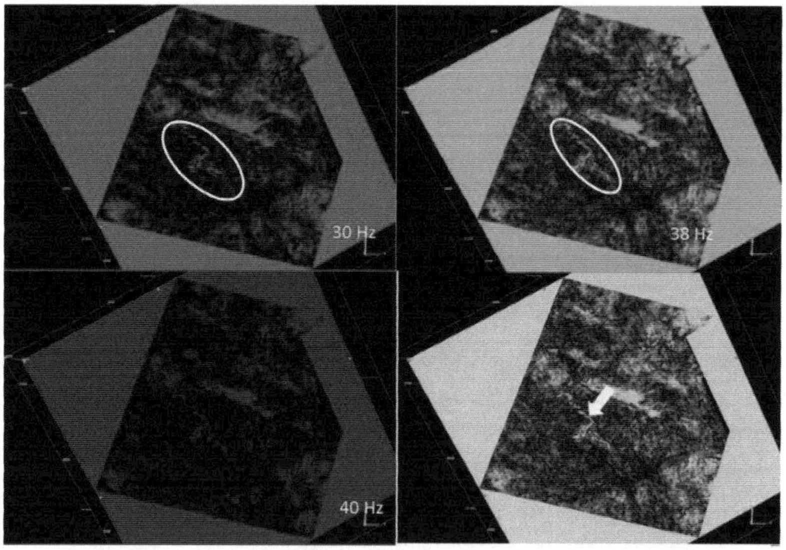

Figure 4: Time slice through spectral magnitude components at 30Hz-38Hz-40Hz with the different geological feature appears in yellow circle (4a and 4b) and yellow arrow for the combination (4d).

Figure 5: The analysis of 10Hz-45Hz-60Hz and 30Hz-38Hz-40Hz show the appearance of black colour and white colour with channel can be detected at 40Hz-4Hz

4.3 Frequency Analysis and Channel

Modulation of the frequency can be referred in figure 5 with the addition of the all three frequencies - red in a lower frequency, green with a middle frequency, and blue with a higher frequency. White colour shows all frequencies are represented while the black colour means the absence of all the frequencies. The interesting features that can be defined by the combination of all colour is a channel which is about 40Hz-45Hz. It might be the abandoned channel during synrift.

5. Discussion and conclusion

Spectral decomposition reveals the thickness, sharp edge and also discontinuity of the geological features and also channel as shown in the figure 1 to 5. As a conclusion, there are four geological features for analysis 1 and three geological features can be determined during analysis 2. The variety of thickness appearance of the geological features which is based on frequency can also be determined based on the spectral decomposition analysis. The appearance of sharp edge, discontinuity of the geological features which might be fault or fold enhanced the understanding of the study area. Based on the modulation, the abandoned channel can be detected in the southeast part of the study area.

6. Acknowledgement

We would like to thank to Dr. Nguyen Huy Ngoc and Dr. Eko Widi Purnomo for the guidance and Centre of Excellence in Subsurface Seismic Imaging & Hydrocarbon Prediction (CSI), PETRONAS University of Technology for research allocation and PETRONAS for allowing the use of the data for testing.

7. References

[1] Ghosh D., Firdaus M., Martin B. and Nasir D 2010 Geophysical issues and challenges in Malay and adjacent basins from an E&P perspective *The Leading Edge*
[2] Ghosh D., Nor A.I., Bernato V. and Hamdan M. 2015 Seismic Attributes adding a new Dimension to Prospect Evaluation & Geomorphology Identification in the Malay and adjacent basins *The Leading Edge*
[3] Annur S, Deva G, Suhaileen S and Nguyen H.N. 2016 Realisation of Mechanically Derived strains Using Geophysical Data (Seismic and Borehole): Fractured Basement Reservoir *ICIPEG 2016*
[4] Kurt J. M. 2015 Techniques and best practices in multiattribute display *Interpretation*
[5] Rocky R, Thomas S and Deborah S 2015 Geologic pattern recognition from seismic attributes: Principal component analysis and self-organizing maps *Interpretation*
[6] Petronas 1999 The Petroleum Geology and Resources of Malaysia
[7] Mansor M., Abdul H., Menier D. and Pubellier 2014 Structural evolution of Malay Basin, its link to Sunda Block tectonics *Marine and Petroleum Geology*

IOP Conf. Series: Earth and Environmental Science **88** (2017) 012003 doi:10.1088/1755-1315/88/1/012003

Petrophysical Analysis of E5 sand group of Sokor Formation, Termit Basin, Niger

Muhammad Naim Nasaruddin[1], Lo Shyh Zung[1], Abdul Ghani Md Rafek[1]

[1]Department of Petroleum Engineering and Geoscience
Universiti Teknologi PETRONAS
Tronoh, Malaysia

E-mail : m.naimn21@yahoo.com

Abstract. The main objective of well logging is to estimate the petrophysical properties of reservoirs which includes porosity and hydrocarbon saturation for hydrocarbon exploration and production purposes. This paper portrays the result of a wireline log study on the E5 sand group of Sokor Formation of the Paleogene in Termit Basin. The objectives of this research are to differentiate the log characteristics of the sediment in the Sokor Formation and provide a preliminary result of the porosity, permeability and hydrocarbon saturation of the study area. A total number of 3 wells were fully evaluated for this research. A compilation of wirelines data sets availability which includes calliper log, gamma ray log (GR), neutron log (NPHI), density log (RHOB) and resistivity logs (RD and RS) were carefully analysed during this study. Petrophysical analysis reveals that the E5 sand group of Sokor Formation has a very high porosity of up to 28.7% and high hydrocarbon saturation up to 88.7%. The suggests future work should include the integration of well log analysis with core analysis.

1. Introduction

The Termit Basin is located in the Western African Rift System (WARS) of Western Central African Rift System (WCARS). It is bounded by the Tenere and Tefidet basin in the north and Bornu basin in the south[1]. It has a total length of approximately 300km and covers an area of approximately 2.7 $x10^4$ km^2. It is composed of several formations known as Donga, Yogou, and Madama (Upper Cretaceous), Sokor-1 (Paleocene-Eocene) and Sokor-2(Oligocene). Existing petroleum system in Termit Basin are present in the Upper Cretaceous and Paleogene specifically the Yogou–Sokor-1 and Sokor-1–Sokor-2 systems with most of the hydrocarbon discoveries were found in Sokor-1 Formation[2].

Previous studies have shown that Sokor Formation has excellent reservoir properties suitable for hydrocarbon accumulation. However, quantitative log analysis for E5 sand group of Sokor Formation to determine its porosity and water saturation has not been fully evaluated. This paper shows the result of the petrophysical analysis of 3 sets of well logs data (A1, A2 and A3) from the Termit Basin area in Western Central African Rift System (WCARS). The aim of the petrophysical analysis is to analyse the log responses and to convert them into petrophysical properties such as porosity, permeability and oil saturation [3]. In this analysis, qualitative interpretation was done by the optimizing the calliper log, gamma ray log (GR), neutron log (NPHI), density log (RHOB) and resistivity logs (RD and RS) to determine the lithology of the rock Formation and distinguishing between reservoir and non-reservoir zones. Quantitative interpretation was conducted to estimate the values of petrophysical properties of the zones of interest (reservoir) such as porosity and hydrocarbon saturation. The petrophysical analysis is aimed to provide a more realistic representation of the petrophysical properties of Sokor Formation which can support the formation evaluation for exploration purposes in the future.

2. Geological Setting and Background

Niger is located in West Africa between the latitude of 12° N and the Tropic of Cancer line. It covers an area of 1.27 x 10^6 km^2. It is bounded by Algeria and Libya to the north, Mali and Burkina Faso to the west, Benin and Nigeria to the south and Chad to the east[4]. Termit basin is one of Mesozoic-Cenozoic rift basins and the largest basin in the West Africa Rift System (WARS) of the Western Central African Rift System (WCARS). Termit Basin is enclosed by Tefidet and Tenere basins on the northwest, and Borau basin at southeast[1]. It has a total length of approximately 300km long measured from north to south. The width of the narrowest area of the Termit Basin which is located at the north end, is about 60km, while the widest area is about 110km across located at the south area of the Termit Basin [5]. It covers an area of approximately 2.7 $\times 10^4$ km^2 and trends NW – SE[2]. The general vertical profile of Termit Basin consists of the Precambrian basement which are made of igneous and metamorphic rock. The basement created a unique trend of faults that act as the trend direction for the Cretaceous-Tertiary rifts in Termit Basin. It is overlain by the Lower Cretaceous, Upper Cretaceous, Paleogene, Neogene and Recent sediments deposited in sedimentary basins[6]. Several kilometers of Lower Cretaceous clastic sediments, primarily lacustrine shale and sandstone, were deposited during rifting. During late Lower Cretaceous, active rifting caused a thermal subsidence to occur. This allowed the progradation of spreading fluvial, deltaic and, shallow marine facies[5].

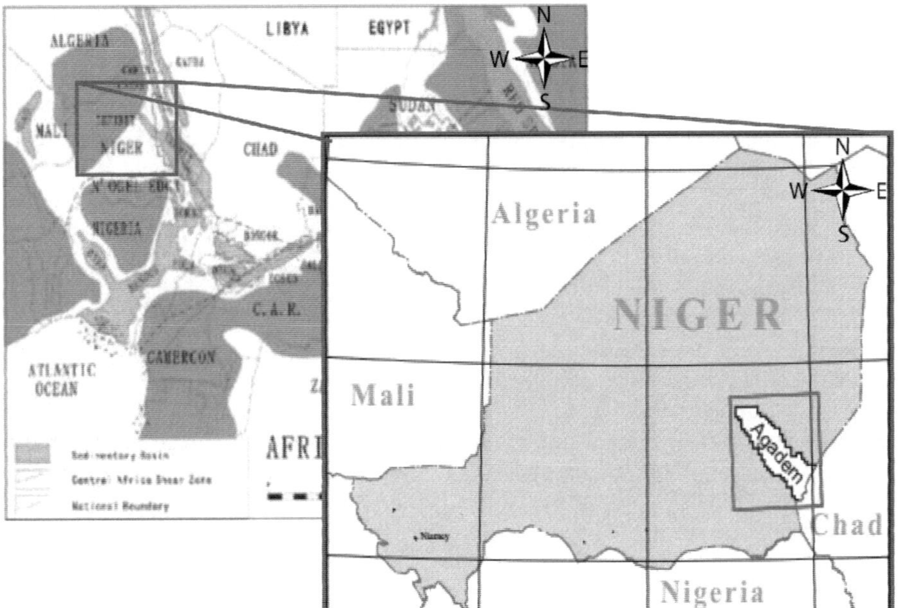

Figure 1. Regional Map of Termit Basin and Agadem Block (Modified from Liu, 2015 and Pan, 2013)

3. Petroleum System

3.1. Source Rock

The hydrocarbon source beds of the Termit Basin are located within the Upper Cretaceous mudstones which is in Donga and Yogou Formation while Paleogene mudstones which are located in the Sokor-1 and Sokor-2 Formation[7]. The Donga Formation varies of grey to dark marine shale, a poor to fair quality source rock, which is dominated by the Type II and III kerogen[1]. The TOC value for the Donga Formation mudstones varies from 0.42% to 1.47%, averaging 0.89%, and S2 yields range from 0.29 to 2.8 mg HC/g rock, averaging 1.1 mg HC/g rock. The Yogou Formation can be separated into two members known as the upper and lower members separated by a conformable boundary. The upper member consists of by interbedded light to medium grey sandstones and light to dark grey mudstones, while the lower member consists of thick dark grey to black mudstones. The TOC within the Yogou Formation varies from 0.36% to 23.32%, with an average of 2.6%. The Rock-Eval pyrolysis, S2 within the Yogou Formation varies from 0.15 to 70.15 mg HC/g rock, which has an average of 5.59 mg HC/g rock. The maximum temperature Tmax vs Hydrocogen index HI analysis reveals that the organic matter of the upper member of the Yogou Formation contains Type I, II, II-III and III kerogen while the lower member is dominated by Type II-III and III material[2]. Gray to dark gray lacustrine shale of Sokor Formation composed of organic shale, that was deposited during the of the second rift cycle. It is a good to excellent source rock which has a high HI values reaching 800 mg HC/g. The TOC content ranges from 0.2% to 23.7% with kerogen Type I and II.

3.2. Reservoir

The reservoirs rock for the Termit Basin are found in the Sokor-1 and Yogou Formations. The reservoir within Sokor-1 and Yogou Formations are sandstones of the fluvial, deltaic and lacustrine depositional environments. The thickness of sandstone in, Sokor-1 Formation varies from less than 1m-50 m, (2% of the sandstone is above 40m), which represent composite bodies of stacked channels. It composed high quality quartz sandstones, have porosity varying from 16% to 36% with mean of 20%-28%, and permeability from 250 md to 1 Darcy. The thickness of sandstones in Yogou Formation varies from less than 1 m- 25 m. The average porosity and permeability varies from 20% to 25% and 100 to 250 mD, respectively[1].

3.3. Seal

The mudstones within the Sokor-2 represent the regional seal of the Termit Basin that are deposited during the second rifting cycles because it is composed of thick of lacustrine mudstones. The average thickness of the seal can reach up to 100m and 500m in the proximal and central basin, respectively. Intra-Formational mudstone and shale that formed within Sokor-1 and Yogou Formations act as local seal rock for individual reservoirs both at top and lateral seals against which reservoir sands are juxtaposed due to faulting. Hydrocarbon is mainly found in structural traps, which were dominated with footwall blocks and horsts, controlled by faults formed during rifting cycle[1].

3.4. Trap

The WCARS hydrocarbon traps are mostly made up of fault blocks[8]. The hydrocarbon stored within the reservoir rock of Termit basin are mainly founded by structural traps. The structural trap is dominated by the footwall blocks and horsts, which are affected by the faults that are formed during the rifting process [3].

4. Objective

To investigate the petrophysical properties of E5 sand group of Sokor Formation.

5. Methodology

The methodology of this research uses both qualitative and quantitative interpretation. Qualitative interpretation is done by identifying the zones of interest which is clean sandstone within the formation. The separation between the clean sandstone and shale is determined by using the gamma ray, density, neutron and resistivity logs. The clean sandstone with hydrocarbon has a characteristics of low GR response, low density response, mid-high neutron log response, high resistivity response, and wide separation between the shallow and deep resistivity response.

The initial step in quantitative interpretation is estimating the shale volume. The shale volume is determined to estimate the fraction of clay in the rock formation for the calculation of effective porosity. The shale volume is calculated using the gamma ray response. The equation for the shale volume Vsh using GR response is:

$$Vsh = \frac{GRlog - GRsand}{GRshale - GRsand} \quad (1)$$

Neutron porosity obtained from the neutron log is known as the total neutron porosity Φ_{TN}. In order to estimate the neutron effective porosity Φ_{EN}, the shale-bound water must be removed from the neutron total porosity. Total density porosity Φ_{TD} can also be calculated using the density logs via equation (2). In order to estimate the density effective porosity Φ_{ED}, the shale-bound water must be removed from the total density porosity. The equation for the correction of effective density porosity from the total density porosity is given in equation (3):

$$\Phi TD = \frac{\rho matrix - \rho log}{\rho matrix - \rho fluid} \quad (2)$$

$$\Phi ED = \Phi TD - (Vsh\Phi sh) \quad (3)$$

Since the analysis of the wells shows the presence of oil, the effective porosity, Φ_E for both density and neutron logs is calculated using the equation:

$$\Phi E = \frac{\Phi ED + \Phi EN}{2} \quad (4)$$

Water saturation, Sw, is calculated using the Archie's equation. Archie's equation is the equation that is derived from the resistivity and porosity of sedimentary rock to estimate water saturation. The Archie's equation is given as

$$Sw = \left(\frac{aRw}{(\Phi^m)Rt}\right)^{\frac{1}{n}} \quad (5)$$

Sw	= Water Saturation	m	= cementation exponent
a	= tortuosity constant	n	= saturation exponent
Rw	= resistivity of Formation water	Rt	= resistivity of true Formation

Hydrocarbon saturation, Shc, is computed using the equation:

$$Shc = 1 - Sw \quad (6)$$

IOP Conf. Series: Earth and Environmental Science **88** (2017) 012003 doi:10.1088/1755-1315/88/1/012003

6. Results and Discussion

6.1. Qualitative Interpretation

Lithology identification is conducted to describe the mineralogy of the formation, thus distinguishing between sandstone (reservoir) and shale (non-reservoir). Gamma ray logs play an important role to determine lithology as it measures the concentration of the clay content in the Formation. Low response of GR logs indicates the presence of sandstone, while high response of GR log indicates the presence of shale.

Determination of formation permeability was done using calliper logs and resistivity logs. Calliper logs records the size of the borehole diameter relative to the bit size. The presence of mudcake represents the presence of sandstone due to the porosity and permeability of the sandstone that act as a filter that allows the mud filtrate to enter the formation, leaving the mud at the borehole wall, hence, reducing its diameter. Separation between deep and shallow resistivity logs represents the permeability of the formation. Wide separation of the deep and shallow resistivity logs represents a permeable formation. Three zones of sandstone intervals were identified to have a good potential hydrocarbon reserves.

Fluid content of the sandstone is determined using the density logs (RHOB) and neutron logs (NPHI). The characteristics of the cross over between the two logs response were used to determine the fluid content. All the log responses from the sandstone zone generate a lower density and middle-high neutron porosity which is associated with the presence of liquid content. Hydrocarbon content is determined by the deep resistivity logs (RD). Indication of the RD which is middle-high (exceeds 20ohm-m) represents hydrocarbon is reserve within the reservoir interval. Figure 3 shows the well log responses of well A2 in Sokor-1 Formation. Three zones of low GR response are labelled as X1, X2 and X3 are identified which indicates the presence of sandstone formation. Figure 2 shows the well response of well A2. All three wells are then correlated to assess the homogeneity distribution of the sandstone interval. Figure 3 shows the well correlation for wells A1, A2 and A3.

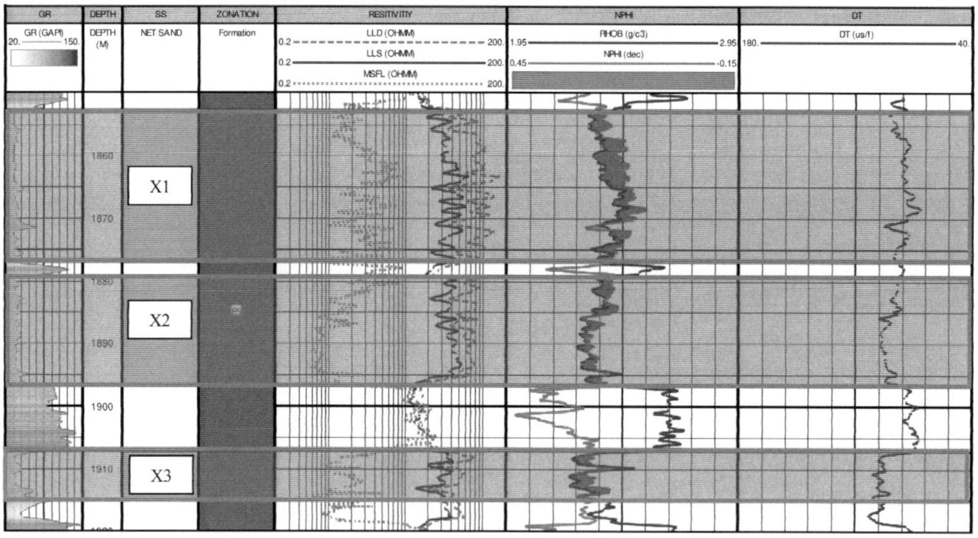

Figure 2. Well log response of well A2 in E5 sand area, area shaded in green indicates the presence of oil.

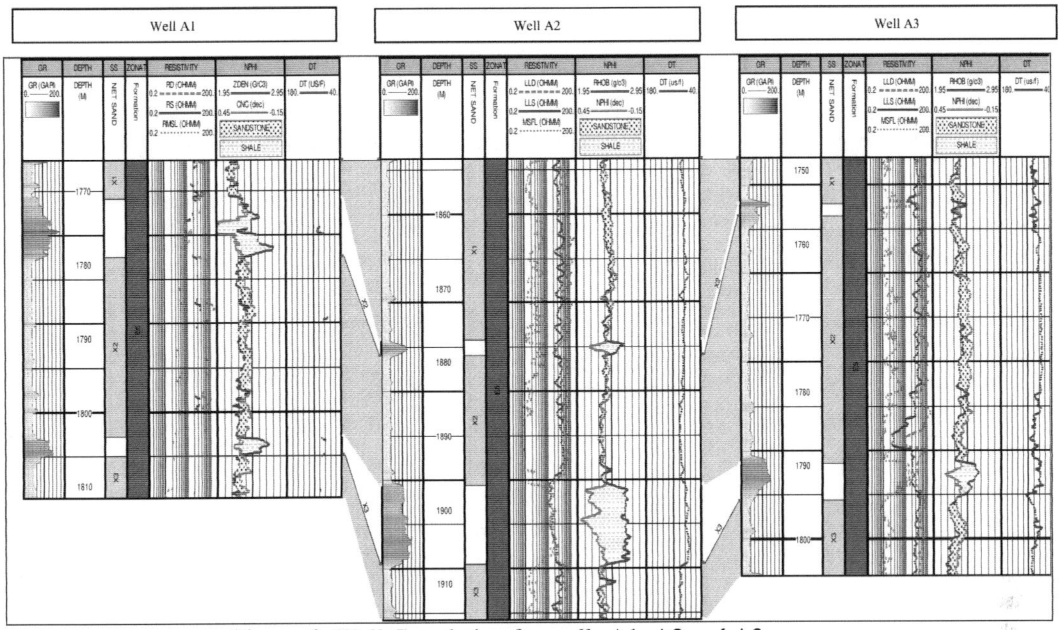

Figure 3. Well Correlation for wells A1, A2 and A3

Table 1, Table 2 and Table 3 represent the summary of results obtained from the qualitative interpretation for E5 member in Sokor-1 Formation of Well A1, A2 and A3 respectively.

Table 1. Results of qualitative interpretation for E5 sand group in Sokor-1 Formation of Well A1

No	Depth Interval(m)	Thickness(m)	Lithology	Shows	Zone
1	1766-1771	5	Sandstone	Oil	X1
2	1779-1803	24	Sandstone	Oil	X2
3	1806-1829	23	Sandstone	Oil	X3

Table 2. Results of qualitative interpretation for E5 sand group in Sokor-1 Formation of Well A2

No	Depth Interval(m)	Thickness(m)	Lithology	Shows	Zone
1	1853-1877	24	Sandstone	Oil	X1
2	1879-1897	18	Sandstone	Oil	X2
3	1907-1915	12	Sandstone	Oil	X3

Table 3. Results of qualitative interpretation for E5 sand group in Sokor-1 Formation of Well A3

No	Depth Interval(m)	Thickness(m)	Lithology	Shows	Zone
1	1749-1755	6	Sandstone	Oil	X1
2	1763-1790	27	Sandstone	Oil	X2
3	1795-1805	10	Sandstone	Oil	X3

6.2. Quantitative Interpretation

Quantitative petrophysical analysis was conducted to determine the petrophysical properties of the formation. Petrophysical properties are calculated using the equation that mentioned earlier in this paper. Quantitative petrophysical analysis was conducted at the zones of interest that was determined using the qualitative interpretation. Table 4 summarizes the results from the petrophysical analysis from all three wells in the area. The petrophysical properties are represented by the values of porosity, Φ, water saturation, Sw, and hydrocarbon saturation, Shc.

Table 4. Results of petrophysical analysis for E5 sand group in Sokor-1 Formation for all wells

No	Well	Zone	Φ (%)	Sw (%)	Shc (%)
1		X1	28.7	30.1	69.9
2	A1	X2	22.5	30.1	69.9
3		X3	25.6	34.3	65.7
4		X1	16.8	32.7	67.3
5	A2	X2	20.2	27.5	72.5
6		X3	22.1	29.5	70.5
7		X1	23.2	35.7	64.3
8	A3	X2	19.8	34.6	65.4
9		X3	26.7	27.0	73

6.3. Discussion

Based on the petrophysical analysis conducted on all wells A1, A2, and A3 shows that sand body E5 in Sokor-1 Formation has an excellent vertically distribution of high thickness of sandstone, up to 27m, because they were deposited in a fluvial environment. Low gamma ray, mudcake, wide separation of deep and shallow resistivity, crossover of low density and middle high neutron, and high resistivity response from the well logs show the favourable characteristics for accumulation of oil. The bodies of sandstone were correlated in 3 wells within the study area labelled as X1, X2 and X3 using beds marker. The porosity of all the interpreted zone has a range between 16.8 to 28.7%. The hydrocarbon saturation show has a range from 64.3% to 88.7%. This result shows that the sand group has the ideal petrophysical properties to allow hydrocarbon accumulation.

7. Conclusion

The objective in petrophysical analysis of E5 sand group for Sokor Formation in Termit Basin, was accomplished. The results show that the Sokor Formation contains adequate properties with high porosity average of 22.51% and up to 28.7% and high hydrocarbon saturation up 88.7%. A more realistic result can be obtained from petrophysical-core integration to validate the parameters calculated, which can be conduct in the future. A wider area within the Termit Basin for petrophysical analysis can also be conducted in the future to obtain a more accurate result. As a conclusion, the methodology and results obtained from this study can be used to improve the petrophysical properties of E5 sand group of Sokor Formation for exploration purposes in the future.

8. Acknowledgment

We would also like to thank the Department of Geoscience, Faculty of Geoscience Petroleum Engineering of University Technology PETRONAS (UTP) for all their helpful contribution.

9. References

[1] L. Wan, J. Liu, F. Mao, M. Lv, and B. Liu, "The petroleum geochemistry of the termit Basin, Eastern Niger," *Marine and Petroleum Geology,* vol. 51, pp. 167-183, 2014.

[2] B. Liu, L. Wan, F. Mao, J. Liu, M. Lü, and Y. Wang, "HYDROCARBON POTENTIAL OF UPPER CRETACEOUS MARINE SOURCE ROCKS IN THE TERMIT BASIN, NIGER," *Journal of Petroleum Geology,* vol. 38, pp. 157-175, 2015.

[3] N. Ishwar and A. Bhardwaj, "Petrophysical Well Log Analysis for Hydrocarbon exploration in parts of Assam Arakan Basin, India."

[4] M. Zanguina, A. Bruneton, and R. Gonnard, "An introduction to the petroleum potential of Niger," *Journal of Petroleum Geology,* vol. 21, pp. 83-103, 1998.

[5] M. J. Warren, "Tectonic inversion and petroleum system implications in the rifts of Central Africa," in *Frontiers and Innovation, CSPG CSEG CWLS Convention*, 2009, pp. 4-8.

[6] A. Eyike, S. C. Werner, J. Ebbing, and E. M. Dicoum, "On the use of global potential field models for regional interpretation of the West and Central African Rift System," *Tectonophysics,* vol. 492, pp. 25-39, 2010.

[7] M. Harouna and R. Philp, "Potential petroleum source rocks in the Termit Basin, Niger," *Journal of Petroleum Geology,* vol. 35, pp. 165-185, 2012.

[8] X.-h. Pan, S.-q. Yuan, Z.-f. Ji, G.-c. Hu, and L. Liu, "Forming Mechanism and Petroleum Geological Features of the Western-Central African Rift Basins (WCARBs)," in *IPTC 2013: International Petroleum Technology Conference*, 2013.

Reducing Uncertainties in Hydrocarbon Prediction through Application of Elastic Domain

S Z Shamsuddin, M Hermana, D P Ghosh and A M A Salim

Center of Seismic Imaging, Universiti Teknologi Petronas, 32610 Bandar Seri Iskandar, Perak, Malaysia.

syazashamsuddin90@gmail.com

Abstract. The application of lithology and fluid indicators has helped the geophysicists to discriminate reservoirs to non-reservoirs from a field. This analysis is conducted to select the most suitable lithology and fluid indicator for the Malaysian basins that could lead to better eliminate pitfalls of amplitude. This paper uses different rock physics analysis such as elastic impedance, Lambda-Mu-Rho, and SQp-SQs attribute. Litho-elastic impedance log is generated by correlating the gamma ray log with extended elastic impedance log. The same application is used for fluid-elastic impedance by correlation of EEI log with water saturation or resistivity. The work is done on several well logging data collected from different fields in Malay basin and its neighbouring basin. There's an excellent separation between hydrocarbon sand and background shale for Well-1 from different cross-plot analysis. Meanwhile, the Well-2 shows good separation in LMR plot. The similar method is done on the Well-3 shows fair separation of silty sand and gas sand using SQp-SQs attribute which can be correlated with well log. Based on the point distribution histogram plot, different lithology and fluid can be separated clearly. Simultaneous seismic inversion results in acoustic impedance, Vp/Vs, SQp, and SQs volumes. There are many attributes available in the industry used to separate the lithology and fluid, however some of the methods are not suitable for the application to the basins in Malaysia.

1. Introduction

Malaysia has been producing hydrocarbons from various fields since the early 1900s, with the first field was found in onshore Miri, Sarawak. The following major discoveries were made offshore in Sarawak, Sabah and Malay Basins. The location of Malay basin is at the eastern side of Peninsular Malaysia. Malay basin is known to be an extensional pull-apart basin with anticlines trending from east to west direction including series of half-grabens. Primarily, the basin formed during the late Eocene to early Oligocene, followed by thermal subsidence leading to sedimentation in the early Miocene. Later in the middle Miocene, regional stress fields changed and the basin inverted forming east-west anticlines trend [3].

Pitfall of amplitudes is one of the many geophysical issues faced by the interpreters that lead to the misinterpretation of hydrocarbon reservoirs for examples are coal, soft shale, and high-porosity brine sand. Coal has been the excellent source rock for Malay basin reservoirs and good horizon markers, however the presence of coal masks interfere the reservoirs underneath by forming strong soft negative impedance. This is due to the acoustic properties of coal that has low density and low velocity [9], which shows similar acoustic impedance response to the gas sand. Another issue is that high quality brine sand shows similar amplitude versus offset (AVO) response to hydrocarbon sand. In the case of poor quality gas sand, the reflectivity shows positive impedance which might be misguided as non-reservoir. In addition, unconsolidated shale or known as soft shale that occur mostly in younger sediments, produces lower acoustic impedance than the surrounding rocks; similar response as hydrocarbon sand. These issues add to the difficulties to differentiate the amplitude anomalies based on acoustic impedance alone, whether they are caused by the lithology or fluid content. Therefore, this analysis is conducted to select the most suitable lithology and fluid indicator for the Malaysian basins that could lead to better eliminate pitfalls of amplitude.

2. Methodology

This paper addresses different application of rock physics analysis such as extended elastic impedance (EEI), Poisson ratio, SQp-SQs attribute and Lambda-Mu-Rho (LMR). The elastic impedance logs can be correlated with any desirable log, in this project are gamma ray and resistivity or water saturation logs to produce EEI_litho and EEI_fluid respectively [8]. Meanwhile, the SQp-SQs attribute is adopted from Hermana et. al which SQp resembles the gamma ray log and SQs imitates resistivity log, thus both attributes are used for lithology and fluid indicator respectively [5].

2.1. Poisson ratio

Poisson's ratio (σ) characterizes the relationship between Vp and Vs of material, which can affect the P-wave reflection coefficients [7]. The magnitudes are used to distinguish between different lithology, which generally brine-saturated sand has higher Poisson's ratio compared to gas-saturated sand.

$$\sigma = \frac{1}{2}\frac{\left(\frac{V_p}{V_p}\right)^2 - 2}{\left[\left(\frac{V_p}{V_p}\right)^2 - 1\right]} \quad (1)$$

2.2. Extended Elastic Impedance

Whitcombe [8] extended the elastic impedance (EI) theory to maximize the separation of fluid or lithology by introducing the angle chi, $\boldsymbol{\chi}$ [1]. The normalization of elastic impedance equation, where the angle $\boldsymbol{\theta}$ is replaced by $\boldsymbol{\chi}$ is shown in Equation 2.

$$EEI(\chi) = V_{p0}\rho_0\left(\frac{V_p}{V_{p0}}\right)^{p}\left(\frac{V_s}{V_{s0}}\right)^{q} \quad (2)$$

Where $p = \cos\chi + \sin\chi$, $q = -8K\sin\chi$, $r = \cos\chi - 4K\sin\chi$. The EEI logs are correlated with gamma ray log to produce EEI_litho log and resistivity or water saturation log to obtain EEI_fluid log.

2.3. Lambda-Mu-Rho

Lame parameters such rigidity, μ and incompressibility, λ are utilized with combination of measured density ρ in order to improve the interpretation of amplitude versus offset (AVO) [4]. These parameters use acoustic and shear impedance to discriminate fluid and lithology by using the equations;

$$\lambda\rho = AI^2 - 2SI^2 \quad (3)$$

$$\mu\rho = SI^2 \quad (4)$$

2.4. SQp-SQs

These attributes are derived based on attenuation concept–rock physics approximation:

$$SQ_p = \frac{5}{6\rho}\frac{\left[\left(\frac{M}{G}\right)-2\right]^2}{\left[\left(\frac{M}{G}\right)-1\right]} \quad (5)$$

$$SQ_s = \frac{10}{3\rho} \frac{\left(\frac{M}{G}\right)}{\left[3\left(\frac{M}{G}\right)-2\right]} \quad (6)$$

SQ_p is defined as Scaled inverse Quality factor of p-wave which is used as lithology indicator, while SQ_s is used as pore fill indicator [5]. From Equation 5 and Equation 6, the M, G and ρ represent bulk modulus, shear modulus and density respectively. The M/G is approximates as $(V_p/V_s)^2$.

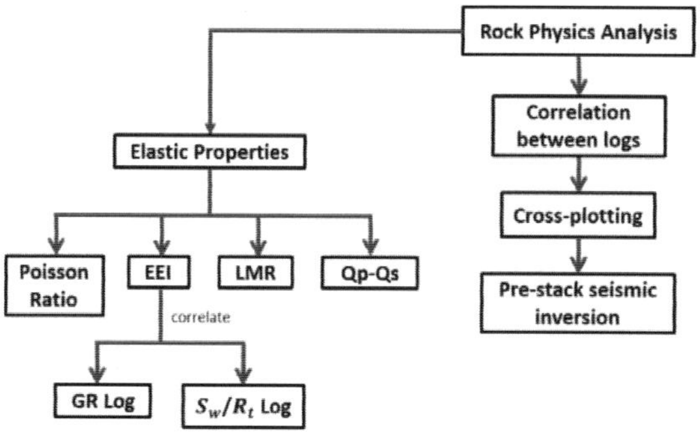

Figure 1. The workflow used for this project.

Initially, the work is done on three well logs from three different fields in Malaysia basins namely Well-1. Well-2 and Well-3. Elastic logs are generated using different rock physics methods for each well. Later, cross-plot analyses are done using logs to observe which attributes can separate lithology and fluid better. The method of cross-plotting has proven to be the best method to observe separation of lithology or fluid [6]. Nevertheless, the cross-plotting analysis and EEI correlation is done at certain depth (including reservoir target) to omit the effect of depth trend.

Among the three wells used, only one field that has pre-stack seismic data which is from Field-3. The analysis is then followed by running simultaneous pre-stack seismic inversion to observe whether the elastic parameters chosen can differentiate fluid and lithology at seismic scale. The inversion is completed based on a low frequency model generated from Well-3 with application of high-cut frequency from 4 Hz to 10 Hz. Later, the inversion analysis acts as the quality control before running the inversion process.

3. Results

3.1. Feasibility Study on Well Log Data

Several elastic logs show good separation of lithology or fluid in well domain. There's an excellent separation between hydrocarbon sand and background shale from several cross-plots generated. In Figure 2, the correlated EEI cross-plot shows good separation of gas sand from shale, which can be difficult to differentiate from logs only. The gas sand cluster is represented by the combination of low EEI_litho and EEI_fluid values. In case of Well-2, LMR cross plot manages to separate clearly the gas sand from the brine sand and shale as shown in Figure 3. The gas sand cluster is represented by the low to high values of Mu-Rho and low values of Lambda-Rho. Well-3 lithology can be distinguished the best using the SQp-SQs attributes, where the hydrocarbon sand cluster is deviated away from the wet trend from the cross-plot in Figure 4.

IOP Conf. Series: Earth and Environmental Science **88** (2017) 012004 doi:10.1088/1755-1315/88/1/012004

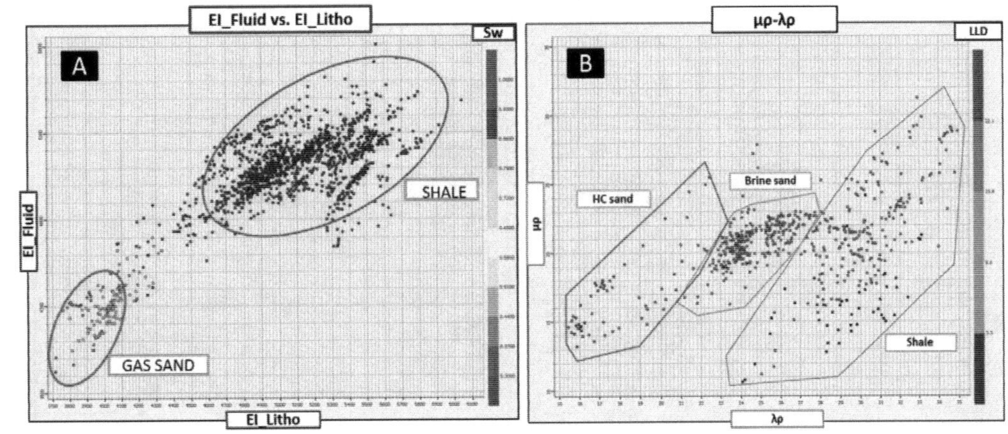

Figure 2. The crossplot between elastic impedance logs for Well-1 in A. The EEI_Litho log is a lithology indicator, while the EEI_fluid corresponds to the fluid indicator. B displays the Lambda-Rho and Mu-Rho crossplot, applied using Well-2. Three clusters of lithology can be separated. The colour key represents the resistivity values.

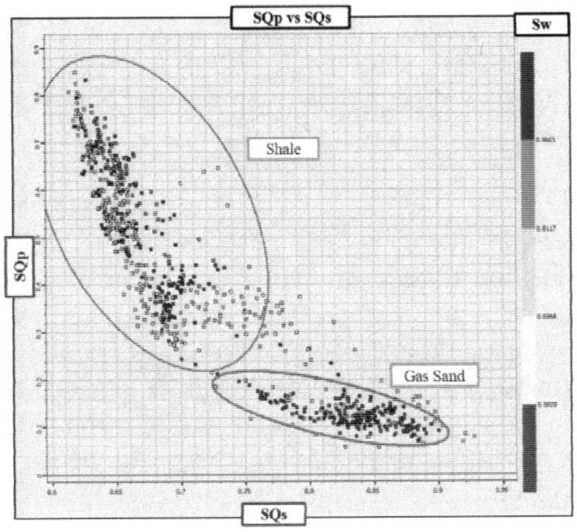

Figure 3. The combination of attenuation attributes separate shale and gas sand at the Well-3 target. The colour key represents water saturation values.

The occurrence of different lithology can be determined much clearer using the histogram plot. Figure 4 shows a summary of the point distribution constructed from four different elastic properties at Well-3. Each lithology can be defined based on the distribution curves trends by assuming the points are normally distributed. In the case of Poisson Ratio, the curves can be grouped into three different lithology; gas sand, silty sand and shale [Figure 4(A)]. This case is applicable for Lambda-Rho (λρ) and elastic impedance (EI) distributions (B and C respectively) with three lithology types can be defined. However, both EI and λρ do not show distinct separation between silty sand (yellow) and shale (green), as silty sand and shale curves might overlap with each other. Hence, it is possible for other interpreters to group the lithology into two classes only. In the meantime, SQ_s attribute presents four clear distribution curves. From the well display on its right, lithology can be categorized separated into shale (grey), poor silty sand (yellow), silty sand (green) and gas sand.

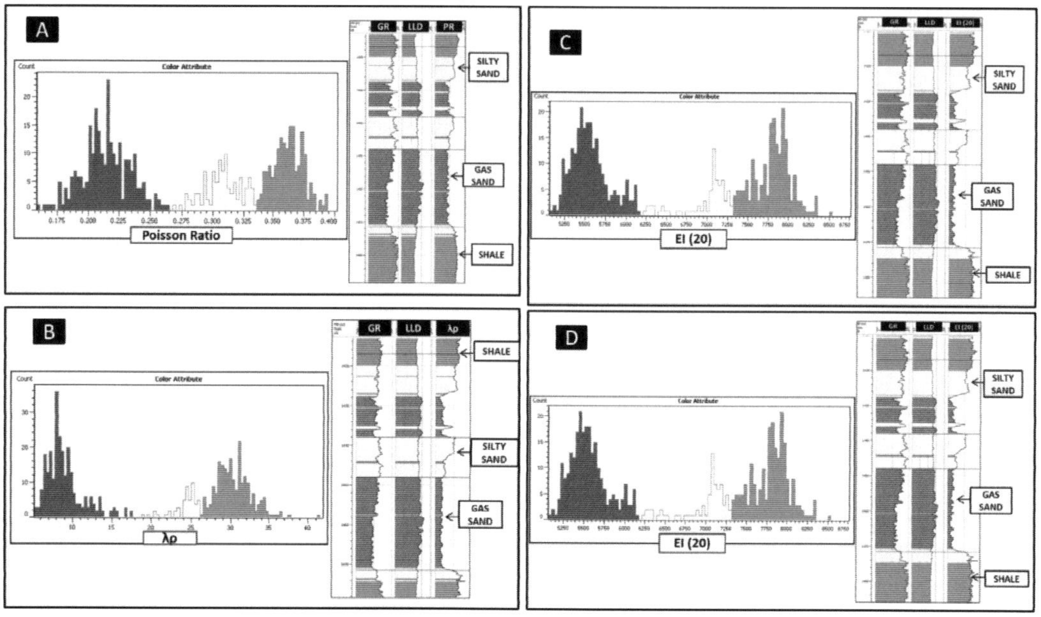

Figure 4. Plot A represents the frequency of Poisson Ratio distribution, B represents Lambda-Rho, C displays the plot of elastic impedance at 20° angle, and D shows distribution of SQs points. On the right side of each histogram shows the lithology type at well location with GR, resistivity and elastic parameters log (from left to right in order).

3.2. Seismic Inversion on Selected Field

The results of the inversion include various data volumes such as AI, SI and ρ from Field-3. The lithology is composed of sand, silty sand (silt-sized or poor sand) and shale. The volumes are then manipulated to form elastic volumes such as Vp/Vs, SQp and SQs and LMR. By reference to the results of feasibility from well data, SQp-SQs could define lithology and fluid based on seismic data as well. In Figure 6, acoustic impedance (AI) volume displays good separation of low and high impedance lithology. However, AI cannot distinguish between low impedance from gas sand and coal, which can be misguided as hydrocarbon reservoir. The sand package is clearly captured within the horizon window in SQp and Vp/Vs volumes as displayed in Figure 5 and 6. Both Vp/Vs and SQp volumes show low values of sand and high value for coal. Meanwhile, the sand filled with hydrocarbon can be seen well in the fluid indicator SQs, see Figure 6. Therefore, if attenuation attributes are combined for interpretation, gas sand can be determined better. LMR is known to be very useful as litho-fluid indicator through combination of high MuRho values and low Lambda-Rho values [2]. Although LMR volumes can differentiate sand similar as other lithology indicators, the fluid can be separated better using mentioned fluid indicator, SQs.

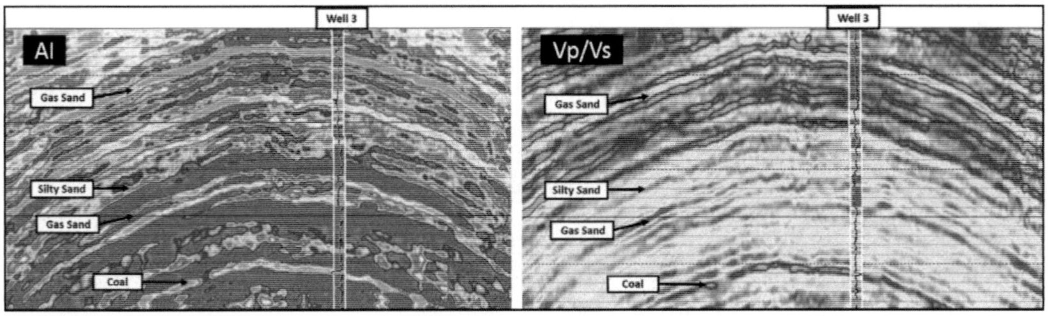

Figure 5. Acoustic impedance (left) and ratio of Vp and Vs (right) volumes extracted from the inversion result.

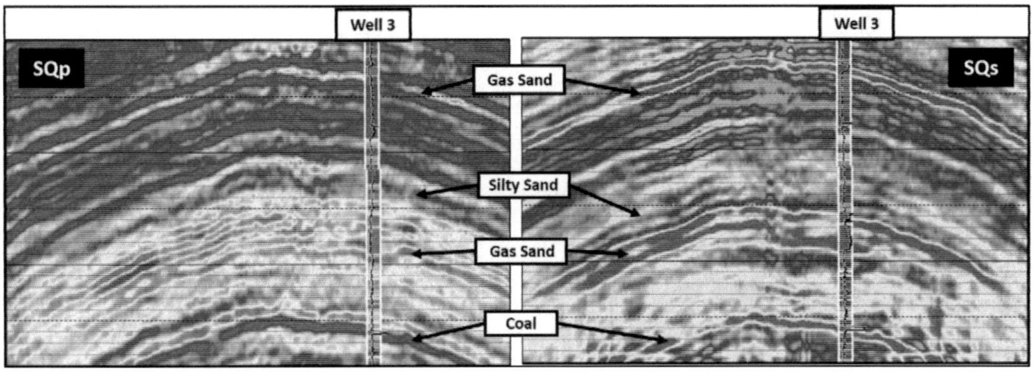

Figure 6. The left figure displays the SQp volume as lithology indicator and the right figure displays the fluid indicator, SQp volume. The combination of both attributes manage to delineate reservoir package.

4. Conclusion

There are various methods available to delineate reservoir from non-reservoir. However, the effectiveness of each method depends on the rock properties of the reservoir, which might be different for different fields. For the sand-dominated well, it would be convenient to use conventional rock physics methods to separate the lithology and fluid. On the other hand, if the well is dominated by silty sand, it is difficult to distinguish the pay sand based on the available rock physics methods only. The analyses suggest that the most suitable method for the silty sand-dominated formation is the SQp-SQs attribute as they manage to display clear separation of hydrocarbon sand from the background trend and different lithology. The implementation of inversion of pre-stack seismic data provides better insight to the extent of the reservoir and application of elastic parameters at the seismic scale. This project can be further explored in determining reservoir properties such as porosity and water saturation for better reservoir volume estimation.

5. Acknowledgements

We would like to express our deepest gratitude to the parties involved in completing this paper especially Yayasan Universiti Teknologi Petronas (YUTP) for funding this project, PETRONAS for providing the well logging data and the Center of Seismic Imaging (CSI) for the complete facilities and to the members for the never-ending support.

6. References

[1] Connolly P 1999 Elastic Impedance *The Leading Edge* 438-452

[2] Ekwe A C, Onuoha K M and Osayande M 2012 Fluid and Lithology Discrimination Using Rock Physics Modelling and Lambdamurho Inversion *AAPG International Conference and Exhibition (Milan)*

[3] Ghosh D, Abdul Halim M F, Brewer M, Viratno B and Darman N 2010 Geophysical issues and challenges in Malay and adjacent basins from an E & P persepctive *The Leading Edge* 436-449

[4] Goodway B, Chen T and Downton J 1997 Improved AVO fluid detection and lithology discrimination using Lamé petrophysical parameters: "λp", "μp", and "λ/μ fluid stack": from P and S inversions *CSEG Recorder* **22**

[5] Hermana M, Ghosh D and Chow W S 2016 Optimizing the Lithology and Pore Fluid Separation using Attenuation Attributes *Offshore Technology Conference Asia (Kuala Lumpur)*

[6] Hicks G J and Francis A M 2006 Extended Elastic Impedance and Its Relation to AVO Crossplotting and Vp/Vs *EAGE 68th Conference & Exhibition (Vienna)*

[7] Ostrander W J 1984 Plane-wave reflection coefficients for gas sands at nonnormal angles of incidence *52nd Annual International SEG Meeting (Texas)* 1637-1647

[8] Whitcombe D N, Connolly P A, Reagan R L and Redshaw T C 2002 Extended elastic impedance for fluid and lithology prediction *Geophysics* **67** 63-67

[9] Yao Q and Han D 2008 Acoustic properties of coal from lab measurement *SEG Las Vegas 2008 Annual Meeting (Nevada)* p 1815-1819

Lacustrine Environment Reservoir Properties on Sandstone Minerals and Hydrocarbon Content: A Case Study on Doba Basin, Southern Chad

N. F. Mohd Sumery[1], S. Z. Lo[2], A. M. A. Salim[3]

[123]Geoscience Department, Universiti Teknologi PETRONAS, Perak, Malaysia.

Email: Fathiha.Ms@gmail.com

Abstract. The contribution of lacustrine environment as the hydrocarbon reservoir has been widely known. However, despite its growing importance, the lacustrine petroleum geology has received far less attention than marine due to its sedimentological complexity. This study therefore aims in developing an understanding of the unique aspects of lacustrine reservoirs which eventually impacts the future exploration decisions. Hydrocarbon production in Doba Basin, particularly the northern boundary, for instance, has not yet succeeded due to the unawareness of its depositional environment. The drilling results show that the problems were due to the: radioactive sand and waxy oil /formation damage, which all are related to the lacustrine depositional environment. Detailed study of geological and petrophysical integration on wireline logs and petrographic thin sections analysis of this environment helps in distinguishing reservoir and non-reservoir areas and determining the possible mechanism causing the failed DST results. The interpretations show that the correlation of all types of logs and *rho* matrix analysis are capable in identifying sand and shale bed despite of the radioactive sand present. The failure of DST results were due to the presence of arkose in sand and waxy oil in reservoir bed. This had been confirmed by the petrographic thin section analysis where the arkose has mineral twinning effect indicate feldspar and waxy oil showing bright colour under fluorescent light. Understanding these special lacustrine environment characteristics and features will lead to a better interpretation of hydrocarbon prospectivity for future exploration.

1. Introduction

Lacustrine dominated basin is a standing water body surrounded by land that mostly feeds and drains by rivers and streams. They are varied in water depth, size, salinity and sedimentation. Some lake deposits can act as both source and reservoir. The lacustrine reservoir may have high porosity and permeability, thus making it one of the prolific areas for hydrocarbon exploration where many fields in the world have been producing from this type of environment.

This study is conducted on Doba Basin that has potential in becoming a good lacustrine reservoir. It is a Late-Mesozoic basin situated in Southern Chad of Central Africa. The appraisal wells have been drilled in the study block of Doba Basin (particularly on northern basin boundary), however no hydrocarbon has been produced yet even though the exploration well proved the presence of it. The drilling results show that the problems were all related to the lacustrine depositional environment. This type of environment has some unique characteristics in wireline log as compared to conventional logs. All the geologically related features have significant impact on the concept measures and the future exploration. To address this issue, an integrated reservoir study on petrophysical and petrographical characteristics are needed to reveal more detailed lacustrine reservoir features particularly on the radioactive sand and waxy oil properties for better hydrocarbon exploration in Doba Basin. This study is therefore conducted with the attempt to solve the problems listed in the Table 1.

Table 1: Problems encountered during well logging process due lacustrine geological features

Geological Features	Problems Encountered	Explanation
Relatively short transport distance of sedimentation (incomplete weathering cycle)	Differentiation of reservoir and non-reservoir (failed to recognize sand and shale)	Radioactive sand always mistaken as shale due to high gamma ray content. This Doba Basin's arkosic sand form due to relatively short distance sedimentation. It is believed that the provenance is coming from the basement high.
	Inconclusive of Drill Stem Test (DST) results	1) The presence of arkose in the formation causes the failure of the cement bond logging (CBL). The poor cementing in-fill causing the failure/inconclusive of DST 2) Waxy oil has tendency to clogs the pore throats thus reducing the permeability, which eventually affects the oil mobility to the surface.
Terrestrial source	Possible mechanism of formation damage	Usually formation damage occurs due to the unintended impedance to the flow of fluids into or out of a wellbore. E.g: waxy oil which originated from terrestrial sources.

2. Geological Setting

2.1. Regional Geology

Regional tectonic studies revealed that the lithosphere of the African continent was affected by the staged expansion of Southern Atlantic Ocean in Early Cretaceous which forms the West and Central Africa Rift System (WCARS) [2]. The strong dextral strike slip tension of WCARS developed two genetically related but physically separated Late Mesozoic – Cenozoic passive rift system; the West African Rift System (WARS) and the Central African Rift System (CARS) [5,6,7,10,13]. CARS has been forming a series of sedimentary basins, consists of Bongor, Doba, Doseo and Salamat Basin in southern part of Republic of Chad. The oil and gas exploration and development activities however have been limited to the Doba Basin where over 50% of the country's wells have been drilled and where all the country's current oil production is coming from [12].

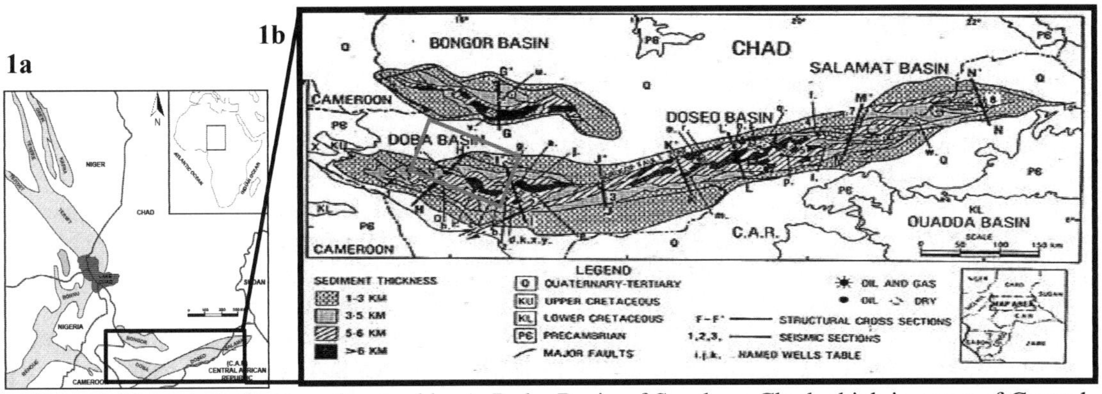

Figure 1:Map of studied location (red box), Doba Basin of Southern Chad which is a part of Central African Rift System, CAS. (Figure 1a: Modified from Genik 1993, Figure 1b: After Aremu, n.d)

Doba Basin located in southern part of Republic of Chad, northwest of CARS as shown in Figure 1. Stratigraphically, Doba basin has a basement composed of Precambrian metamorphic rocks overlain by, at most over, 7,500 meters continental deposits of thinner Lower Cretaceous (Mangara, Kedeni, Doba and Lower Kome) formations, but thick Upper Cretaceous (Upper Kome and Miandoum) formations [6]. The Lower Cretaceous formations mainly contain lacustrine sediments, the sandstone and mudstone interbeds are major reservoir strata, and especially a large set of thick shales are developed at top of formation, acting as the good regional caprocks. Upper Cretaceous is dominated by abundant fluvial sandstones, with alluvial plain mudstones at the top. Cenozoic formations contain fluvial coarse grained clastics sediments. The petroleum Geology of Doba Basin [1] are made of:

- **Reservoir rock -** Lower Cretaceous of Lower Kome, Kedeni and Mangara Formation and Upper Cretaceous of Miandoum Formation of varyingly arkosic sandstones.
- **Source Rock -** Lower Cretaceous of Doba Shale; early rift lacustrine shales with TOC content of typically 1-4% (type I and III organic matter)
- **Seal -** Lower Cretaceous (lacustrine) Mangara Shale and Upper Cretaceous (flood plain and inter-distributary to lacustrine) Miandoum shale.

2.2. Lacustrine Reservoir

2.2.1. Lacustrine Radioactive Reservoir

The hydrocarbon reservoir potential within lacustrine basins is partially impacted by overall tectonic setting [8 9] where within extensional settings there is limited transport distance. The radioactive sand, resulted from limited transport distance, is coarse-grained detrital rocks that have high affinity of radioactive minerals. In principal, sand composes of quartz which gives low Gamma Ray (GR) reading but the high concentration of radioactive minerals in clean sand causes the rise of GR values. Radioactive sand can be classified into 6 types which are [4]; i) Feldspathic sandstone or arkose, ii) Micaceous sandstones, iii) Mixed feldspathic-micaceous sandstone, iv) Greensand or glauconitic sandstones, v) Heavy minerals within sandstones, and vi) Shaly sand and sandstone [4]. This radioactive sand may originate from basement high within the basin itself, implying that the detrital materials from the sedimentary source have not undergone sufficient transportation and weathering cycle, then the parent minerals like feldspar (K), micas, glauconite and heavy minerals would be retained in the sedimentary rock. In this case, the sands and gravels may be highly arkosic (high feldspar content), with high content of K^{40} which was deposited in lacustrine environment particularly in Kedeni and Mangara Formation of Lower Cretaceous age.

IOP Conf. Series: Earth and Environmental Science **88** (2017) 012005 doi:10.1088/1755-1315/88/1/012005

2.2.2. Waxy Oil

The lacustrine environment has been recognized to produce source rock of Type I and III organic matter [11] where:

i) Type I formed from algal remains deposited under anoxic conditions in deep lakes and tend to generate waxy crude oils when submitted to thermal stress during deep burial high.

ii) Type III formed from terrestrial plant material that has been decomposed by bacteria and fungi under oxic or sub-oxic conditions. Petroleum geochemists tend to consider long-chain n-alkanes as being typical of higher plants (from wax coatings of leaves), and often use them as potential indicators of land plant contribution to the source rock of an oil.

In Doba Basin, the hydrocarbon quality had shown that the oils are paraffinic with wax content up to 25% [1] which indicate that the source is highly Type III as it is more waxy-prone than Type I.

3. Methodology

The methodology used in this study comprises of integrated study of wireline logs, mud logs, rock matrixes, DST and CBL results analysis and petrographic thin section analysis as shown in Table 2.

Table 2:Methodolody applied in this study

Methods	Descriptions	
Wireline Logs Response	Sand Shale Recognition	• Natural Gamma Ray Spectroscopy and Wireline Logs Integration • Rho matrix analysis
	CBL and DST Results Analysis	• CBL and DST Results analysis integration with wireline logs
Petrographic Thin Section (Characterisation of Mineral)	Normal Polarized Light	• Detailed information about mineralogy of detrital grains matrix and cement composition can be done
	Fluorescent Light	• The occurrence of waxy oil within the rock matrix can be easily distinguished

4. Results and Discussions

4.1. Sand Shale Recognition

4.1.1. Natural Gamma Ray Spectroscopy and Wireline Logs Integration

By using natural gamma ray spectroscopy, the individual gamma ray sources can be determined thus help in interpreting the types of sands in the reservoir. In general, in shales, high potassium indicates micaceous clay while high uranium indicates organic shale. As for the sandstones, high potassium means arkosic sands, high thorium and uranium suggest more heavy minerals within the sandstone matrix. In Doba Basin, one of the wells show high gamma ray (>90) value but there is formation of mudcakes which are usually associated with sand intervals (Refer Figure 2). Moreover, the amount of potassium in that interval is the highest compared to the uranium and thorium and this concludes that the interval has arkosic sands.

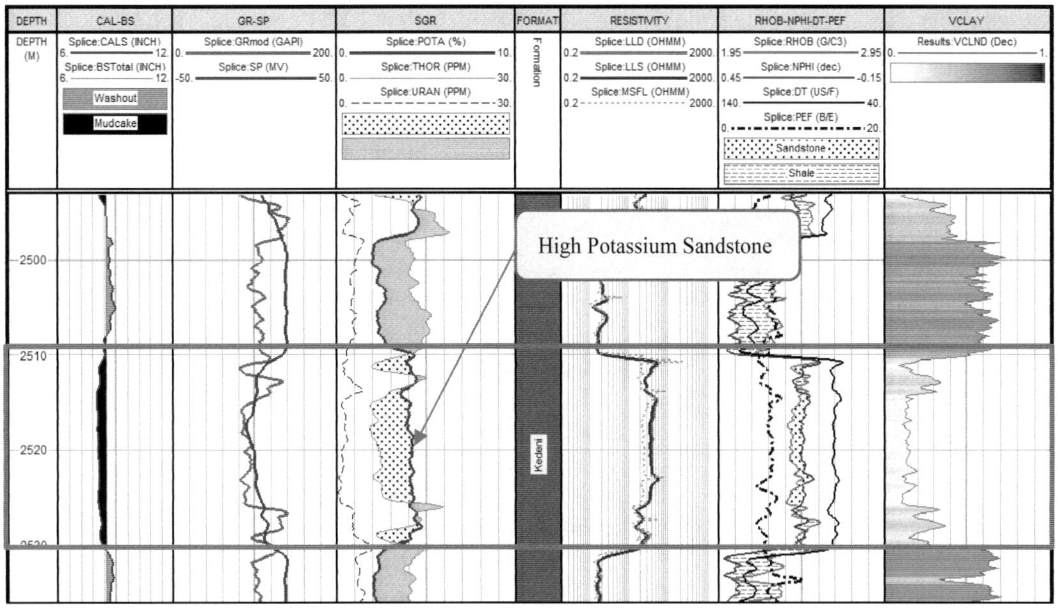

Figure 2: Conventional well log data (Caliper, Bit Size, Gamma Ray, Resistivity, Density-Neutron, Sonic and Photoelectric) with spectral gamma ray value. At depth of 2510m to 2530m, the potassium (blue line) content is higher than uranium and thorium.

The goal of formation microimager (FMI) interpretation is to characterize formation properties to assist sedimentological interpretation such as structure and lithology. Shale and sandstone for example can be easily defined by comparing the colour contrast as shown in Figure 3. Shale appears to be darker compared to sandstone. Here, by integrating conventional log with FMI data, it is clear, at depth 2619m, is defined as sandstone bed, even though the gamma ray displays a high value.

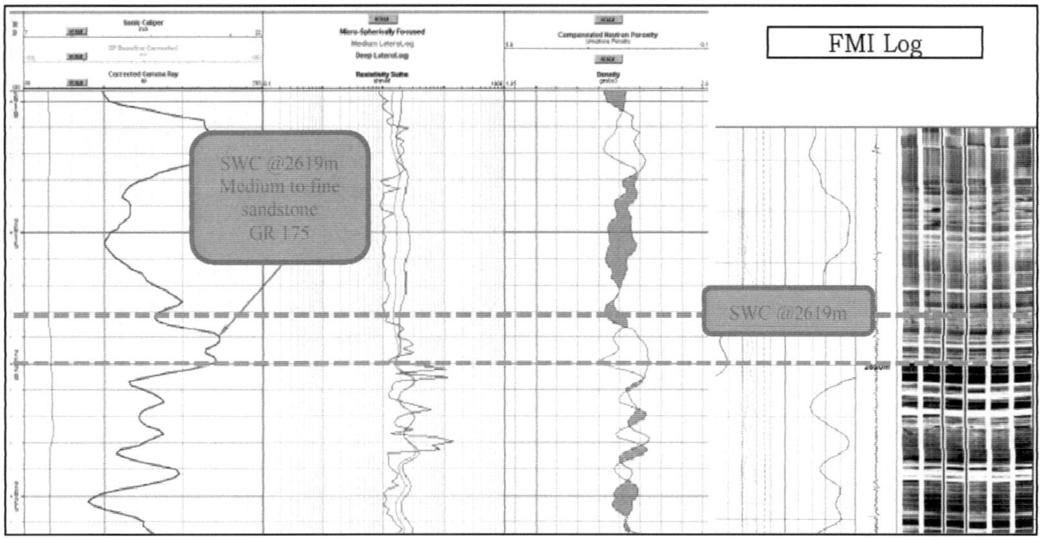

Figure 3:FMI data shows the presence of sandstones at high gamma ray interval

IOP Conf. Series: Earth and Environmental Science **88** (2017) 012005 doi:10.1088/1755-1315/88/1/012005

Rho matrix analysis is another method of differentiating sandstone and shale bed. Based on the GR log, there is no distinct change in its value (which probably means same lithology), however, by using the *rho* matrix technique, the sand and shale bed indicates a totally different value as shown in Figure 4.

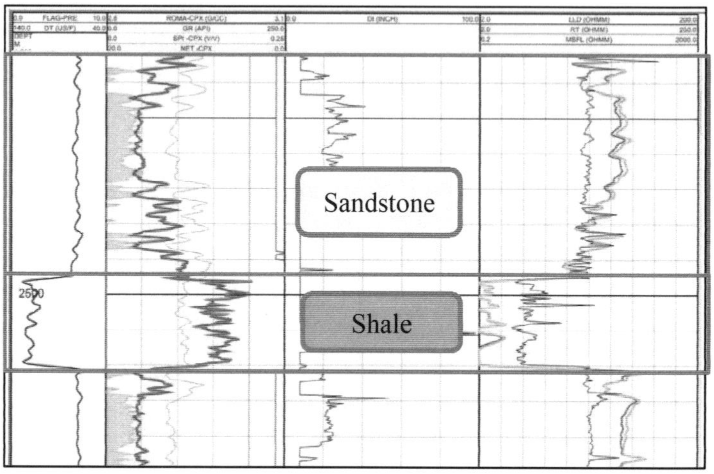

Figure 4: Sand and Shale can be easily identified using rho matrix technique as it shows distinct different value even at the constant (slightly different) Gamma Ray value.

4.1.2. *Petrographic Thin Section Analysis*

Petrographic analysis of thin sections provides detailed information on mineralogy of detrital grains matrix and cement composition. Under microscopic images using normal polarized light, quartz can be easily distinguished as a common mineral in sandstones. It shows no twinning in contrast to images shown in Figure 5 (pictures were taken during analysis). From quick-look petrographic interpretation, the minerals show obvious twinning (lamellar albite, perthite and microcline tartan twinning) which is usually associated with feldspar, a high gamma ray mineral.

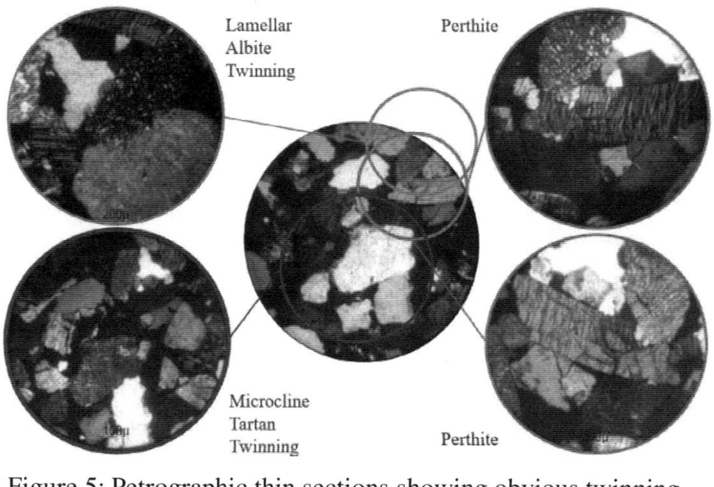

Figure 5: Petrographic thin sections showing obvious twinning that indicates feldspar. All the feldspar twinnings were found in one thin section. High amount of feldspar grains cause increase in gamma ray value

4.2. *Failed DST Results*

4.2.1. *Poor Cement Bond Logging (CBL) Leading to Failed DST Results*

Cement bond logs were run to determine the quality of the cement bond to the production casing, and to evaluate cement fill-up between the casing and the reservoir rock. Poor fill-up of cement leaves large channels behind the pipe that, likewise, allows the flow of unwanted fluids, such as gas or water into an oil well. Both poor bond and poor fill-up problems can also allow fluids to flow to other reservoirs behind the casing. This causes failed or inconclusive DST which leads to serious loss of potential oil and gas reserves. The interpretation in Figure 6 shows that poor cementation happens at high radioactive sandstones. The arkosic sandstones cause the cement to be non-solidified and leave spaces between the borehole and casing.

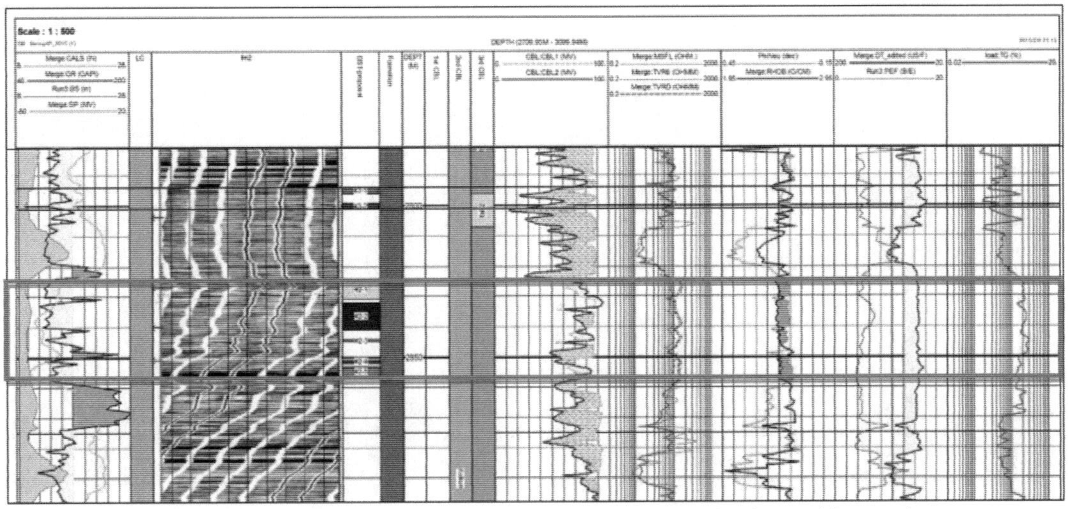

Figure 6: Bad cementing (orange bar) occurred at arkosic sand interval.

4.2.2. Formation of Waxy Oil Leading to Failed DST Results

For the case of waxy oil formation, Figure 7 shows the portion where 10m core was taken from sandstone and basement rock. There is oil drips from the sandstone part which indicates the presence of oil. However, when DST was performed, no oil was produced. This waxy oil formation is one of the major problems in production as it clogs the pores thus preventing oil mobility to surface.

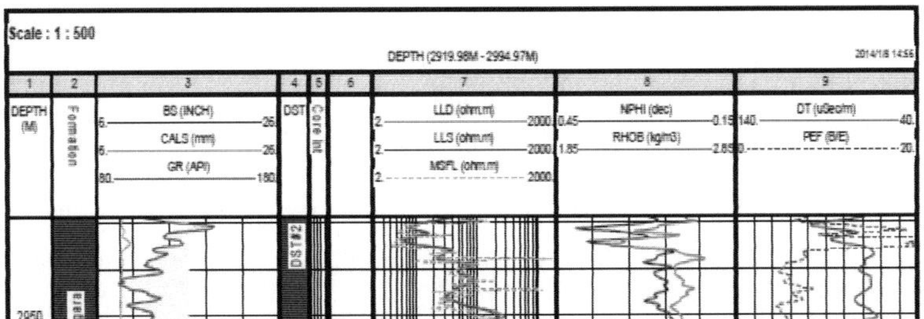

Figure 7: DST was done just above the basement

Petrographic thin section also helps in distinguishing and proving the occurrence of waxy oil. Under the normal polarized light, the sample do not show any distinct feature (oil absent) as shown in Figure 8 (top section). However, under fluorescent light (Figure 8, bottom section), there are obvious bright colours which indicate the presence of wax within the matrix. The existence of waxy oil is believed to be generated from the transported terrestrial organic matter that contributes to the mixed Type I-III [3] but more on Type III kerogens.

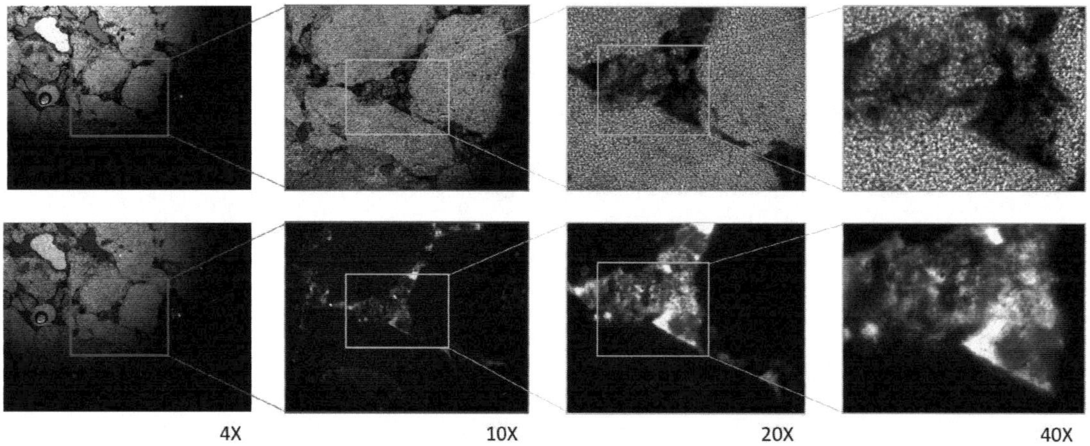

Figure 8: Under fluorescent light (right), the bright colour indicates wax

5. Conclusions

Based on this study, a few conclusions can be made;

1. The reliability in distinguishing reservoir and non-reservoir areas based on the gamma ray log alone will lead to the misinterpretation of the lithology due to the well presence of radioactive sand in the area. Therefore, use of other types of logging tools is necessary.
2. Application of rho matrix technique can help in identifying sand and shale from different density values.
3. The arkosic sandstones are the cause of bad cementation in borehole which affect the DST results. The presence of waxy oil in the formation also leads to DST failures due to formation damage.
4. The petrographic analysis is useful in defining the minerals (e.g: feldspar show twinning features) present in the samples. Not only that, by using fluorescent light, wax is easily distinguished since it gives out a bright colour.

The integration study between geology and petrophysics data will lead to the understanding of the lacustrine reservoir characteristics. It is expected that after the evaluation, a better volumetric calculation of the reservoir and pay zone can be achieved, resulting in more accurate prospectivity of the studied area and also help in solving the problems in other similar lacustrine environment basin.

6. Acknowledgement

The author would like to acknowledge the co-authors for the continuous support in preparing and presentating this research. A special thanks to URIF for the funding and Geoscience Department of Universiti Teknologi PETRONAS for giving authors the opportunity to do this research.

7. References

[1] Aremu, M. (n.d) Chad Prepares to Be an Oil Producers. Oil and Gas Online. Retrieved on September, 29th 2015 from http://www.oilandgasonline.com/doc/chad-prepares-to-be-an-oil-producer-0001

[2] Binks, R.M. and Fairhead, J.D. (1992). A Plate Tectonic Setting for Mesozoic Rifts of West and Central Africa. Department of Earth Sciences, University of Leeds, UK. Elsevier Science Publishers B.V.

[3] Caroll, A.R. and Bohacs, K.M. (2001). Lake-type Controls on Petroleum Source Rock Potential in Nonmarine Basins. The AAPG Bulletin, Volume 85, No. 6, PP. 1033-1053.

[4] Chudi, O., Simon, Roya., SPE, and Warri, SPDC. (2012) Petrophysical Characterisation of Radioactive Sands – Intergrating Well Logs and core Information: A Case Study in Niger Delta. Nigeria Annual Internatiolanl Conference and Exhibition , Abuja, Nigeria, August 6-8, 2012.

[5] Fairhead, J. D. (2009) The Mesozoic West and Central African Rift System: Qualitative Evaluation. AAPG International Conference and Exhibition, Cape Town, South Africa, October 26-29, 2008. Search and Discovery Article #30077.

[6] Fairhead, J. D., Green, C. M., Masterton, S. M., and Guiraud, R. (2013) The Role That Plate Tectonics, Inferred Stress Changes and Stratigraphic Unconformities Have on The Evolution of The West and Central African Rift System and The Atlantic Continental Margins. Science Direct, Tectonophysics 594, 118–127.

[7] Genik, G.J. (1993) Petroleum Geology of Cretaceous-Tertiary Rift Basins in Niger, Chad, and Central African Republic. The AAPG Bulletin, Volume, No.8, PP.1405-1434.

[8] Katz, B. J. (1995) Factors Controlling The Development of Lacustrine Petroleum Source Rocks - An Update. In Huc, A. Y. (ed.), Paleogeography, Paleoclimate, and Source Rocks. Am. Assoc. Petrol. Geol. Studies in Geology 40: PP. 61–79.

[9] Katz, B. J. (2001) Lacustrine Basin Hydrocarbon Exploration – Current Thoughts. Journal of Paleolimnology 26: 161–179, Kluwer Academic Publishers. Netherlands.

[10] Pan, X.H., Yuan, S.Q., Ji, Z.F., Hu, G.C., and Liu, L. (2013). Forming Mechanism and Petroleum Geological Features of Western-Central African Rift Basins (WCARBs). IPTC 17116, Paper Presentation.

[11] Paul Farrimond (2016) Waxy Oills Form Lacustrine Source Rokcs, Integrated Geochemical Interpretation. Retrieved on September, 30th 2016 from https://www.igiltd.com/news/technical-note-12116

[12] United Hydrocarbon International Corp. (2013) Geology. Why Chad. Retrieved on September, 30th 2015 from http://unitedhydrocarbon.com/cms/why-chad/

[13] Warren, M.J. (2009). Tectonic Inversion and Petroleum System Implications in the Rifts of Central Africa. Jenner GeoConsulting, Suite 2008, 1235 17th Ave SW, Calgary, Alberta, Canada, T2T 0C2. Frontiers+Innovation, CSPG SCEG CWLS Convention.

Improving Vintage Seismic Data Quality through Implementation of Advance Processing Techniques

A.H. Abdul Latiff[1], P.G. Boon Hong[2], S.N.F. Jamaludin[3]

[1, 2, 3] Centre for Seismic Imaging, Department of Geosciences, Universiti Teknologi PETRONAS, Malaysia

Email: abdulhalim.alatiff@utp.edu.my

Abstract: It is essential in petroleum exploration to have high resolution subsurface images, both vertically and horizontally, in uncovering new geological and geophysical aspects of our subsurface. The lack of success may have been from the poor imaging quality which led to inaccurate analysis and interpretation. In this work, we re-processed the existing seismic dataset with an emphasis on two objectives. Firstly, to produce a better 3D seismic data quality with full retention of relative amplitudes and significantly reduce seismic and structural uncertainty. Secondly, to facilitate further prospect delineation through enhanced data resolution, fault definitions and events continuity, particularly in syn-rift section and basement cover contacts and in turn, better understand the geology of the subsurface especially in regard to the distribution of the fluvial and channel sands. By adding recent, state-of-the-art broadband processing techniques such as source and receiver de-ghosting, high density velocity analysis and shallow water de-multiple, the final results produced a better overall reflection detail and frequency in specific target zones, particularly in the deeper section.

Keywords: deghosting, shallow water demultiple, high density velocity analysis

1. Introduction

Seismic imaging, sometimes called "reflection seismology" is a primary exploration method that estimates the seismic characteristics of the Earth's subsurface. The technology measures reflected acoustic energy waves, and is used mostly for coal, oil and gas and geothermal exploration. Seismic images provide geophysicists and geologists a method of mapping the subsurface structure of rock formations. This data tells geophysicists and geologists where rock properties change, thus helping find oil and gas deposits. When geological objectives are not achieved in the processed data, the exploration team were encouraged to re-process the data rather than re-acquiring a new seismic data because of time and cost constraint, where new processing work saves a lot of time and is cost-effective. In addition, the technological advances and improved methods of data processing were made over time when the data was first processed compared to current processing algorithms.

The field of study is located in Penyu Basin, offshore Peninsular Malaysia (Figure 1a). The reprocessing of the seismic data is aimed at obtaining a more concise geological image using modern imaging technologies of data processing to overcome limitations of the previous processing which were conducted in 2004 and 2009. The Kirchhoff Pre-Stack Time Migration (PSTM) workflow is applied

[1] Corresponding author

IOP Conf. Series: Earth and Environmental Science **88** (2017) 012006 doi:10.1088/1755-1315/88/1/012006

alongside enhanced 3D processing technologies to attain improved results. Previously in 1991 and 2011, Texaco Exploration Penyu Inc. and Lundin Petroleum had discovered hydrocarbon reserves in Penyu Basin, but eventually discover that the basin to be marginal and no commercial discoveries were ever made. It is said that the poor coverage and imaging quality especially at the basement faults and across the syn-rift half-grabens had contributed to the lack of success [1] (Figure 1b).

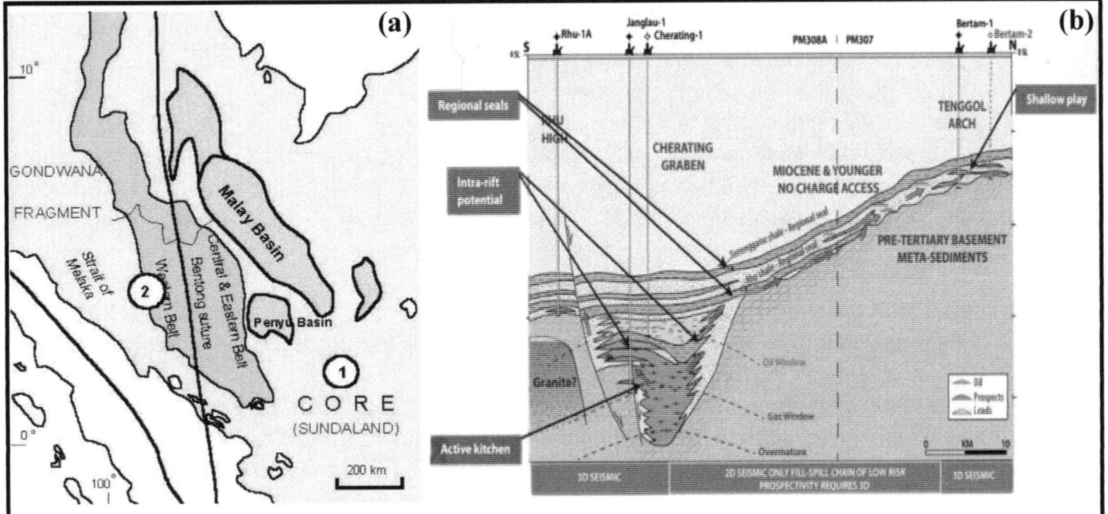

Figure 1: Survey location map and schematic cross-section of Rhu play concept [2].

This paper will mainly focus on application of an enhanced Pre-Stack Time Migration (PSTM) processing workflow to improve on the previous imaging of the seismic data located within the Penyu Basin. Through re-processing the available seismic data, we aim to obtain a better quality seismic section with full retention of relative amplitudes and significantly reduce seismic, structural and stratigraphic uncertainty. In addition, it is hope that re-processing will help to facilitate new prospect discovery through enhanced data resolution, fault definitions and events continuity, particularly in syn-rift section and basement cover contacts. The improved dataset will be subsequently used for future exploration wells placement, and enable AVO and inversion techniques to be applied in a more quantitative manner. In this region, our zone of interest is range in between 1.0 to 4.0 seconds of two way time.

2. Advance Processing Techniques

a. De-multiple

Multiples is a type of energy which has more than one reflection in its travel path. It may disguise, hide or falsely assumed to be primaries which cause data analysis and processing images will be more difficult than normal. The primary wavelet may be affected and distortion in quality of imaging and obstructs primary reflections. A wide range of techniques have been derived and applied to solve for multiple attenuation. Overall, these techniques differ based on the type of multiples – long period or short period. Predictive deconvolution (τ-p deconvolution), surface related multiple elimination (SRME)[3], and shallow water de-multiple (SWD) focuses on short period multiples [4], where high resolution linear & parabolic radon de-multiple [5] is for long period multiples. De-multiples techniques mentioned before are generally time consuming, expensive, while accurately removing the multiples masking the important events.

b. De-ghosting

In marine seismic, there is always the unavoidable ghost effect from the result of reflections of waves off the sea surface. The energy from a marine source emanates in all directions, such as downwards into the earth, while upward travelling energy is also reflected downwards by the free surface causing a secondary signal. These ghost reflections can either be source ghost, receiver ghost or both, where all three of reflections meet the sea surface before being recorded at the geophone. They pose a threat because of the constructive or destructive interference on primary reflections and in some cases mistaken as primaries.

The process of source and receiver de-ghosting upgrades the conventional data to the appearance of a broadband data. This broadens the bandwidth and improve visibility of impedance contrasts for enhanced imaging. classified There are two possible approaches to de-ghosting in processing; a statistical approach based on the classical minimum phase deconvolution algorithm, and a deterministic approach which entails computation of the analytic inverse of the ghost operator [6]. While statistical deconvolution has remained the workhorse in wavelet processing of land data, it has been eschewed in modern marine processing flows in favour of deterministic techniques.

c. Kirchhoff Migration

Kirchhoff migration is the most flexible migration algorithm and can be implemented in 2D and 3D, pre- and post-stack and as a time or depth migration. The principal parameters for Kirchhoff migration are trace spacing, summation aperture and dip limit [7]. These can all be tested on test sections before production migration is carried out. Some implementations also request frequency limits. It consists of searching the input data in x-t space for energy that would have resulted if a diffracting secondary source were located at a particular point. It is summation along the hyperbolic trajectory that is governed by the velocity function.

3. Re-processing Results

a. Receiver and Source De-ghosting

In conventional processing, receiver de-ghosting is usually applied before any de-multiple processing. The aim is to remove receiver and source side of ghost event (including ghost caused by event and multiples) and to produce data with broader bandwidth and higher resolution using a pre-migration receiver de-ghosting algorithm. Receiver and source de-ghosting was performed in common shot domain to derive the ghost filter using a bootstrap approach. A few iterations were executed using the bootstrap approach and the original data to refine the initial ghost filter to account for time shift, noise and receiver depth uncertainty. Finally the inverse of the resulting ghost filter was applied on the original shot gather to obtain a ghost-free output.

After zero phasing, the ghost reflections are distributed across each wavelet whereby it can be seen on both side lobes of the primary reflection and overlapping the primary reflections itself. Thus after application, the amplitudes are reduced in the side lobes and the primary signature. From the stack sections, the data appears to be much cleaner (Figure 2a and Figure 2b) and notches reduced significantly in the amplitude spectrum. Furthermore, the bandwidth of the processed data is broadened while significant improvement obtained in low frequency section.

b. Shallow Water Demultiple (SWD)

Shallow Water Demultiple (SWD) is used to attenuate the water bottom multiple in shallow water areas where the water bottom reflector is not recorded or is unclear due to interference from the direct arrival. SWD was tested with a low frequency limit applied to the SWD multiple model prior to the adaptive subtraction process to preserve the low frequency content in the data. SWD involves 3 steps; 1) Predict water bottom wavelet: In the shallow water area, the water bottom wavelet is not recorded due to the fact that the near offset is much bigger than water depth, even though the water bottom wavelet can be reconstructed from water bottom related multiple; 2) Construct water-bottom related multiple models by similar method as SRME; 3) Subtract water-bottom related multiple by adaptive subtraction. One quick way to evaluate the effectiveness of SWD is by examining the first bounce water bottom multiple

IOP Conf. Series: Earth and Environmental Science **88** (2017) 012006 doi:10.1088/1755-1315/88/1/012006

removal after application of SWD (Figure 3). In addition, the resultant amplitude spectrum saw a reduction of multiple notches at low frequency while preserving low frequency data content. In order to generate a better multiple model at the upper part of the data, the hybrid SWD method is used. The non-hybrid SWD model usually degrades in upper part. This mainly caused by limited angle of estimated operator. The hybrid SWD using model base approach for upper part gives a better model than SWD.

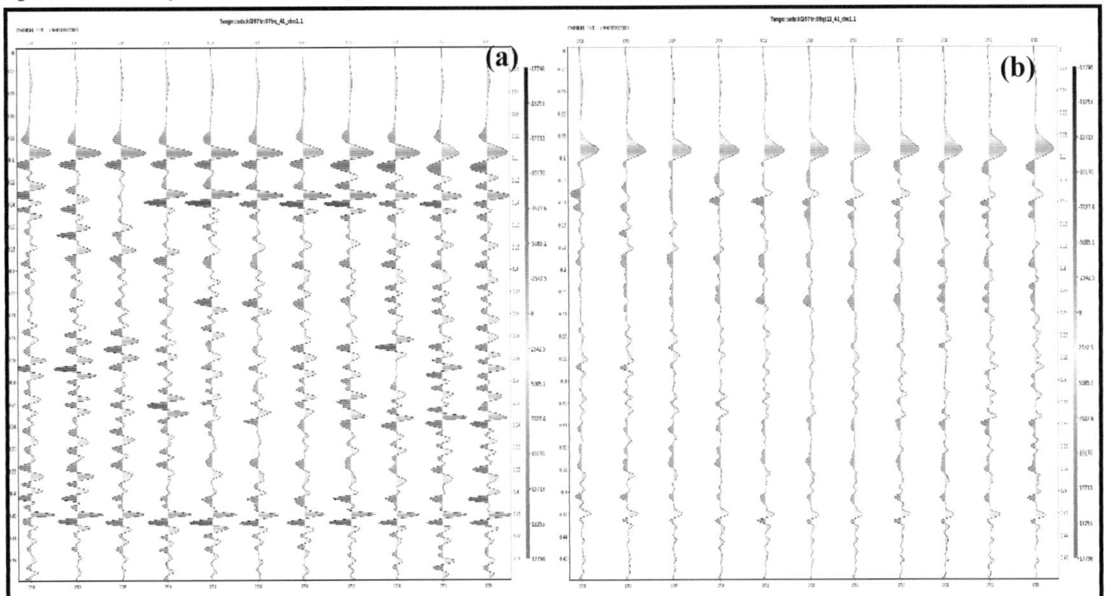

Figure 2: Near trace gather in wiggle display before (a) and after (b) receiver and source de-ghosting.

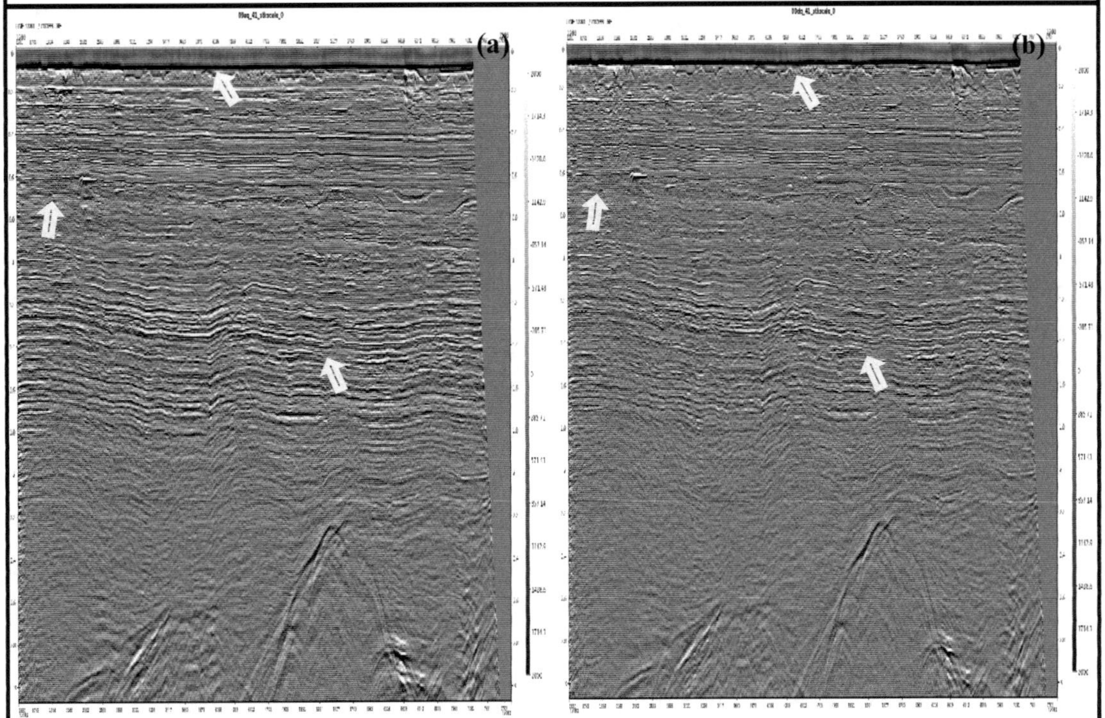

Figure 3: Stack before (a) and after (b) shallow water de-multiple process.

c. High Density Velocity Analysis

After PSTM, the smoothed PSTM velocity field was removed and a high density velocity analysis was conducted using amplitude oriented kinematics program. The program is a robust and continuous velocity analysis program which uses the concept of fully flatten gathers that combine with AVO intercept and gradient will be in phase (Figure 4).

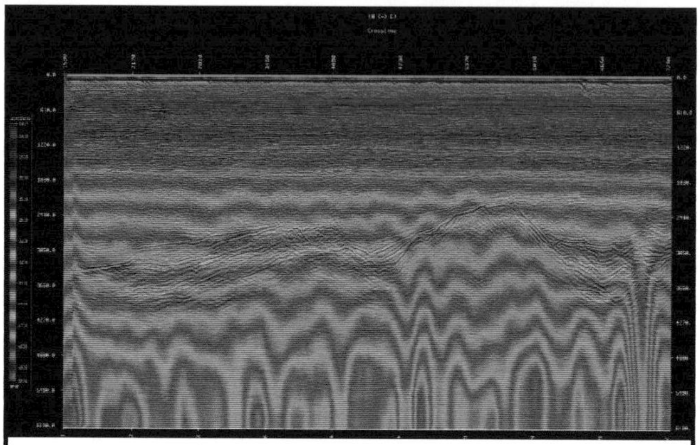

Figure 4: High density velocity field overlaid on stack.

4. Comparison with Vintage Data

Figure 5 represent the selected seismic section from the survey processed from two separate years. The target area we are interested in, is improving at the deeper section (around 750 m – 3km in depth or 1 sec to 4 second in time). The basement area has more structural complexity (e.g. syn-rift half grabens, post-rift structure) as well as it is heavily faulted, which makes it challenging to image accurately. However, there are several improvements after modifying and applying a few added steps to the vintage PSTM workflow.

Firstly, with SWD, source and receiver de-ghosting incorporated into the sequence, the vintage data which was initially severely contaminated with multiples, now appears to be much more geologically accurate. With a significant reduction of surface multiples and ghost reflections, we can be almost certain that the events present are only primary reflections. Thus this allows more confident interpretation. The effective noise and multiples attenuation also allowed for good quality velocity analyses and the subsequent migration.

There is also notable improvement in event continuity and impedance contrast in the re-processed data. The enhancement is credited to the use of high density velocity analysis, and new post-stack processing techniques. The events are more defined and less disrupted across the strata, making it appear cleaner. This in turn highlights the faults, fractures and any sort of discontinuity particularly in the basement region. Overall, the comparison between 2009 vintage data and new re-process image is obvious. It proves that with advanced tools and methods in seismic data processing, a superior imaging result can be acquired.

5. Conclusions

The re-processing work conducted has generally achieved the objectives, delivering a good quality pre-stack time migration dataset, as shown in Figure 5. In previous exploration based on the vintage seismic dataset, the hydrocarbon potential of the basin is considered marginal with no commercial discoveries due to the poor coverage and imaging quality especially at the basement faults and across the syn-rift half-grabens. However, the results discussed above have significantly eliminate structural and stratigraphic uncertainty while retaining the signal to noise ratio of the data. The enhanced resolution, fault definitions and events continuity allow more confident mapping and interpretation to be conducted for further prospect delineation. Essentially the revised processing PSTM work flow has proven that with state-of-the-art tools and methods, appropriate parameterization, innovate approach and careful quality control, we can obtain a geologically accurate results. Therefore, reprocessing seismic data is still a more cost-effective approach to achieve better-quality results.

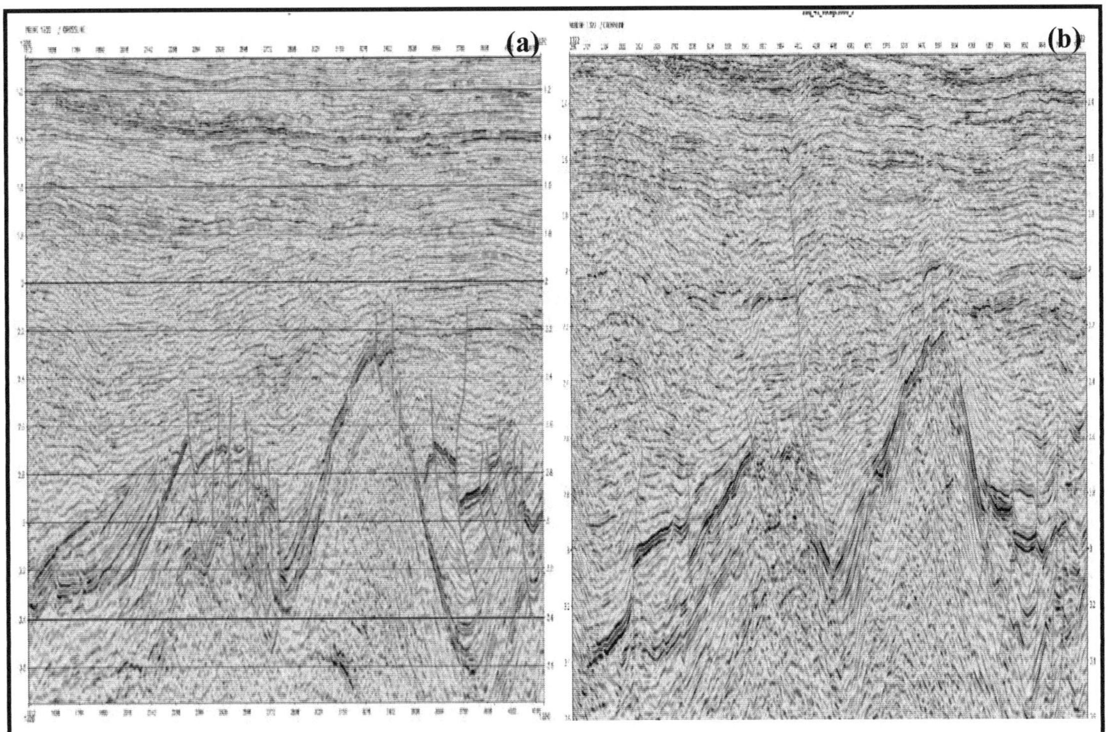

Figure 5: (a) 2009 vintage seismic section and (b) new re-process seismic section.

Acknowledgement

We would also like to express our gratitude to PETRONAS for providing us the seismic data, and Universiti Teknologi PETRONAS for the opportunity to work on this project.

References

[1] Madon, M. & Anuar, A. (1999). The Petroleum Geology and Resources of Malaysia: The Penyu Basin. Petroliam Nasional Berhad (PETRONAS).
[2] Tan, D. N. (2009). Malay and Penyu Basins. In C. S. Hutchison, C. S. Hutchison, & D. N. Tan (Eds.), Geology of Peninsular Malaysia (pp. 175-196). Kuala Lumpur: University of Malaya and Geological Society of Malaysia.
[3] Biersteker J., 2001. MAGIC: Shell's surface multiple attenuation technique: 71st Meeting, SEG, Expanded Abstracts, 1301-1304.
[4] Barnes*, S. R., et al. (2015). "Improved shallow water demultiple with 3D multi-model subtraction." 4460-4464.
[5] Claerbout, J. F., 1976, Fundamentals of geophysical data processing: McGraw-Hill Book Co.
[6] Perz, M.J. and Masoomzadeh, H. (2014). Deterministic marine deghosting: tutorial and recent advances.
[7] Yilmaz, O. (2001). Seismic Data Analysis. Society of Exploration Geophysics.

Importance of Seismic Diffractions for Fractures Imaging

Y. Bashir[1, 2], D. P. Ghosh[1], S. Y. Moussavi Alashloo[1], and C. W. Sum[1]
[1] Centre for Seismic Imaging (CSI), Department of Petroleum Geoscience, Universiti Teknologi PETRONAS, 32610 Seri Iskandar, Perak, Malaysia

E-mail: Yasir.bashir@utp.edu.my

Abstract. The use of diffraction imaging is important and rapidly gaining momentum in the oil and gas industry as the need of the industry moves toward exploiting smaller and more complex structures to find hydrocarbon. Illumination of these small scale discrete transmissivity structures such as faults or fracture zones prior to exploration offers substantial benefits for all phases of field development. We have established a new way of seismic imaging through diffractions study using reflection seismology. A subsurface geological model is developed from one of highly productive field located in the southern part of the Malay Basin. To study the seismic component in reflection seismology, a Finite difference modelling is used to generate synthetic seismic data. Both velocity and density models are used for the explanation of wave propagation, intimating the subsurface from side to side in both one-way and two-wave wave propagation. One–way wave equation migration methods are tested to image the faults in the synthetic seismic section of the Malay basin and we demonstrated the need of diffraction imaging for highly dipping faults and complex structure.

1. Introduction

The Malay Basin is a mature Tertiary extensional basin with a later inversion regime in the late Miocene. The overall geology is a simple "layer cake" seismically, with compressive anticlinal inversion structures [1]. Extreme tectonic activity is a cause of a compressional structure like complex faulting with steep dips and over-thrust.

The spectral practices of migration are well known to have a Wavefield extrapolation achieved in the Fourier domain (frequency and wave number)[2]. Seismic data is recorded in the space-time domain (x-t) and transformed into the Fourier domain *(k-w)* are undergoing converted namesake. The algorithms which belong to the categories of fundamental migration are Stolt [3], Phase Shift [4], Split Step [5], Phase Shift Plus Interpolation -PSPI [6], and the Fourier Finite Difference –FFD [7]. We describe the migration in the form of the wave equation in figure 1 that is an enlightenment of all migration algorithms which are being rummage-sale by the processors in all over the world.

[2] Yasir Bashir (Yasir.bashir@live.com)

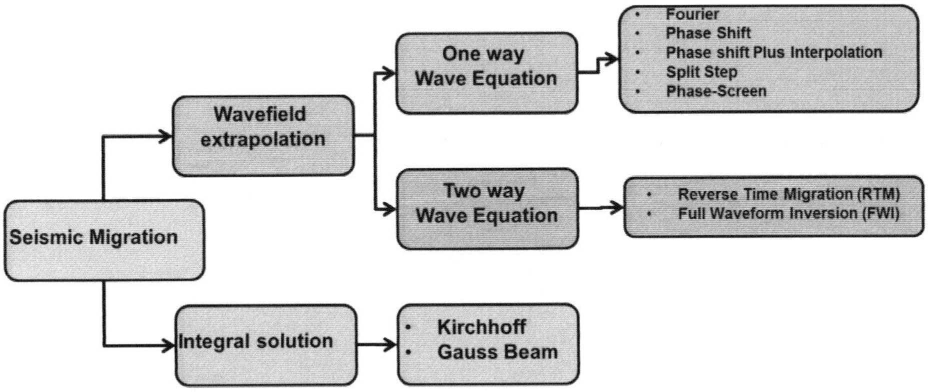

Figure 1: Classification of Seismic migration on the basis of wave equation techniques, from simple to advanced methods [8].

These methods mentioned in Figure 1 are reasonably vary from low computational power to high computational power, and implement the Wavefield extrapolation based on the derivative of the Fourier transformation and properties of the translation in time[2]. However, every algorithm has some inherent inadequacies by way of its own, which prevent the imaging of small-scale and high-sloped structures, whose reflections have trajectories with reflections on more than one interface of the seismic data.

2. Methodology

2.1. Geological Model

The model presented in this paper is taken from one of the fields in the Malay Basin. The model comprises four superimposed layers of different velocities, with major faulting because of uplifting of the sub layers. The synthetic seismic data of this model was obtained from a well-known seismic modeling technique as given below.

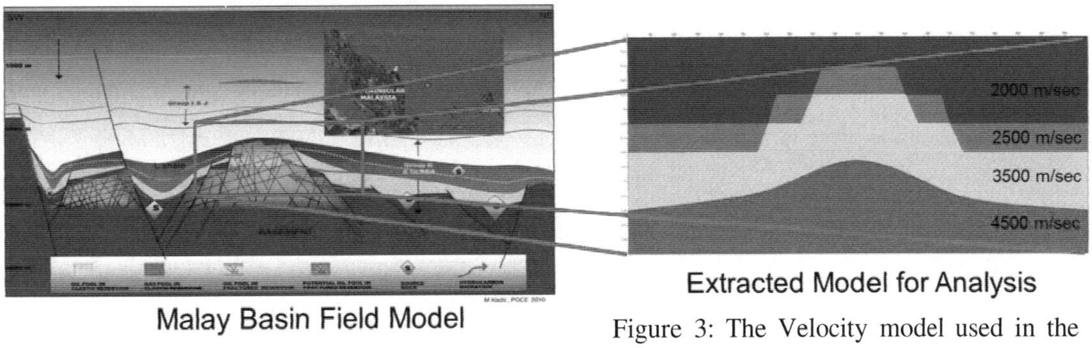

Figure 2: A geological model from the Malay Basin [12].

Figure 3: The Velocity model used in the study, representing number of highly dipping faults and an anticline bellow[11].

The finite difference methods (FDM) have been widely used in seismic modeling and migration. One of the disadvantages of FDMs is inadequate computational cost. The FD approximations give a numerical solution of a partial differential equation. The precision of this solution is reliant on the order of approximation. Eventually, all the finite difference methods are based on Taylor Series approximation [9]. Therefore, a discrete version of wave equation that includes the Initial situation or preliminary point, where the wave field to starts propagate, is required.

IOP Conf. Series: Earth and Environmental Science **88** (2017) 012007 doi:10.1088/1755-1315/88/1/012007

The modeling was conducted in a grid of 10x2 m (dx=10, dz=2) and with Ricker wavelet, which is defined by its dominant frequency of 50 Hz. These constraints were respectable enough to avoid numerical dispersion and unpredictability in finite difference modeling. The modeling was performed by ignoring multiples and absorption edge in the physical limit of the model.

The modeling was performed using the same model with different velocity and density parameters.

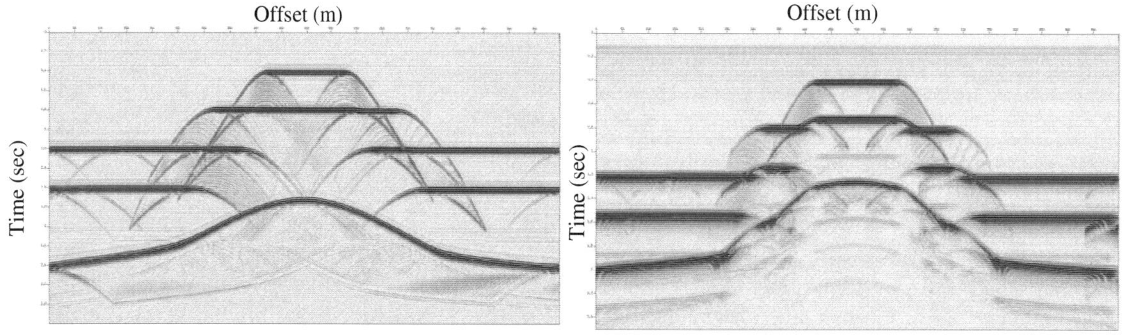

Figure 4: Zero-offset synthetic seismic section using finite difference modelling, showing alterations in parameters of velocity (left) and a constant velocity (right).

3. Discussions and obtained Results

The extrapolation of the Wavefield in the frequency domain is computationally highly efficient and numerically stable, which is contributing to make the use of spectral techniques attractive in the migration of seismic data[2]. However, ignoring the down-going Wavefield, the most important limitation in the imaging of high dipping structures are imposed to these algorithms. The result obtained with the imaging of data from the above geological faulted model clearly demonstrates the limitations of conventional spectral techniques imaging and the competency of two-way wave migration.
The comparison of dissimilar wave equation migration techniques reveals data with similar structural imaging; however, these techniques consider that using the same processing capability and faster implementation provides optimum imaging results.

One-way-FK-Reverse-Time migration can be distributed into FK, Phase shift methods, and phase shift plus interpolation method.

We started the solution for the scaler wave equation for the zero-offset Wavefield. If the medium velocity is constant, migration can be expressed as direct mapping form temporal frequency *w* to vertical wavenumber *kz*. In mid-1970's, R. H. Stolt invented what has become known as Stolt or FK migration.

$$k_z = \mp \sqrt{\frac{w^2}{v^2} - k_x^2} \quad (1)$$

$$P(k_x, k_z, t = 0) = \left[\frac{v}{2} \frac{k_z}{\sqrt{k_x^2 + k_z^2}}\right] P\left[k_x, 0, \omega = \frac{v}{2}\sqrt{k_y^2 + k_z^2}\right] \quad (2)$$

where P (k_x, z=0, ω) is zero-offset section and P (k_x, kz, t=0) is the migrated section in the frequency-wavenumber domain.

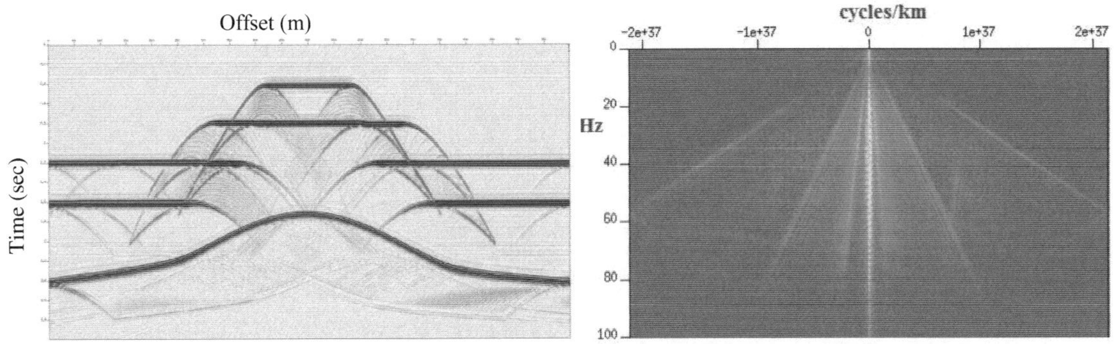

Figure 5: A zero-offset section that contains a set of faults and reflectors (left), F-k spectrums of un-migrated data, showing high amplitude in the central part and low amplitudes away from the sides (right).

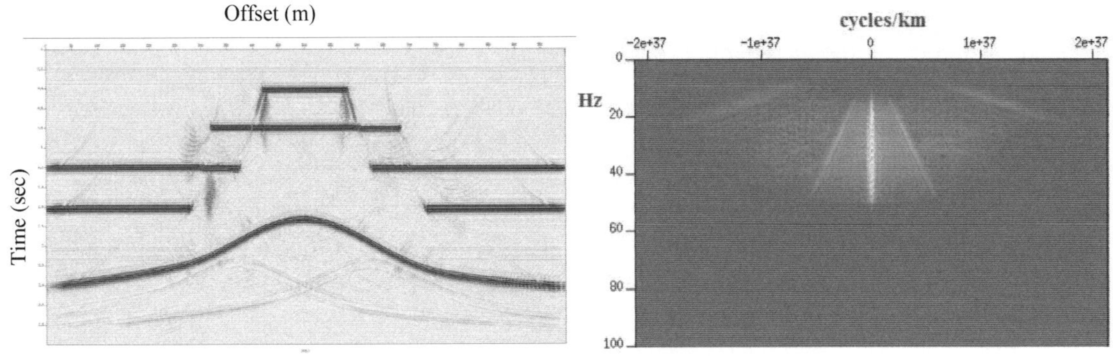

Figure 6: migrated data using Stolt's constant velocity algorithm (left) and f-k spectrum of the migrated data which showing low amplitude distributed on the true position of reflection (right).

Phase shift method was based purely on the reverse-time modeling approach. In this technique, the approach of constant velocity criteria was successfully removed, as which was imposed by the Stolt method. It was also the earliest and, conceivably until very recently, the only method that was capable of imaging steeply dipping reflectors. The phase shift approach was developed by Jeno Gazdag shortly after the Stolt method appeared [4]. The starting PDE is the Helmholtz equation given by:

$$\nabla^2 P(x.z.\omega) + \frac{\omega^2}{v^2 P(x,z,\omega)} = 0 \qquad (3)$$

Since the starting exploding reflection is at t=0, the location of the reflector is therefore given by inverse transforming the solution in k_x and w and evaluating at t=0:

$$r(x,z) \approx p(x,z,t=0) = \sum_{k_x}\sum_{\omega} P^{\sim}(k_z,0,\omega)e^{-i(k_x x + k_z z)} \qquad (4)$$

The above condition of r(x, z) =p(x, z, t=0) is known as the zero-offset migration imaging condition. Phase Shift Plus Interpolation –PSPI method is a phase-shift like method for dealing with strong lateral velocity variation [6]. The initial impression of PSPI is to familiarize several reference velocities in order to describe the lateral velocity in each extrapolation step, and obtain the multi-reference Wavefield in the f-k domain. Based on the relationship of the local velocity and reference velocity, the final migration results are obtained by interpolating the reference Wavefield in the f-k domain. The basic formulas are:

$$P_0(x, y, z, \omega) = P(x, y, z, \omega).e^{i\frac{\omega}{v(x,y,z)}dz} \quad (5)$$

and

$$P'\left(k_x, k_y, z + dz, \omega\right) = P_0\left(k_x, k_y, z, \omega\right)e^{i\left(k'_z - \frac{\omega}{v'_{ref}}\right)} \quad (6)$$

where k'_z is obtained using the reference velocity. After the reference wavefield is Fourier-transformed back to the frequency-space domain, the final migration result is obtained by linear interpolation[6].

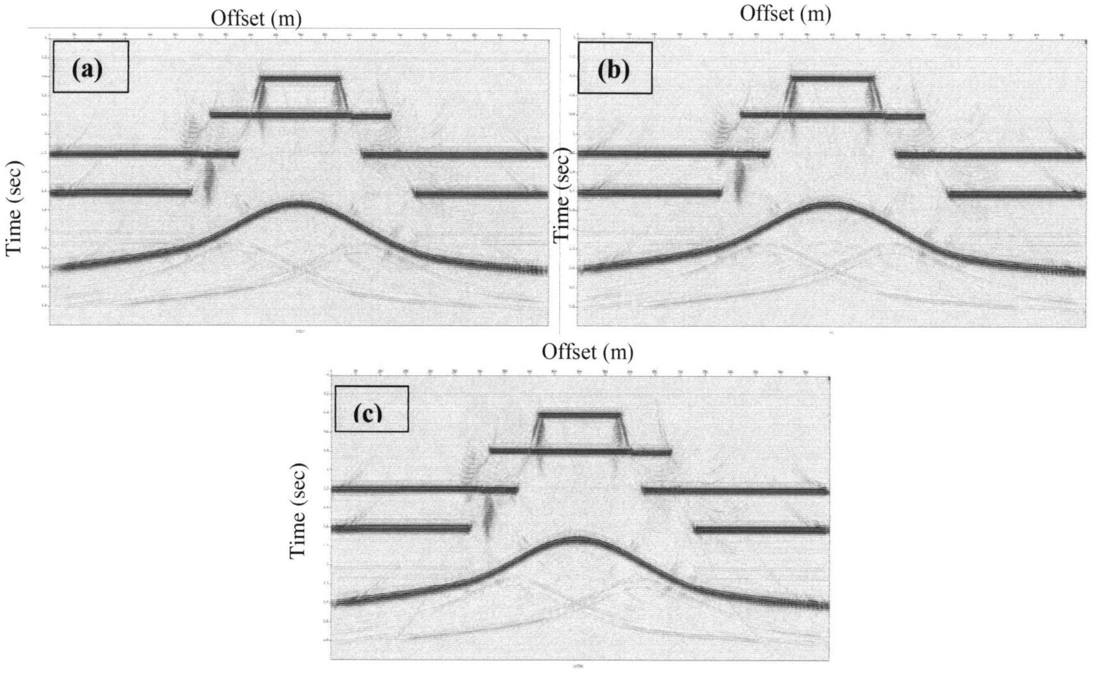

Figure 7: Comparison of migration algorithm operators. (a) Gazdag (b) Phase Shift and (c) Stolt migration.

Figure 7 shows the recovery of all the seismic reflection properly and illuminates the horizontal reflections, but none of these migration algorithms are capable of compensating for all the component of faults as shown above. For these types of structure we need diffraction imaging by to separate diffracted and reflected waves and image it separately [10]. Figure 8 shows the separated diffraction using plane-wave destruction filter, which calculates the slope of the data on the basis of reflection and then separates the diffracted event.

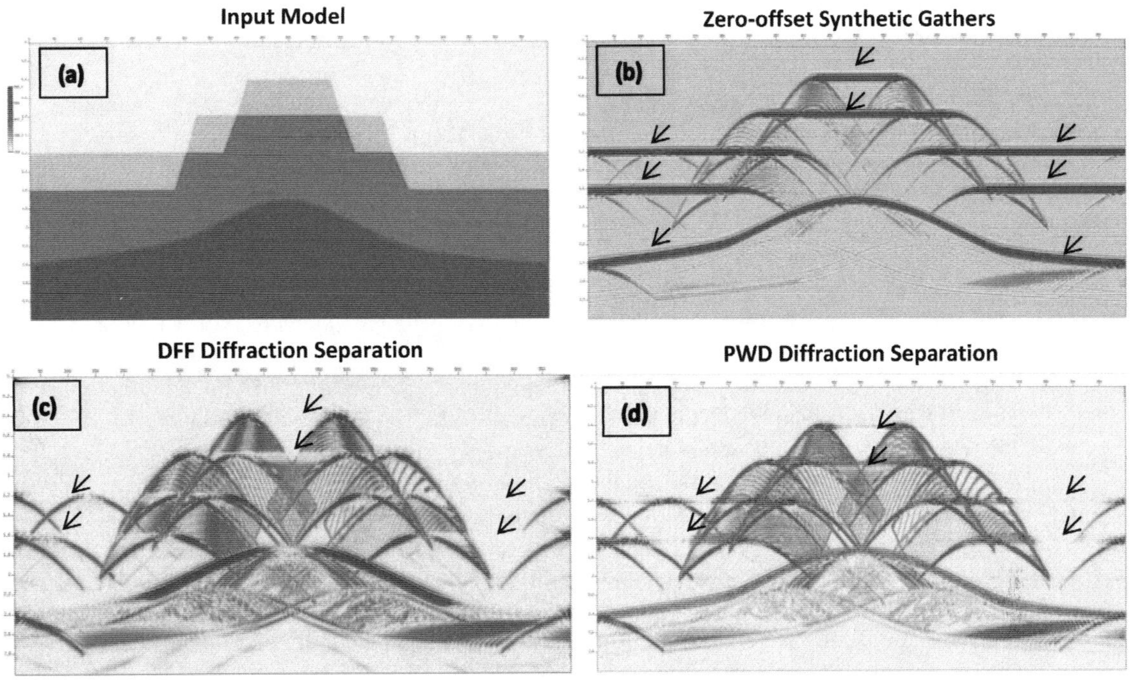

Figure 8: Diffraction separation process (a) input model; (b) zero-offset synthetic shot gather data; (c) diffraction separation using dip frequency filtering; and (d) separated diffraction using plane-wave destruction filtering.

These highly dipping faults have an illumination problem in the reflection data imaging. Figure 8(b) is the zero-offset section of the model data after finite difference modelling. Diffractions are caused by the irregular fault boundaries and preserved in the section. Figure 8 (c) shows separated diffractions using DFF and Figure 8(d) is the separated diffractions section using PWD, which shows better results for the diffraction separation. The arrows clarify the reflections which are suppressed from the given data. Both filtering techniques are able to eliminate upper straight reflections, but if we look at the flanks below of the anticline, our DFF is eliminating those reflections from the data which do not originally belong to the diffraction data. On the other hand, the PWD filtering technique is sophisticated enough to remove the reflection seismic data and preserve the diffractions.

4. Conclusions

Diffractions give us valuable information about subsurface structures; the objective of this paper is to show that, when we migrate our data, most of the diffracted events cannot collapse correctly. Consequently, we lose information from the diffracted waves. In order to prove the importance of diffractions, dissimilar wave equation techniques are applied to the synthetic data for imaging faults. Application of one-way wave equation migration using F-K, PS, and Gazdaz migration is used to image the data but none of them were able to image the dipping components of fault. Therefore, the best way to image these events is to preserve these diffractions by separating the reflection and diffraction. We have proposed two methods for diffraction separation: Dip Frequency Filtering (DFF) and Plane-wave Destruction filtering (PWD). The possibilities of using these methods are dependent on the complexity of the subsurface structure. The results demonstrate the effectiveness of the desired objective to preserve the diffraction for Imaging. Future work can be extended to apply these methods on real seismic data, especially carbonate fields for Karsts imaging.

Acknowledgments

We are thankful to Universiti Teknologi PETRONAS (UTP) and the Centre for Seismic Imaging (CSI) to provide the facility for this research. Authors are thankful to UTP & PETRONAS staff for their suggestions & contribution. In this work, I have used Seismic Unix and Madagascar.

Reference

[1] D. Ghosh, M. Sajid, N. A. Ibrahim, and B. Viratno, "Seismic attributes add a new dimension to prospect evaluation and geomorphology offshore Malaysia," *Lead. Edge*, vol. 33, no. 5, pp. 536–545, 2014.
[2] A. R. L. Santos, R. da Cruz Pestana, and G. C. Aldunate, "SEISMIC IMAGING OF HIGH SLOPE STRUCTURES USING ONE-WAY WAVE EQUATION MIGRATION TECHNIQUES," *Rev. Bras. Geofísica*, vol. 33, no. 1, pp. 71–78, 2015.
[3] R. H. Stolt, "Migration by Fourier transform," *Geophysic*, vol. 43, pp. 23–48, 1978.
[4] J. Gazdag, "Wave equation migration with the phase-shift method:," *Geophysic*, vol. 43, pp. 1342–1351, 1978.
[5] R. M. L. Freire, "Migração por mudança de fase em duas etapas," *Migração por mudança fase em duas etapas*, 1988.
[6] J. Gazdag and P. Sguazzero, "Migration of seismic data by phase shift plus interpolation," *Geophysics*, vol. 49, no. 2, pp. 124–131, 1984.
[7] D. Ristow and T. Rühl, "Fourier finite-difference migration," *Geophysics*, vol. 59, no. 12, pp. 1882–1893, 1994.
[8] S. H. Gray, Etgen, John;, Dellinger, Joe;, and D. Whitmore, "Seismic migration problems and solutions," *Geophysic*, vol. 66, no. 5, p. 1622.
[9] J. A. Scales, *Theory of seismic imaging*, vol. 2. Springer-Verlag Berlin, 1995.
[10] Y. Bashir, D. Ghosh, and C. Sum, "Preservation of seismic diffraction to enhance the resolution of seismic data," in *SEG Technical Program Expanded Abstracts 2017*, Society of Exploration Geophysicists, 2017, pp. 1038–1043.
[11] B. Yasir, D. P. Ghosh, S. Y. Moussavi Alashloo, and C. W. Sum, "Seismic Modeling and Imaging Using Wave Theory for Fault and Fracture Identification," in *Offshore Technology Conference Asia*, 2016.
[12] M. B. Kadir, "Fractured Basement Exploration Case Study in Malay Basin," in *PGCE 2010*, 2010.

IOP Conf. Series: Earth and Environmental Science **88** (2017) 012008 doi:10.1088/1755-1315/88/1/012008

Structural Analysis: Folds Classification of metasedimentary rock in the Peninsular Malaysia

A Shamsuddin[1]
PETRONAS University of Technology (Universiti Teknologi Petronas – UTP), 31750, Tronoh, Perak, Malaysia[1]
E-mail: anassash@gmail.com
Abstract. Understanding shear zone characteristics of deformation are a crucial part in the oil and gas industry as it might increase the knowledge of the fracture characteristics and lead to the prediction of the location of fracture zones or fracture swarms. This zone might give high influence on reservoir performance. There are four general types of shear zones which are brittle, ductile, semibrittle and brittle-ductile transition zones. The objective of this study is to study and observe the structural geometry of the shear zones and its implication as there is a lack of understanding, especially in the subsurface area because of the limitation of seismic resolution. A field study was conducted on the metasedimentary rocks (shear zone) which are exposed along the coastal part of the Peninsular Malaysia as this type of rock resembles the types of rock in the subsurface. The analysis in this area shows three main types of rock which are non-foliated metaquartzite and foliated rock which can be divided into slate and phyllite. Two different fold classification can be determined in this study. Layer 1 with phyllite as the main type of rock can be classified in class 1C and layer 2 with slate as the main type of rock can be classified in class 1A. This study will benefit in predicting the characteristics of the fracture and fracture zones.

1. Introduction

Metasedimentary rocks are highly correlated with the temperature and also pressure which might be caused by regional metamorphism or contact metamorphism. These two different conditions influence the rock properties. Different types of rocks with different properties will form because of this relation and can be related to the shear zones. According to Haakon F (2010), a shear zone is a tabular zone with strain rate is higher compared to the surrounding rocks. Shear zones can be divided into four zones which are brittle, ductile, semibrittle and brittle-ductile zones (George H D et. al 2012). The characteristics of shear zones allow us to divide them into these types of zone. Figure 1 shows a simple model that might help in understanding the metamorphic facies together with its characteristics might help in the study that might influence the fracture.

Different properties of the rock were highlighted because they might influence rock characteristic especially structures. There is a lack of understanding of the rock characteristic especially in the subsurface as the resolution of the seismic is limited. The limited resolution of the seismic gives limitation in understanding the rock characteristic in the subsurface. Thus, an outcrop study was conducted in order to enhance the understanding of the rock characteristic on the surface and to improve the understanding of the rock characteristic in the subsurface. The similarity of the rock types from the outcrop-surface and in the subsurface were highlighted by Ngoc H N (n.d.) and Suhaileen S (2005). Based on this study, the possibilities to locate the fracture zone, which might be an element that can increase reservoir performance can be better understood.

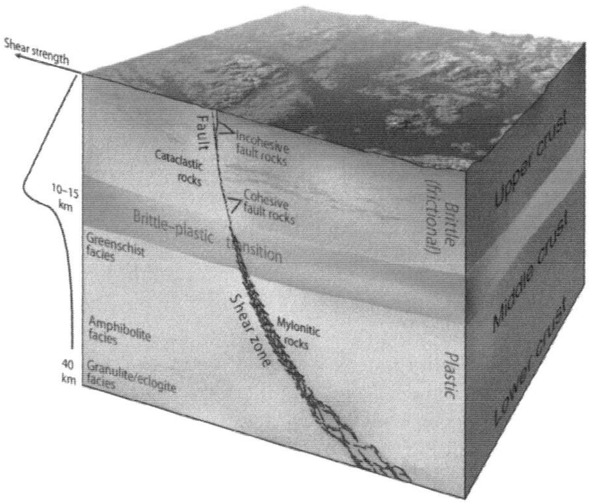

Figure 1: Simple model for shear zones that were illustrated by Haakon F 2010

2. Geological setting of the study area

A field study was conducted on the outcrops that are well exposed along the coastal part of the Peninsular Malaysia. It is located in the eastern belt of the Peninsular Malaysia. The study area might be related to the fault zones which are Lepar Fault zone, Mersing Fault zone and Balau Murau Fault zone. Based on the previous studies, this study area is filled with the distribution of phyllite, slate, shale and sandstone, argillaceous carbonaceous with local limestone as well as felsic to intermediate volcanic. The study was conducted on the metasedimentary rock as this type rock shows high similarity with type of rocks in the subsurface.

3. Methodology

A few structural analysis was conducted in order to understand the characteristic of the structure in the subsurface. Other types of rock which is granite was analysed by Annur S (2013). In this paper, a structural analysis for metasediment was conducted in order to understand the structural characteristic especially on geometry of the rock. The study was conducted based on the structural characteristics (fold, faults and fracture) that were exposed in the study area as the weathering affects the exposure of the rock. Strike and dip orientations of rocks were collected. Fold geometry was studied based on the fold classifications layer thickness variations (Figure 2.0) by Ramsay (1967) and also by using the combination of the dip of the axial plane and the plunge of hinge line (Figure 3.0)

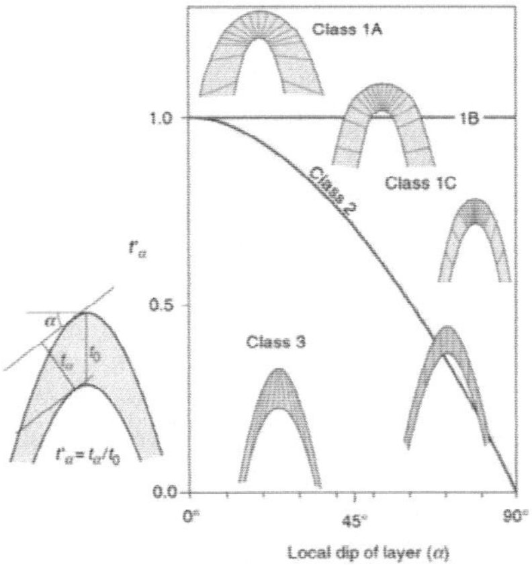

Figure 2: Fold classifications layer thickness variations based on Ramsay, 1967 from (Haakon F 2010)

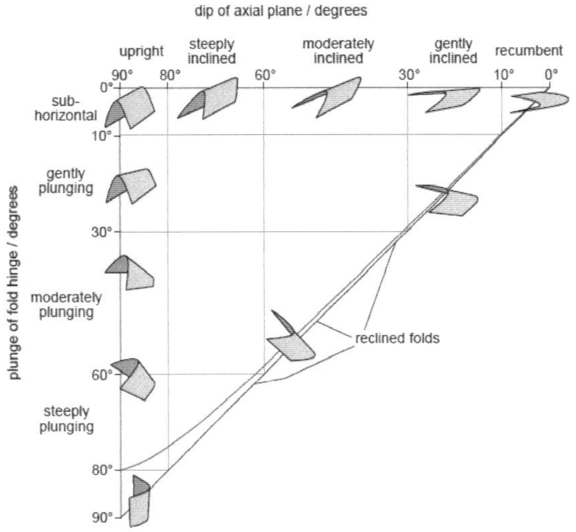

Figure 3: Fold type classifications using the dip and the axial plane and the plunge of hinge line based on Ramsay, 1967 from (Angela L C 2010)

4. Metasedimentary rocks

Three main types of metasediment which are metaquartzite, slate and phyllite were determined in the study area. The characteristics of the rock types can be determined as in the following table (Table 1). Figure 4 and figure 6 show the type of rocks in the study area.

No	Rocks	Characteristics
1	Metaquarzite Colour: white to beige	1) Type of rock: Non foliated metamorphic rock 2) Others: Mineral - Quartz grain (Size: Medium to coarse)
2	Slate Colour: Black	1) Type of rock: Foliated metamorphic rock - Parent rock: shale, mudstone or siltstone 2) Grain size: Very fine grain 3) Others: Slaty cleavage with quartz veins in the joint
3	Phyllite Colour: Dark gray	1) Type of rock: Foliated metamorphic rock - Parent rock: slate 2) Others: Quartz veins in the joint

Table 1: Metasediment rock types (metaquarzite, slate and phyllite) based on colour, type of rock and other characteristic that can be observed in the study area

Figure 4: Different types of metasediment rocks with non-foliated, white to beige colour metaquartzite and foliated, gray phyllite can be observed in this location

5. Structural Geometry

Structures that are exposed in this field can be classified into a few types of geometry. The highlighted structure in this field area is fold with structural development that might relate to the formation of the fracture or can be known as fold related fracture. Small scale structures were determined in a fold and the analysis was divided into layer 1 and layer 2 which can be determined in figure 5.0. Based on the analysis, the plunging of the fold hinge about more than 30° oriented from South to North. Based on the classification in figure 3, which was classified by Ramsay 1967, the fold in this location can be classified as moderately plunging.

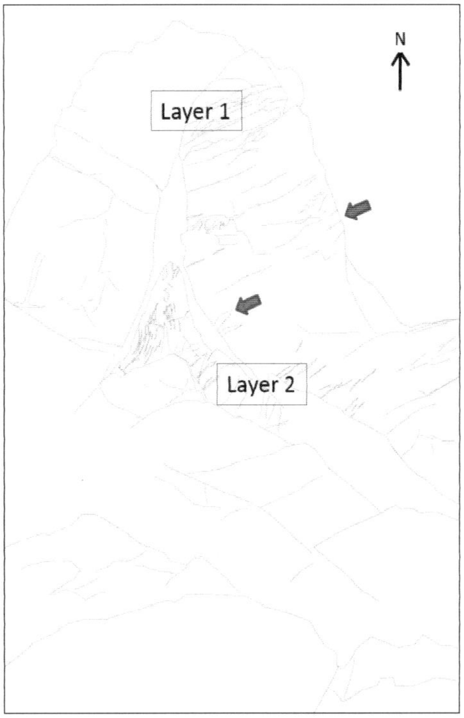

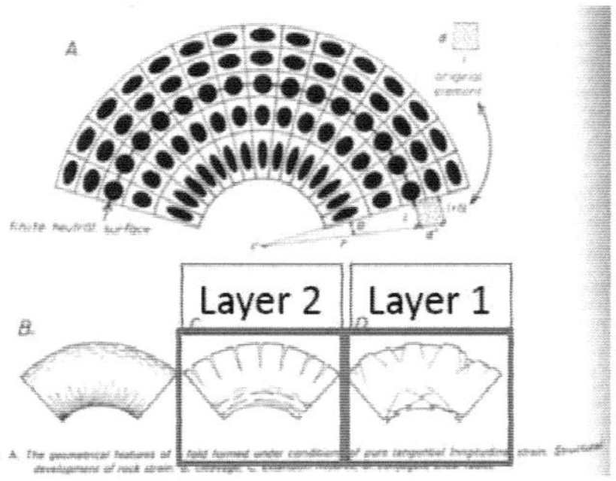

Figure 5: Left figure shows two layers which were divided into layer 1 and layer 2 for this study area. The characteristic of the layer can be compared with the right figure which based on strain and small scale structures in folds based on Ramsay (1967).

IOP Conf. Series: Earth and Environmental Science **88** (2017) 012008 doi:10.1088/1755-1315/88/1/012008

Based on the fold classification layer thickness variations by Ramsay (1967) in figure 2, the fold in this area were analysed. Ramsay (1967) classified the fold based on t'α=tα/t₀ as in figure 6 left side (bottom). In this case, the calculation is t'= tα/ t₀ with t' is similar to t'α in Ramsay (1967). Table in figure 6 shows detail analysis of the fold. Based on the analysis, the fold in the study area can be classified as Class 1A for Layer 2 and Class 1C for Layer 1 (Figure 6.0: Fold classifications layer thickness variations). Other interesting features in this fold is tension gashes as well as pinch and swell that can be defined in the second layer.

Layer 1

α	tα	t'
10	40	0.97
20	40	0.97
30	36	0.87
40	30	0.74
50	22	0.53
60	21	0.50
70	15	0.34
80	22	0.53

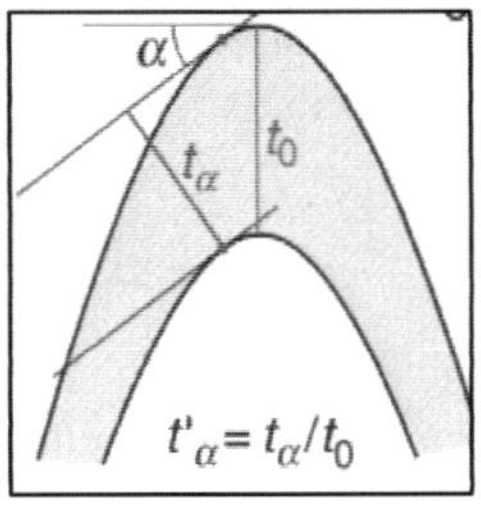

Layer 2

α	tα	t'
10	93	1.00
20	90	0.97
30	88	0.95
40	87	0.94
50	90	0.97
60	87	0.94
70	99	1.06
80	107	1.15

Figure 6: The analysis of fold in the study area (middle figure) was classified by using layer thickness variations based classification from Ramsay (1967). Two layers for fold were classified as in the figure (left-bottom). Layer 1- phyllite can be classified as Class 1C and Layer 2 - slate as Class 1A.

6. Discussion

The main structural that can be observed in this study area is fold which can be classified as Class 1C and Class 1A. This classification is based on the layer as there are two main types of rock which is slate (Layer 2) and phyllite (Layer 1). Based on the analysis, different types of rocks show different classification of fold geometry. The analysis of this study is crucial as it can enhance the understanding of the rock as different types of rocks with different properties will form different. Understanding different types of the rock with different properties is important because the analysis in this study shows different characteristics of the rock can be determined. The formation of the fracture in the layering of the fold can be the evidence of the relation in between fold and fracture. The characteristic of structural analysis especially fold in the outcrop is highlighted as it can enhance the understanding of the relation of fold with the fracture which cannot be determined in the seismic data which is low in resolution.

7. Conclusion

Fractures are the major paths as it gives some effects on increasing or decreasing the reservoir performance. Based on the structural identifications on this research, factors that might be related to the fracture zones can be determined based on the rock characteristic that are highly correlated with the temperature and also pressure. Fractures that were observed can be fracture related faults and also fracture related fold. In this study, based on the classification from Ramsay (1967), the fold classification can be divided into two classes which are layer 1 with phyllite in Class 1C and layer 2 with slate as the main rock in class 1A. The fracture between the layering of the rock can be observed and it shows the relation in between fold and fracture. This study might help in predicting the characteristics of the fracture and fracture zones. Currently, I am still working on predicting the fractures especially in the subsurface which is still in progress.

8. Acknowledgement

I would like to thank to Suhaileen Shahar for the guidance and Centre of Excellence in Subsurface Seismic Imaging & Hydrocarbon Prediction (CSI), PETRONAS University of Technology for research allocation.

9. References

[1] George H D, Stephen J R and Charles F K, 2012. *Structural Geology of Rocks and Regions*
[2] Ngoc H N, n.d. Seismic attributes for study Pre-Tertiary fractured basement in Malay Basin
[3] Suhaileen S, 2005 The prospectivity of Fractured Basement Play of the Malay Basin
[4] Haakon F, 2010. *Structural Geology*
[5] Annur S & Deva G 2013 Fracture Analysis of basement rock: A case example of the Eastern Part of the Peninsular Malaysia
[6] Ramsay J G, 1967. *Folding and fracturing the rocks*
[7] Angela L C, 2010. *Geological Field Techniques*

STUDY OF ROCK PHYSICS AND SEISMIC ATTRIBUTES OF HYDROCARBON RESERVOIRS IN SABAH BASIN

Adelynna Shirley anak Penguang[1], Luluan Almanna Lubis[1,2], Maman Hermana[2], Prof. Dr. Deva Prasad Ghosh[2]

[1]Petroleum Geoscience Department,
[2]Centre of Excellence in Subsurface Seismic Imaging & Hydrocarbon Prediction,
Universiti Teknologi PETRONAS,
Perak, Malaysia.

Email: adelynna23@gmail.com

Abstract. Elastic moduli are known to discriminate fluid and lithology for effective reservoir characterization. However, in some cases, the elastic modulus can also show the same response for different type of fluid and lithology. In our study area, a few elastic moduli were used to evaluate the hydrocarbon sands, water sand and shale as identified based on well log response. Seismic attributes such as Relative Acoustic Impedance attribute and RMS amplitude attribute are also applied to confirm the well log analysis and investigate the seismic response. Based on the results, the effective elastic moduli for our study area is the Lambda-Mu-Rho, Velocity Ratio (Vp/Vs), Poisson's ratio and Scaled Inverse Quality (Q) Ratio. The seismic attributes such as the Relative Acoustic Impedance highlights the impedance difference between the formations while RMS amplitude shows the hydrocarbon presence. The workflow of this study is applicable for other areas for effective fluid and lithology delineation.

1. Introduction

Acoustic impedance (AI), which is the the product of velocity and density [1], is affected by different lithology type. Seismic inversion of acoustic impedance is usually done to get the 3-D elastic modulus volume that can be used for interpretation of a hydrocarbon field. High acoustic impedance interval is interpreted as shale and the low acoustic impedance as hydrocarbon-bearing sand in most clastic reservoirs. The acoustic impedances' differences are also used as an expression for the reflection coefficient to make the synthetic seismograms that can be seen on the seismic cross-section [1]. Other elastic modulus such as velocity ratio and Lambda-Mu-Rho are also commonly used to distinguish different fluid and lithology. Elastic properties therefore are very important in ensuring an effective interpretation of reservoir.

2. Study Area and Objectives

Our study area has two wells in the Sabah basin which are Well UTP-1 and Well UTP-2. Well UTP-1 has oil, gas and water sand while Well UTP-2 has gas sand. The Sabah basin is known to have soft shale which complicates the seismic interpretation as they have the same seismic amplitude response [2]. Therefore, this paper aims to compare the response of these two wells and determine the best parameter used for 1) well that penetrates all hydrocarbon and water sand, and 2) well that penetrates gas sand only with possibility of soft shale.

3. Methodology

Figure 1 shows the whole workflow of this study. Literature review was first conducted to understand the background of the elastic modulus and seismic attributes followed by Petrel and Rockdoc software familiarization. The well log and seismic data were loaded into Petrel and Rokdoc software. Petrophysical analysis were then carried out using Rokdoc software on the well log data such as generating volume of shale, porosity, and water saturation. Elastic modulus such as acoustic impedance (AI), velocity ratio (Vp/Vs), Lambda-Mu-Rho (LMR), and scaled inverse quality (Q) ratio were also generated during well log analysis. Based on the well log response and the elastic modulus, the hydrocarbon sands and shale cap rock intervals were identified and cross-plotted.

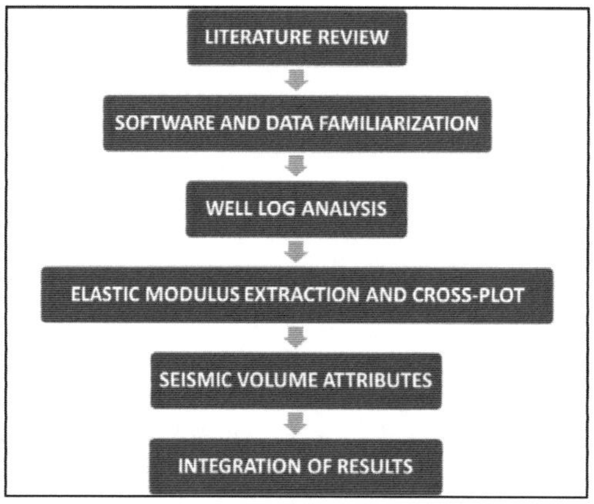

Figure 1. Workflow of Study

The wells were then correlated with the seismic volume of the area using Petrel software. Seismic volume attributes such as Relative Acoustic Impedance and RMS Amplitude attributes were applied to the seismic. Relative Acoustic Impedance attribute is chosen to evaluate the impedance of the area while RMS Amplitude attribute highlights the hydrocarbon anomaly at the well location and further from the well. Results were interpreted, compared and discussed for both wells.

4. Results and Discussion

Figure 2 shows the interpretation of Well UTP-1 and Well UTP-2. Based on gamma ray, resistivity and neutron-density well log response, the hydrocarbon, water and shale intervals were identified and then cross-plotted. The thickness of the investigated intervals are shown in Figure 2.

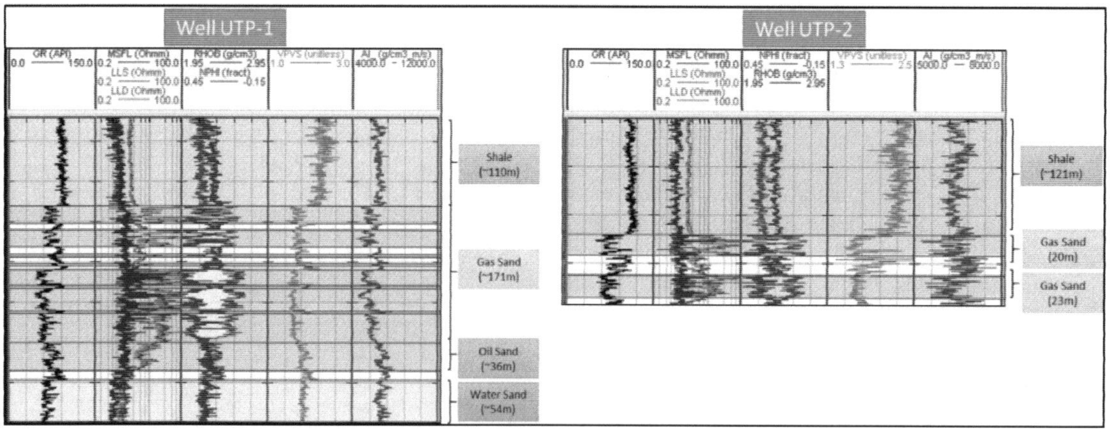

Figure 2. Interpretation of Shale and Gas Sand of Well UTP-1 and Well UTP-2

4.1. Elastic Properties Cross-plot Analysis

4.1.1. Acoustic Impedance (AI) vs. Velocity Ratio (Vp/Vs) cross-plot [3]. Based on the acoustic impedance vs. velocity ratio (Vp/Vs) cross-plot for Well UTP-1 in Figure 3a, the acoustic impedance of the shale and the reservoir rocks are slightly similar. The acoustic impedance values between the oil sand and the water sand overlap. The acoustic impedance value of the gas sand and the shale in Figure 3b which is Well UTP-2 is almost similar which attribute to the case of soft shale. The AI vs. Vp/Vs cross-plot thus is not efficient in separating the fluid and lithology as the acoustic impedance value for the soft shale overlaps with the acoustic impedance value of hydrocarbon sands.

4.1.2. Lambda-Rho vs. Mu-Rho cross-plot [4]. Based on the Lambda-Rho vs. Mu-Rho cross-plot in Figure 4, the Lambda-Rho values for gas sand in both wells are the lowest while the shale has the highest Lambda-Rho values. This is because the gas sand is more compressible than shale. The oil sand and water sand in Figure 4a shows that the oil sand and water sand for Well UTP-1 overlaps. Thus, in our study area, the Lambda-Rho and Mu-Rho parameters are not advisable to use for wells with oil and water sand but more effective for well with gas sand reservoir only.

4.1.3. Velocity Ratio (Vp/Vs) vs. Poisson's ratio cross-plot [5]. Based on the velocity ratio (Vp/Vs) vs. Poisson's ratio cross-plot in Figure 5a, the gas sand has the lowest values of velocity ratio (Vp/Vs) and Poisson's ratio values, followed by the oil, water sand and the shale.

4.1.4. Scaled Inverse Qs (SQs) vs. Scaled Inverse Qp (SQp) [6]. Based on the scaled inverse Qs (SQs) vs. scaled inverse Qp (SQp) cross-plot in Figure 6, the SQs values of the gas sand and the shale for both wells are different. The gas sand has the lowest SQp values. The values of SQs values for the oil and water sand can be seen overlapping. However, the SQp can discriminate the oil and water sand for Well UTP-1.

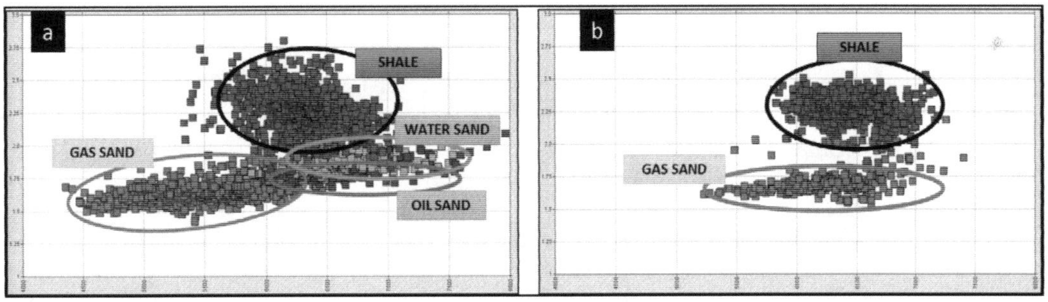

Figure 3. a) AI vs. Vp/Vs cross-plot for Well UTP-1 and b) AI vs. Vp/Vs cross-plot for Well UTP-2

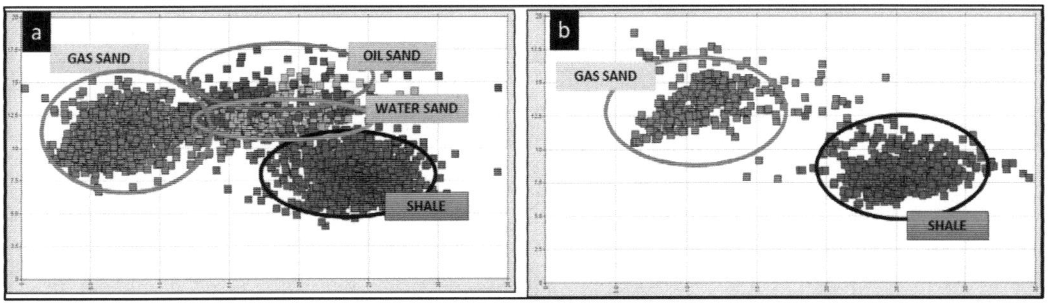

Figure 4. a) Lambda-Rho vs. Mu-Rho cross-plot for Well UTP-1 and b) Lambda-Rho vs. Mu-Rho cross-plot for Well UTP-2

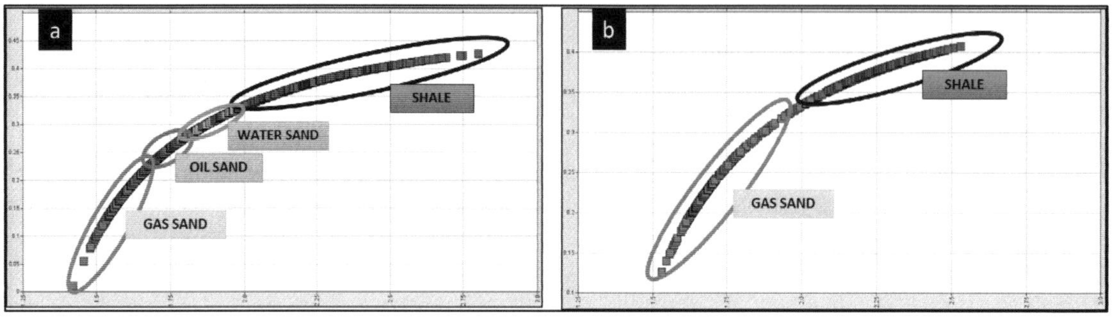

Figure 5. a) Vp/Vs vs. Poisson's Ratio cross-plot for Well UTP-1 and b) Vp/Vs vs. Poisson's Ratio cross-plot for Well UTP-2

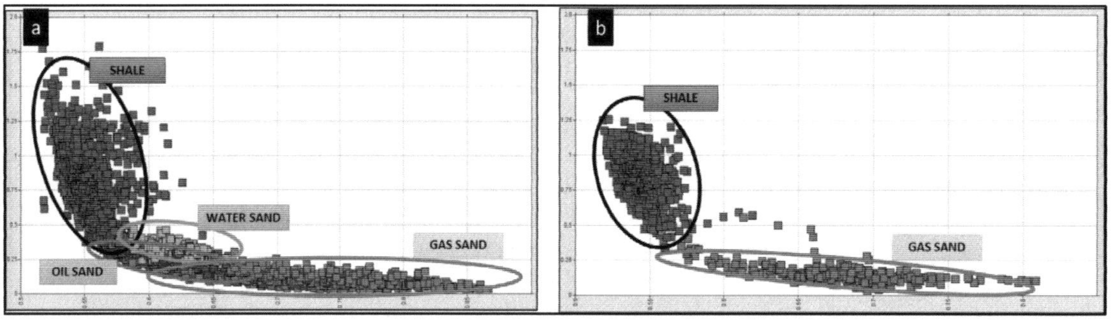

Figure 6. a) SQs vs. SQp cross-plot for Well UTP-1 and b) SQs vs. SQp cross-plot for Well UTP-2

4.1.5. The summarized values of the elastic modulus. Based on the cross-plots, the elastic modulus range of values for each fluid and lithology were summarized and shown on Table 1.

Table 1. Elastic Modulus Values for Well UTP-1 and Well UTP-2

Fluid and Lithology	Acoustic Impedance ((g/cm3)/ (m/s))	Velocity Ratio	Poisson's Ratio	Lambda-Rho ((GPa)/ (g/cm3))	Mu-Rho ((GPa)/ (g/cm3))	Scaled Inverse Qs	Scaled Inverse Qp
Well UTP-1 Gas Sand	4355 - 6936	1.4 - 2.0	0.0 - 0.35	0-22	6 -15	0.55 - 0.86	0.0 - 0.55
Well UTP-1 Oil Sand	5990 - 7962	1.6 - 2.0	0.20 - 0.35	11-34	10-17	0.53 - 0.66	0.14 - 0.56
Well UTP-1 Water Sand	5841 - 7543	1.7 - 2.0	0.25 - 0.35	12-28	10-16	0.56 - 0.66	0.20 - 0.46
Well UTP-1 Shale	5333 - 7159	1.7 - 2.8	0.25 - 0.42	14-34	4 -13	0.50 - 0.64	0.22- 1.78
Well UTP-2 Gas Sand	5220 - 7410	1.5 - 2.3	0.10 - 0.38	6-30	7-18	0.54 - 0.80	0.0 - 0.90
Well UTP-2 Shale	5810 - 7200	1.8 - 2.5	0.30 - 0.4	18-34	5-13	0.50 - 0.62	0.25- 1.25

Based on Table 1, the values of the acoustic impedance for different fluid and lithology in both wells are almost in the same ranges. This would complicate the interpretation of the inversion product if seismic inversion of the modulus is carried out for the area as similar values of acoustic impedance represent different formation. The values of velocity ratio discriminate the gas sand and shale better than the acoustic impedance values for both wells, which suggest that it would be an effective modulus for a gas field. The Poisson's ratio values are slightly overlapped, but the trend is clear between reservoir and shale interval where shale has a higher Poisson's ratio value than reservoir rocks. For Lambda-Rho modulus, the values of different intervals for both wells almost overlap and thus it advised to be coupled with Mu-Rho modulus for better interpretation. The SQs values for the formation in both wells almost overlap, but is more effective when combined with SQp. These values and the trend of the elastic moduli, such as low or high values for certain fluid and lithology, can be the reference if seismic inversion is carried out on our study area.

4.2. Seismic Attributes Analysis

Seismic volume attributes were applied to our study area by using Petrel software in order to evaluate the formation penetrated by the wells and the formation further from the wells. The results of the attributes were compared with the realized seismic amplitude attribute.

4.2.1. Relative Acoustic Impedance Attribute [7]. Based on Figure 7b, 7e and 7f, the yellow and red colours show the highest impedance change between the formations while the blue colour shows the lowest impedance changes. Figure 7b shows the blue outline of the accretionary prism and the impedance difference inside the prisms. Figure 7e displays the blue outline of the prism shape and internal formation penetrated by Well UTP-1 which can't be seen clearly on the realized seismic amplitude section. Figure 7f shows the red and yellowish outline of the formation penetrated by Well UTP-2.

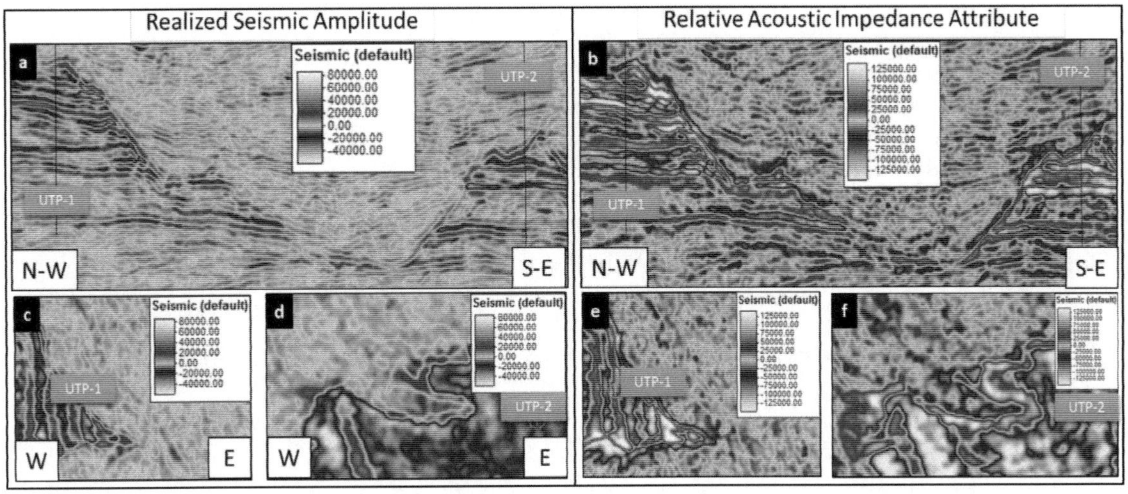

Figure 7. a) Cross-section of realized seismic amplitude, b) Cross-section of Relative Acoustic Impedance, c) Time-slice of realized seismic amplitude for Well UTP-1, d) Time-slice of realized seismic amplitude for Well UTP-2, e) Time-slice of Relative Acoustic Impedance for Well UTP-1, f) Time-slice of Relative Acoustic Impedance for Well UTP-2

4.2.2. RMS Amplitude Attribute. Based on Figure 8b, 8e, and 8f, the 'warm' red yellowish colour indicates hydrocarbon and both wells has penetrated hydrocarbon sands. The RMS amplitude shows the anomaly which can't be seen on the realized seismic amplitude alone. Figure 8b shows hydrocarbon sands, which are displayed as the red yellowish colour, penetrated by Well UTP-1 and Well UTP-2 and thus confirms the well log interpretation. Based on the result, most of the hydrocarbon anomalies are found to be inside the accretionary prisms.

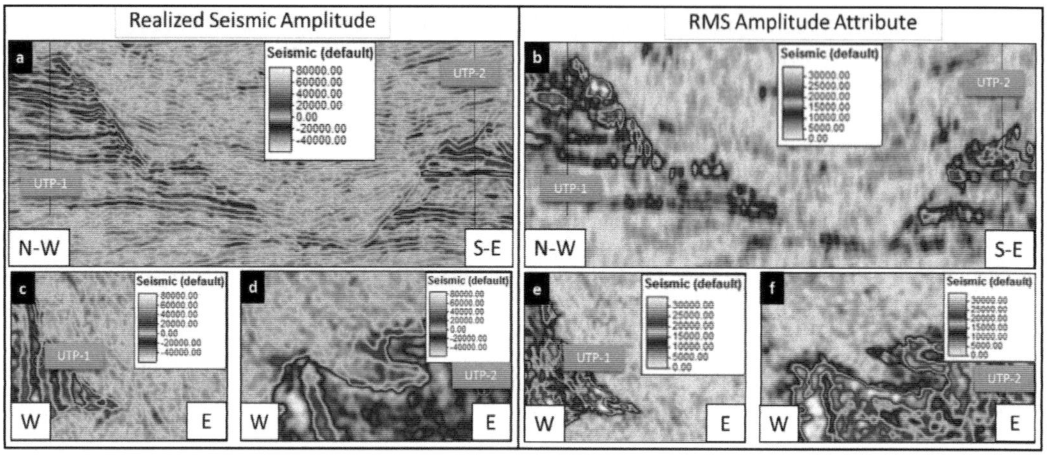

Figure 8. a) Cross-section of realized seismic amplitude, b) Cross-section of RMS Amplitude, c) Time-slice of realized seismic amplitude for Well UTP-1, d) Time-slice of realized seismic amplitude for Well UTP-2, e) Time-slice of RMS Amplitude for Well UTP-1, f) Time-slice of RMS Amplitude for Well UTP-2

5. Conclusion

In conclusion, for our study area, the effective elastic moduli for well with oil, gas and water sand and shale are Velocity Ratio (Vp/Vs), Poisson's ratio and Scaled Inverse Quality (Q) Ratio while the effective elastic moduli for well with soft shale and gas sand is the Lambda-Mu-Rho (LMR), Vp/Vs, Poisson's ratio and Scaled Inverse Quality (Q). The Relative Acoustic Impedance seismic attribute displayed the outline of formation which cannot be seen on the realized seismic amplitude section while the RMS amplitude confirms the hydrocarbon presence. For future study, seismic inversion can be conducted to further study about the formation. This study workflow is applicable to other areas and can help in fluid and lithology discrimination.

6. Acknowledgement

The authors would like to thank PETRONAS for the data provided in this study.

References

[1] Bjorklykke, K. (2010). Petroleum Geoscience: From Sedimentary Environments to Rock Physics. London, New York: Springer.

[2] Tarang, A. and Singh, Y. (2011). "Soft Shale Complication in AVO Interpretation in Sabah Basin". Petroleum Geology Conference & Exhibition (PGCE) 2011

[3] Ødegaard E. and Avseth P., 2004 "Well log and seismic data analysis using rock physics templates" *First Break* 22 37-43

[4] Goodway, W., T. Chen, and J. Downton (1997). "Improved AVO fluid detection and lithology discrimination using Lame petrophysical parameters; "Lambda-Rho", "Mu-Rho", & "Lambda/Mu fluid stack", from P- and S- inversions". 67th Annual International Meeting, SEG, Expanded Abstracts, 183–186.

[5] Fanka, W. R. T. (2012) "Well Log and Seismic Interpretation: Rock Physics Study of Poorly Consolidated Sandstones in the North Sea."

[6] Hermana, M., Lubis, L. A., Ghosh, D., & Sum, S. W. (2016). New Rock Physics Template for Better Hydrocarbon Prediction.

[7] Subrahmanyam, D. and Rao, P. H. (2008). "Seismic Attributes-A review". 7th International Conference & Exposition on Petroleum Geophysics.

IOP Conf. Series: Earth and Environmental Science **88** (2017) 012010 doi:10.1088/1755-1315/88/1/012010

Application of FK Filtering for Coherent Noise Removal in High Frequency Shallow Marine Data

Maman Herman[1], Hazwan Syahmi Hashim[1], Abdul Halim Abdul Latif[1] and Deva P. Ghosh[1]

[1]Centre of Seismic Imaging and Hydrocarbon Prediction, Universiti Teknologi PETRONAS

Email: maman.hermana@utp.edu.my, mherjody@yahoo.com

Abstract. The use of seismic reflection to study earth subsurface is commonly troubled by the presence of the coherent noise in the seismic data. This unwanted signal were generated by various sources and therefore had their own noise characteristic. The acquisition of the shallow marine seismic survey were done using a sparker source to obtain a high frequency seismic data. This however results in a stronger noise level in the recorded data. The application of FK filter were used to eliminate this unwanted noise using the data obtained from the Malacca Strait and Terengganu Bay. The result obtained prove to be successful in reducing the coherent noise while preserving the frequency content of the seismic data.

1. Introduction

Coherent noise is a common problem especially in shallow water area. As this area are consist of water-bottom which are mostly hallow and hard, the primary signal that were acquired are usually mixed with various type of seismic noise which results in a poor quality seismic data. Moreover, as the seismic data were acquired in a high resolution using a high frequency sources, it results in a much higher level of noise at the receiver compare to the conventional acquisition [1].

By observing the field record, we can determine the type of coherent noise [2]. This unwanted signal can comes from various sources such as scattered noise, guided waves, cable noise, ship's propeller and also multiples. Knowing the noise mechanism are important as it helps to determine the best possible way to eliminate this unwanted signal based on their specific characteristics.

In current technology, *F-K* filtering is used as an effective tool to separate noise from a main signal. However this technique is applied on pre-stack seismic data. In the other hand, the zero offset data is commonly used in shallow marine survey to observe the subsurface. Hence it is difficult to apply this technique into zero offset seismic data for same purposes. Some modifications in the processing workflow need to be done to make the FK-filtering is applicable for this data condition.

We enhanced the processing workflow by involving the FK filter to be applicable for our zero offset of shallow marine seismic survey. By applying the FK filter, the frequency content of zero offset seismic data can be maintained, while the coherent noise embedded in this data can be eliminated. Therefore, this technique can be used as an enhancement for the common processing workflow of zero offset data which in current technology is still not inserted.

The main purpose of shallow marine seismic survey is to obtain the shallow marine data of the Terengganu Bay and Strait of Malacca to study the seismic stratigraphy of the Quaternary depositional and paleo-channel. To achieve the objective, we required a high frequency seismic data in order to obtain the highest resolutions possible which were done by using the sparker source. However, the drawback of acquiring the seismic data in a high frequency results in a more prominent coherent noise cause by the ship movement. Therefore a proper noise removal technique must be choosing to eliminate this noise while preserving the frequency bandwidth of the seismic data.

2. Methodology

The algorithm was tested using the shallow marine data acquired at the Malacca Straits and Terengganu Bay. The main target of this survey was to image the shallow marine subsurface of the Malacca Straits and Terengganu Bay.

A zero offset marine survey was later carried out through the collaboration between Universiti Teknologi PETRONAS (UTP) and Universiti Malaysia Terengganu (UMT) to evaluate the seismic stratigraphy of the Quaternary depositional system and paleo-channels of Terengganu Bay and Straits of Malacca in 2013 and 2014 respectively. These data were collected using a high frequency source and high frequency sampling. The sampling intervals were set at 250 micro seconds and the number of samples per trace were 1200. A sparker source was used as a marine seismic impulsive source to produce high resolution seismic data [3][4]. The location of this survey area is shown in Figure 1.

Figure 1: Location map of marine seismic survey of Malacca Strait and Terengganu [5].

Information about the subsurface of shallow marine were crucial for geo-acoustical modelling studies and engineering applications, such as foundation design for offshore structures and also as a preliminary study for shallow gas delineation studies [6]. However, the recorded seismic data were contaminated with random and coherent noise which directly reduce the quality of the data recorded. These noises were generated by the waves and current, as well by the vortexes of the ship's propeller which appear to be more dominant than the primary signal.

Several basic processing steps were applied to remove this unwanted signal from the seismic data. The steps of the processing applied to this seismic data includes: header manipulation, conducting the transformation from time domain to frequency-wavelength (F-K) domain, conducting the filtering in the F-K domain, Q-compensation and conducting the transformation from FK back to the time domain.

Seismic noise can be properly removed by looking at the characteristic of the noise. For example, coherent noise that was cause by the ship movement can be identified by its high frequency range usually more than 80 Hz. The common method used to remove this noise are by using the frequency filtering, where the noise are removed by designing a low pass filter which remove any signal that were above the normal frequency content of a seismic signal which are around 10 – 80 Hz. However, as we need the data in high frequency, this method cannot be implemented as we will also reduce the bandwidth frequency therefore reducing the resolution of the seismic image. The important signal which were needed appeared in the same frequency range with the coherent noise and thus it will be mixed up. Therefore we must find a proper way to eliminate the coherent noise while preserving the frequency bandwidth.

Frequency-wavenumber filtering or better known as *F-K* filter is a process of noise attenuation in the *F-K* domain. This method was applied by transforming the seismic data in the t-x domain into the *F-K* domain using the Fourier transform. As the seismic data were transformed in the *F-K* domain, the linear event will remain the same in both domains, however the signal and noise can now be separated based on the dipping angle [7]. This way we can properly separate this two different signal and later mute the unwanted seismic noise. Although it was more computational extensive compare to the frequency filtering, this method allows the elimination of the coherent noise without reducing the frequency content of the seismic data. This method also gave better results as we can now identify the coherent noise and therefore gave a better noise attenuation without damaging the primary signal.

3. Results and Discussions

The examples of the seismic data from the Malacca Straits in the time domain, *F-K* domain and the filter design in *F-K* domain are shown in Figure 2**Error! Reference source not found.**. In Figure 2a, the recorded seismic data are dominated with coherent noise. The reflectors of the subsurface is masked by this noise, hence the interpretation of this data to pick the boundary of small basin structure is mostly very hard to be done. the coherent noises appear in all traces of zero offset data from beginning to the end of recording. Figure 2b shows the processed data with our workflow. The results showed that the algorithm was effectively removing the coherent noise. The quality of seismic section was improved, now the reflectors of subsurface can easily be interpreted.

In F-K domain, the original seismic data is shown in Figure 2c, the signal was concentrated in the centre of the K domain (K=0). The filter was designed to keep the signal and to remove the noise. Hence, a symmetrical parallelogram filter in FK domain was created. The frequency ranged from around 100 Hz to 2000 Hz and kept the K from -500 to 500 and everything outside that filter was muted. The FK filter design is shown in Figure 2d and the final result is shown in Figure 2b. From that figure, it can be seen clearly that the noise is successfully removed.

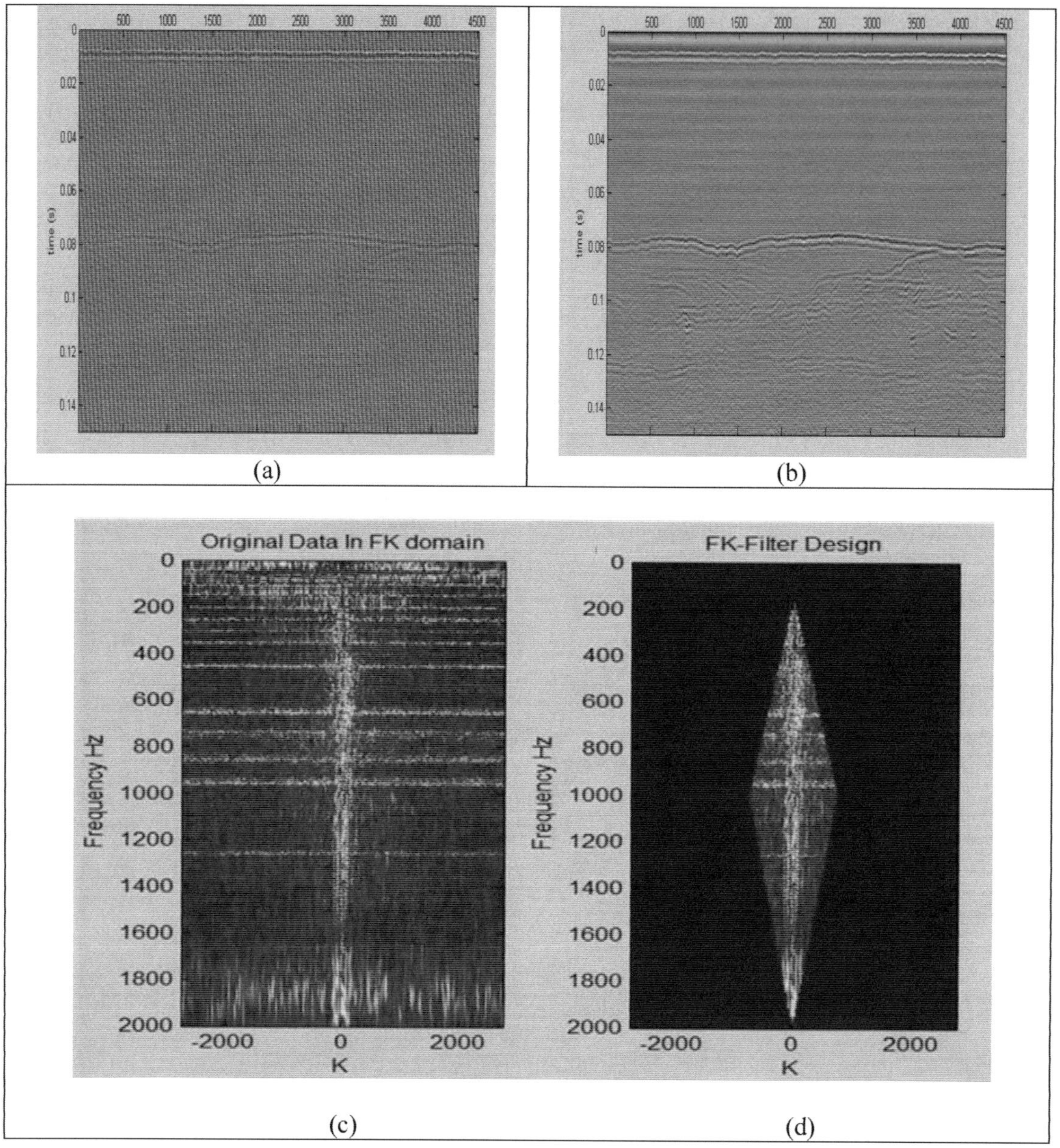

Figure 2. Seismic section from Malacca Strait survey, a) original seismic data, b) seismic data after processing, c) original seismic data in FK domain, and d) Filter design in FK domain.

Another example is coming from Terengganu survey, where all the survey parameters during collecting data are similar as Malacca survey. The characteristic of coherent noise resulted from ship's propeller and current are dominant in the recorded data as shown in Figure 3a.

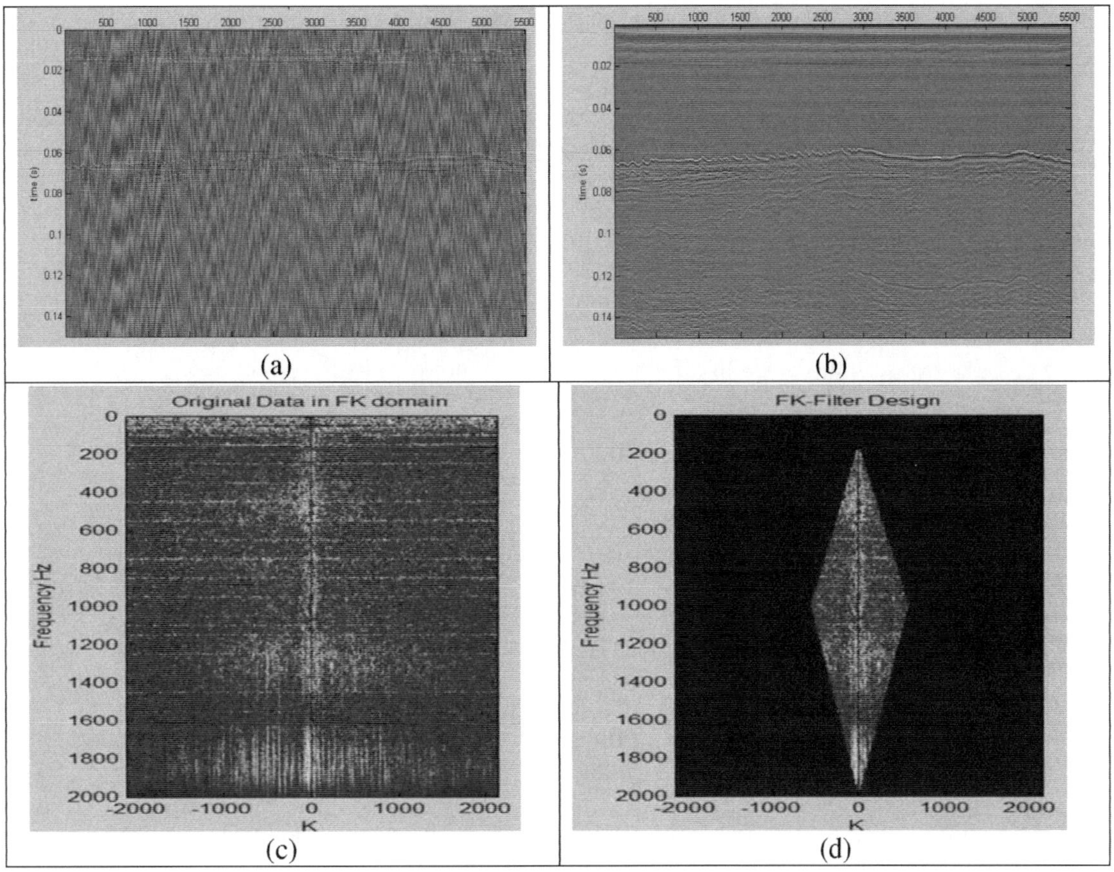

Figure 3. a) The original Seismic section from Terengganu survey, b) seismic section after applying FK filtering and Q-compensation , c) seismic data in FK domain, and d) Filter design in FK domain.

In Figure 3a, the character of random noise in the Terengganu survey is the same as the Malacca survey although the wavelength of this noise was lower. Hence, the picking of subsurface reflectors are mostly impossible especially in the area below see bottom. In FK domain, this seismic data is shown in Figure 3c. To preserve the signal energy around central of K axis, the FK filter as shown in Figure 3d was applied. In this filter, the frequency was kept from 200 Hz to 2000 Hz and the wavelength from -500 to 500. The final result of our workflow is shown in Figure 3b. This figure shows that noise has been removed successfully from the data and the detail of subsurface reflectors below seabed can be identified clearly. The channel or small basin of this area can be recognized nicely.

4. Conclusion

Coherent noise resulted from ship's propeller usually embedded in the recorded seismic data especially in the high frequency shallow marine survey. This noise affects the quality of seismic data that cause several difficulties during the interpretation of the subsurface, which usually mask the geological structure and stratigraphy channel of the subsurface. Since the marine data were required in a highest resolution possible, we must find a proper noise removal technique which does not reduced the seismic bandwidth.

Implementation of FK filter design on high frequency shallow marine data was successfully removing the coherent noise resulted from ship's propeller while preserving the frequency content of the marine seismic data. Reflectors of subsurface in Malacca and Terengganu surveys can be improved. Picking horizon and interpretation of subsurface of this area therefore can be performed without any further difficulties.

With the success of implementation of this work flow in high frequency data, it is therefore possible for it to be practised in the common seismic survey with some adjustment. However, further testing on other real seismic data with different condition of noise will still be required.

Acknowledgment

We thank UTP for funding this research work and PETRONAS for providing the data.

References

[1] D. C. Mosher and P. G. Simpkin, "Status and trends of marine high-resolution seismic reflection profiling: data acquisition," Geoscience Canada, v. 26, pp. 174-188. 1999.
[2] K. Larner, R. Chambers, M. Yang, W. Lynn, and W. Wai, " Coherent noise in marine seismic data," Geophysics., vol 48, No. 7, pp. 854-886, July. 1983.
[3] M. J. Duchesne, G. Bellefleur, M. Galbraith, R. Kolesar, and R. Kuzmiski, "Strategies for waveform processing in sparker data," Mar. Geophys. Res., vol. 28, no. 2, pp. 153–164, Jun. 2007.
[4] M. J. Duchesne and G. Bellefleur, "Processing of Single-chanel, high-resolution seismic data collected in the St. Lawrence estuary,Quebec," Geol. Surv. Canada, 2007.
[5] https://www.google.com.my/maps/@4.30093,97.5385451,1020678m/data=!3m1!1e3?hl=en.
[6] T. Missiaen, S. Murphy, L. Loncke, and J. P. Henriet, "Very high-resolution seismic mapping of shallow gas in the Belgian coastal zone," Cont. Shelf Res., vol. 22, pp. 2291–2301, 2002.
[7] H. W. Zhou, "Frequency-wavenumber (f-k) Filtering." Practical Seismic Data Analysis, pp. 156-159, 2014

Imaging of karst on buried Miocene Carbonate Platform

Siti Nur Fathiyah Jamaludin[1,2] Mukhriz Mubin[1] Abdul Halim Abdul Latiff[1]

[1]Department of Geosciences, Faculty of Geosciences and Petroleum Engineering, Universiti Teknologi PETRONAS, 32610 Bandar Seri Iskandar, Perak.

[2]fathiyah.jamaludin@utp.edu.my

Abstract: Imaging of carbonate rocks in the subsurface through seismic method is always challenging due to its heterogeneity and fast velocity compared to the other rock types. Existence of karst features on carbonate rocks makes it more complicated to interpret the reflectors. Utilization of modern interpretation softwares such as PETREL and GeoTeric® to image karst morphology make it possible to model the karst network within a buried carbonate platform. Using a combination of different seismic attributes such as Variance, Conformance, Continuity, Amplitude, Frequency and Edge attributes, it is able to image the karsts features that are available in the proven gas-field in Central Luconia Province, Malaysia. The mentioned attributes are excellent in visualizing and imaging the stratigraphic features based on the difference in their acoustic impedance as well as structural features, which include karst. 2D & 3D Karst Models were developed to give a better understanding on the characteristics of the identified karsts. From the models, it is found that the karsts are concentrated in the top part and middle part of the carbonate reservoir with some of them becoming extensive and creating karst networks, either laterally or vertically. Most of the vertical network karst are related to the existence of faults that displaced all the horizons in the carbonate platform.

Keywords: Seismic Interpretation, Subsurface Karst, Carbonate Imaging, Seismic Attributes, Karst Models

1. Introduction

Carbonate rocks are very heterogeneous in nature as they are originally composed of living things such as coral, algae, shell fragments and other bioherms. Due to this, carbonate rocks are prone to chemical weathering and with the rise and fall of sea level, carbonates are often exposed subaerially, where they undergo further diagenesis and karstification (Palaz & Marfurt, 1997). Karstification process involve long term effects of weathering from the atmosphere and rain water (meteoric water) that percolates into the carbonate rocks and prolong sub-aerial exposure. The CO_2 in the rain water releases a weak carbonic acid (H_2CO_3) that reacts readily with the limestone and form cavities and voids (Marshak, 2005). Figures 1A and 1B are the examples of carbonate rocks as viewed on a cave and on seismic data. Note the difference in scale. The difference in scale from seismic viewpoint, as stated by Palaz & Marfurt, 1997, is because carbonate rocks create fast velocities and greater densities compared to most siliciclastic rocks, hence contributing to lower vertical and spatial resolution in the carbonate strata for the same thickness of lithology.

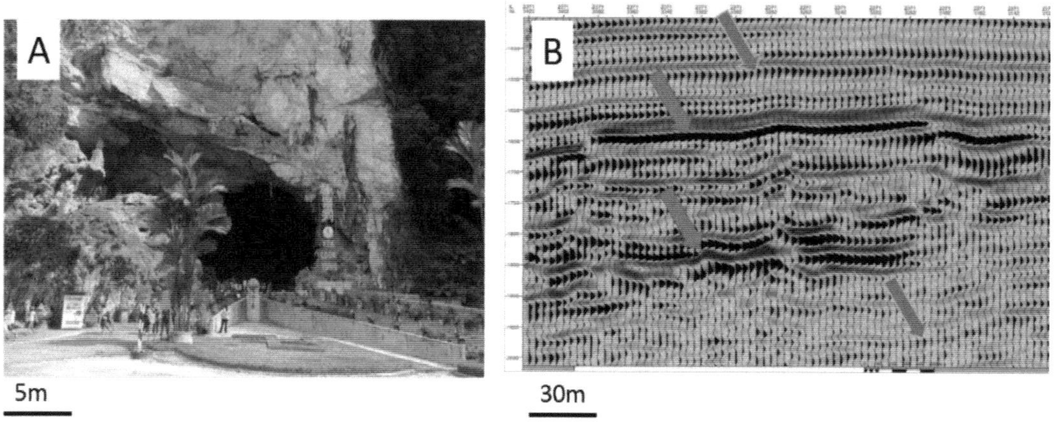

Figure 1: A: Example of a cave (example of karst) in carbonate bedrock from Kek Look Tong Cave in Ipoh, Malaysia. Image is from the second author's personal collection. B: Red arrows are pointing to seismic wavelets that imaged the carbonate rocks while the green arrows are pointing to siliciclastic rocks.

In the context of hydrocarbon exploration, karsts can be considered as possible voids that store the hydrocarbon or can be a potential geohazard to the exploration, especially during the drilling and exploration processes (Chung et al., 2016) as karst networks are understood to have high permeability faults and fractures. As such, recognition of karst features in carbonate reservoirs is the key to increase hydrocarbon production especially in matured field like Central Luconia Province in offshore Malaysia. Apart from that, cognizance of karsts in the carbonate reservoir will increase the safety measure during the exploration to avoid loss while drilling. For that reason, this paper aims to delineate karsts geomorphology in carbonate reservoirs to further understand the karst network within the reservoir. This study also aims to suggest suitable methods for the imaging of karsts features on buried/drowned carbonate platform in the subsurface.

2. Data and Methodology

3D seismic volume from one carbonate platform in offshore Malaysia was used throughout this study. The data were acquired in the late 1990s and are bound for research works within Petronas and UTP. The data used are filtered, scaled and migrated seismic sections with inline intervals of 25m and crossline intervals at 12.5m. The data were kindly provided by Malaysian Petroleum Management (MPM) for research purpose.

2.1 Horizon and Structural Interpretation

Horizons are picked to distinguish geologic intervals such as flooding surfaces and unconformities. The horizons are picked based on reflectors that show an increase in acoustic impedance and are distinct in the amplitude value. The interpretation of the seismic data was focused at two-way-time (TWT) 1480ms to 1800ms only because this is where the carbonate reservoirs are found. The estimated depth for these carbonate reservoirs are around 2.8km to

3.2km in depth. The total length of the carbonate reservoir is estimated to be 11km from east to west and width of 6km from north to south. Faults are interpreted based on the obvious displacement seen on the reflectors and abrupt changes in the seismic amplitude along the horizons. Horizon and structural interpretation are crucial in differentiating the different rock layers, especially the clastic and carbonate reservoirs. The detailed method of interpretation and results for this carbonate platform can be found in Jamaludin et al., 2014.

2.2 Seismic Attributes

Seismic attributes are able to boost and elevate the visibility of karst features to be interpreted. In this project, the seismic attributes are applied using two different softwares which are PETREL and GeoTeric®. Both softwares are compatible for seismic interpretation and attribute visualization and are widely used in the industry worldwide. PETREL is used to perform stratigraphic and structural interpretation as mentioned in Section 2.1 in this paper as well as application of several attributes as described in Section 2.2.1 in this paper.

2.2.1 *Seismic Attributes Utilization in PETREL*

Among the seismic attributes that have been meaningful in the visualization of the karsts in PETREL are Variance, Curvature and Amplitude. The variance attribute represents the trace-to-trace variability over a particular sample interval and therefore produces interpretable lateral changes in acoustic impedance (Chopra & Marfurt, 2007). It is able to highlight the different type of layers and unconformities in the data and will support the interpretation of the user especially while doing horizon picking. Curvature attribute measures the structural deformation and deduces stress regimes that are correlatable with the density and orientation of fractures either in the deep-water complexes or in carbonate build-ups environment. Amplitude attribute on the other hand defines how the amplitude stands out against the surrounding reflectors and background events. These attributes help to make karst features stand out from the other carbonate rock.

2.2.2 Seismic Attributes Utilization in GeoTeric®

GeoTeric® software allows its user to see the geology before interpreting the study area. The Carbonate Evaluation Workflow developed by GeoTeric® focuses on the following carbonate and associated features: Karsts; Clinoforms, Patch Reefs and Mounds, and Chalk. In addition, Fault Density and Combination Attributes are used, and features of interest can be extracted from an attribute and embedded in the Noise Reduced 3D volume. In this section, several attributes that were used such as Semblance, Conformance and Frequency attributes are discussed. Figure 2 below summarizes the workflow used in GeoTeric® Software in this study to successfully image the karsts. Semblance attribute is categorised in edge attributes. It is a form of cross correlation that produces a single coefficient in the range 0-1. In geological terms, semblance represents the measure of coherence or similarity of two traces of seismic data from multiple channels that is equal to the energy of a stacked trace divided by the energy of all traces that make up the stack (GeoTeric®, 2015). Conformance attribute compares the orientation around our target with its surrounding wavelets response. As conformance works in a structurally oriented manner, this attribute is suitable to delineate karst features as it calculates the value based on Dip & Azimuth volume. High Definition Frequency Decomposition (HDFD) on the other hand is able to highlight the dissolution features in the study area. HDFD uses a modified matching pursuit algorithm bringing out a much better vertical localisation of the events.

3. Results & Discussion

3.1 2D Karst Model

Two 2D karst models (Models A and B in Figure 3) were generated from the observed seismic sections in the northern and southern part of the seismic data. These two sections were picked because they showed the most prominent possible karst structures, as well as structural features present in the area. Possible karsts in the area were labelled by the continuity of the seismic lines within each stratigraphic layer. Generation of 2D karst models helps to give a detailed description for the identified karst within the carbonate platform.

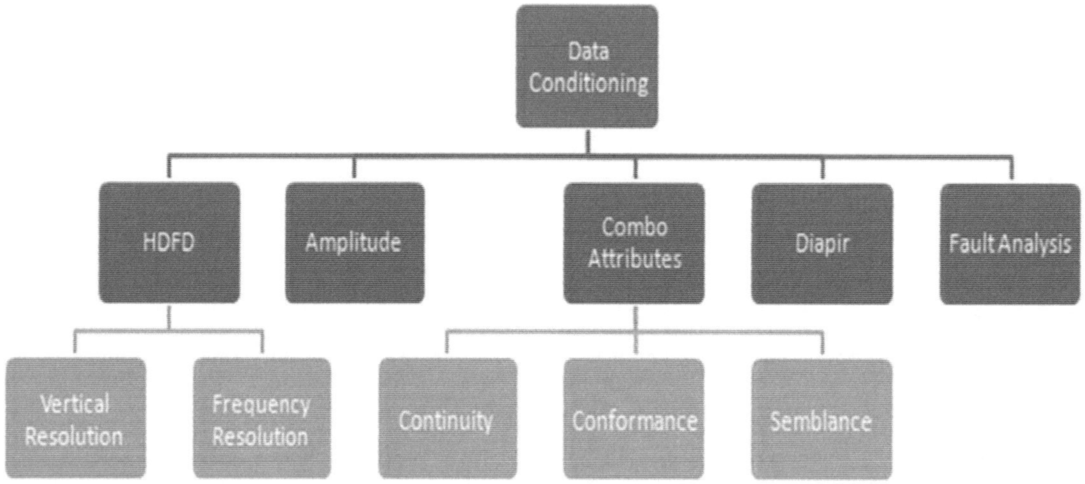

Figure 2: The seismic attributes used in GeoTeric® for this study.

Based on 2D karst models generated in Figure 3, karst structures were most frequently present along conjugate faults, which could be good and possible hydrocarbon trap in the area. Karst also can be found in abundance along Horizon 2 and Horizon 3 which are the Top and Middle Carbonate zones. The shapes of the karst structures were not uniform throughout the area. Based on the karst model generated in Figure 3B, karst structures were seen to be less in continuity among their formation, compared to karst model in Figure 3A. Two young faults can be found at western part of Model B in Figure 3, indicating that it was formed after the deposition of all facies in the study area. Large scale of uncertain structure can be found at the eastern part of the karst Model B in Figure 3, indicating that it could be caused by gas seepage, facies intrusion, or likely possible karst structure in the area.

IOP Conf. Series: Earth and Environmental Science **88** (2017) 012011 doi:10.1088/1755-1315/88/1/012011

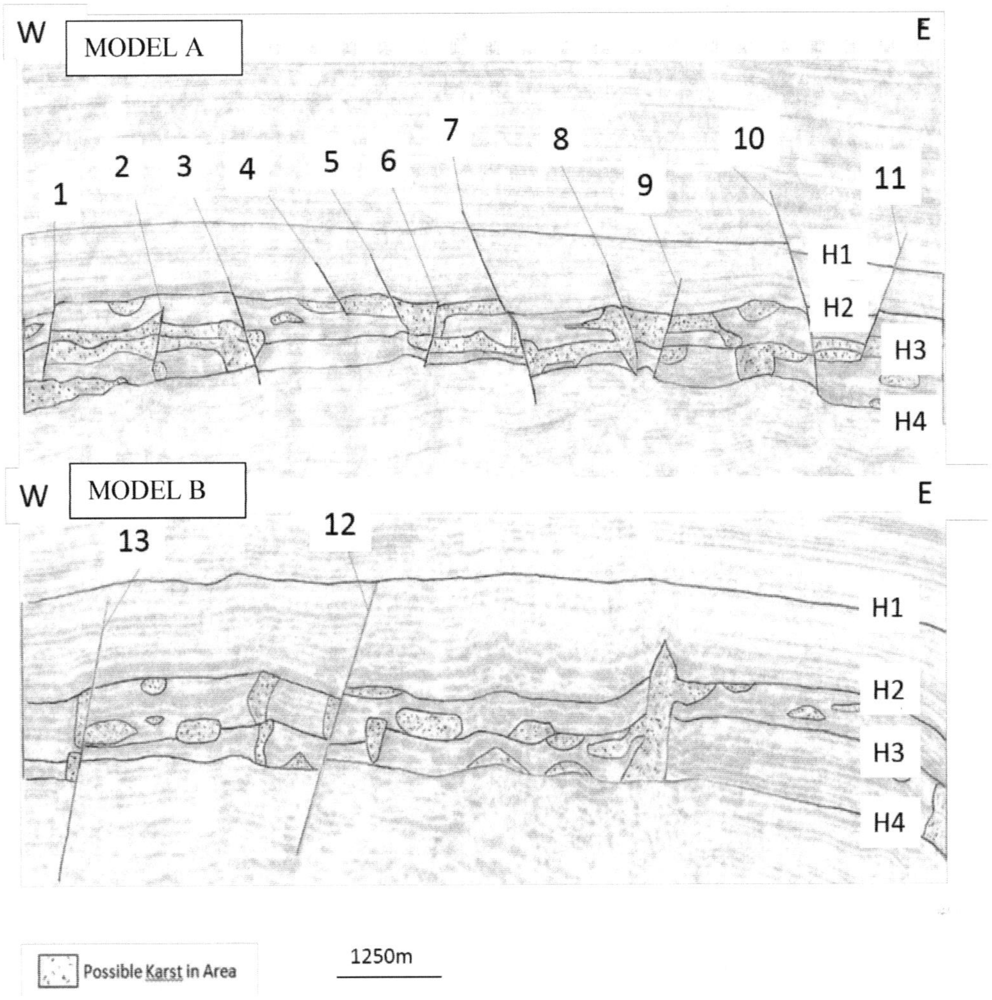

Figure 3: Two 2D Karst Models. Model A shows karst features which are highly associated with faults and tend to develop laterally throughout the interpreted horizons, creating a karst network. Model B shows scattered karst features which are mainly distributed in the Horizon 3 and few of them showing vertical distribution.

3.2 3D Karst Models

A few 3D Karst Models were created using GeoTeric® software. Only the most probable karst in the area was selected for the construction of 3D karst model. The reason for choosing more than one attribute for the construction of the karst model is to compare on which attribute is able to give out the most ideal output on determining the karst geomorphology in the area.

Combo Attribute highlights objects most effectively as it is often necessary to combine a few attributes. This has proven to be successful with carbonates when combining a structural and a stratigraphic attribute. The components that were used in this section were the Semblance attribute, Continuity attribute and Conformance attribute (Figure 4A) along with manipulation of the frequency.

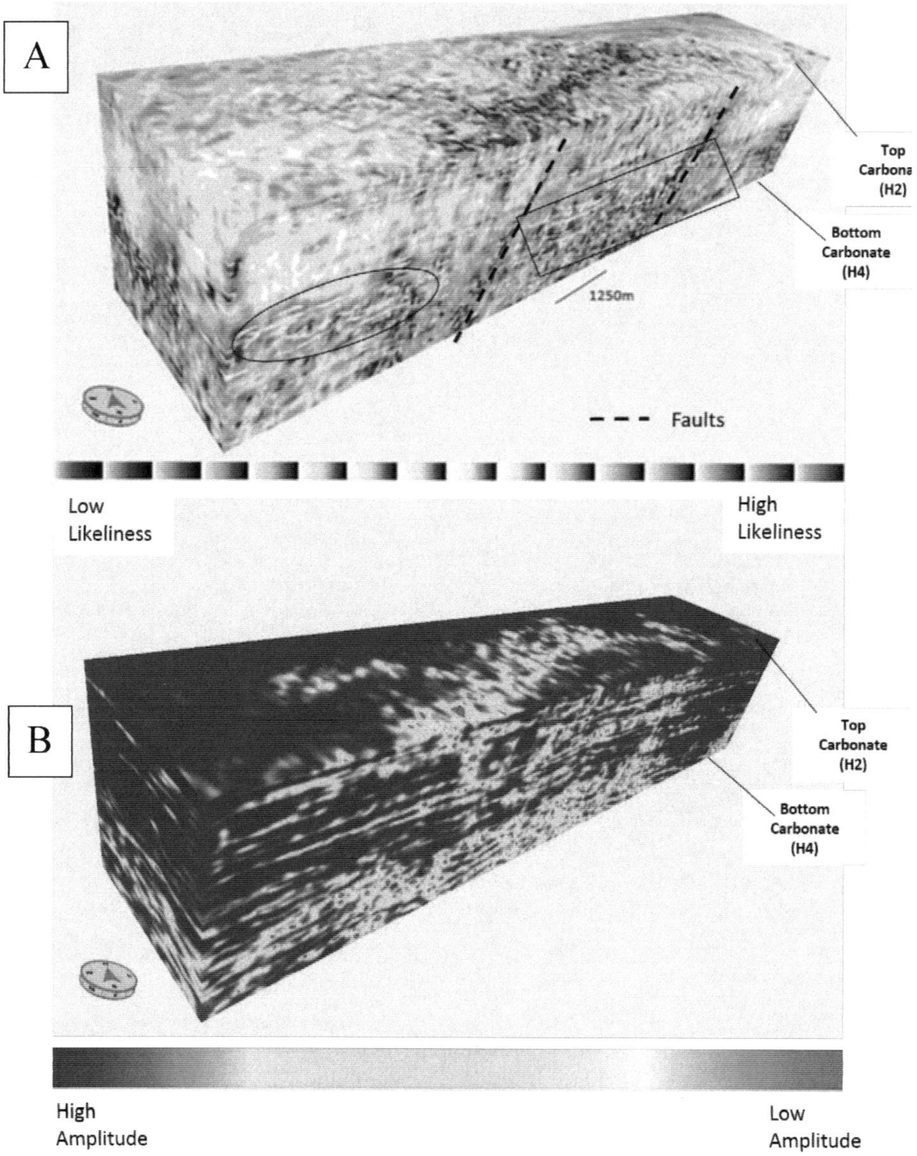

Figure 4: A: Model based on Continuity+Conformance attribute applied with RGB colourmap, to brighten up the features present within the area. The greenish colour on the 3D volume is referring to karst and they are mainly concentrated in the centre part of the reservoir and associated with faults. B: Seismic cube applied with diapir attribute. Area with red colour indicates possible karst structres within the cube.

A diapir attribute (Figure 4B) that was designed for Gas Zone was also applied in this study as both gas and karst are showing chaotic features on seismic reflectors. Carbonate facies shows high amplitude, but cavities within the area shows low amplitude, and these cavities are related to karst features. Each of the combination attributes shows different clarity and uses the respective indicators to highlight features within the area. The colour scale of the attribute plays an important role in determining and highlighting structures that are present within the area. The scale for Combo Attributes were flipped and adjusted to bring out more details on structures present and the indicator for the flipped scaled cube would be different from the original combination. From the developed 3D Models, the karst seemed to be more prominent at the centre of the studied carbonate reservoir. Note that from the fault analysis, most faults occurred in the centre part and the eastern part of the study area. Therefore, karst structures in the area may be related to the formation of faults cutting through the facies. Models developed from Continuity+Conformance attributes (Figure 4A) and Diapir Attribute (Figure 4B) showed that the size of each karst structure observed is roughly around 50-100m in width and height. The diapir attribute gives out the most possible size of the karst structure in the area as it differentiates clearly with red and green colours the most obvious karst features.

4. Conclusion

Referring to the aims of this paper, the geomorphologies exerted on karst features in this studied carbonate platform ranged from circular shape with a possible depression which might be related to a cavern karst. It is observed that the possible karst features are creating karst networks that extend laterally. Some of the karst network follows the faulting zone and create a vertical network. The workflow suggested (Figure 2) which is using Combo attributes consist of Semblance, Continuity and Conformance together with Diapir attributes should be used to give better imaging and visualization of karstification on seismic data. PETREL and GeoTeric® software were used in highlighting the karst geomorphology throughout the study area. PETREL provides good resolution while doing horizon and structural picking but users could not see where the picking is heading and it has longer turnaround time for manual interpretation. Even for automatic picking, a user's interpretation is still required. GeoTeric® on the other hand, enable user to see the features first in whole volume before interpreting and creating attributes volumes which are faster compared to PETREL. This work should continue in the near future by having calculations of the total volume for the area affected with karst. The calculation of total volume would give a hint on the amount of possible voids created by the karst. Apart from that, detailed correlation study with wells data such as wireline and cores would be helpful to improve the developed models.

Acknowledgement

We would like to thank PETRONAS's Malaysian Petroleum Management (MPM) for providing the seismic data, Schlumberger for providing academic license for PETREL and Foster Findlay Associates (FFA) for providing license for GeoTeric® and technical support from GeoTeric® Malaysia Team for this work. Lastly, Thank you to Universiti Teknologi PETRONAS for the funding of this research. This is a preliminary work from a research grant 0153AB-D40.

References

[1] Ibrahim Palaz & Kurt.J.Marfurt. (1997). Carbonate Seismology. Geophysical Development Series, No 6, Society of Exploration Geophysicsts. ISBN: 978-1-56080-038-5, Tulsa, Oklahama, United States of America.

[2] Jamaludin, S., Pubellier, M., & Menier, D. (2014) Relationship between syn-depositional faulting and carbonate growth in Central Luconia Province, Malaysia. Bulletin of the Geological Society of Malaysia, Volume 60, December 2014, pp. 77 – 83

[3] Kiat, C. W., Menier, D., Jamaludin, S. F., & Ghosh, D. (2016). Geomorphology and Karstification of the Southern Field High Carbonates in Central Luconia Province. Offshore Technology Conference Asia. doi:10.4043/26650-ms

[4] No Author., GEOTeric Training Manual, GeoTeric® 2015.

[5] S. Chopra and K. J. Marfurt, *Seismic Attributes for Prospect Identification and Reservoir Characterization*: Society of Exploration Geophysicists, 2007.

[6] Stephen, Marshak. (2005). Earth: Potrait of a Planet (Second Edition), W.W. Norton & Company, Inc. ISBN: 0-393-92502-1, Illinois, United States of America.

Sedimentology and geomorphology analysis of coastal area along pantai penarik, terengganu before and during northeast monsoon season

Tengku Ahmad Imran Ku Yusoff [1], Mohamad Shaufi Sokiman [1]

[1] Geoscience Department, Universiti Teknologi PETRONAS, Bandar Seri Iskandar, 32610 Perak, Malaysia

Abstract. This research is conducted to understand the sedimentology and morphological change before and during the northeast monsoon at the east coast of peninsular Malaysia. The increase in wind speed, wave energy and rainfall during the northeast monsoon are believed to causes the coastal erosion to increase during the season. Rapid development along the east coast area might disrupt the sediments distribution which can increase the coastal erosion rate every year. The understanding on the sediments distribution, erosion and deposition as well as the morphological change can help to figure out if the coastal erosion can affect the infrastructure in the future. The result of the study can show the necessity to perform mitigation or any required action toward the problem that might happen

1. Introduction

Development activities that have been done on the east coast area may have causes the sediment circulation disrupted while the aggressive wave erosion during Northeast monsoon that come from November until March have made the condition worsen [1] [2] [3]. Numerous studies have been done at this coastline, mostly on the coastal erosion at the beach. Among the study that have been done are the Open Sandy Beach Morphology and Morphodynamic as Response to Seasonal Monsoon in Kuala Terengganu, Malaysia, Beach Erosion Variability during a Northeast Monsoon: The Kuala Setiu Coastline, Terengganu, Malaysia and Variation of beach sand in relation to littoral drift direction along the Kuala Terengganu coast.

The local newspaper, The New Strait Times have released an article on February 2016 regarding the bad condition of Terengganu coastline. The article stated that the beach protections made at the coastline are not enough and need to be stretch. One problem arises from recent the monsoon is that the beaches sediments are not fully recovered after the monsoon [4]. Since the issue arises, the understanding about the sediment transport and morphological changes at the Terengganu coastline and the factor affecting it need to be done to find a suitable solution to minimize the erosion effect of Northeast monsoon to the beaches [5] [6].

The objectives of this study are to analyses the sediments distribution and morphological change at the coastal area along Pantai Penarik, Terengganu before and during the Northeast monsoon. The map of study area before and after the Northeast monsoon will be produce to determine the net shore direction.

2. Study area

The study area located at Pantai Penarik at Setiu, Terengganu as shown in Figure 1. It is located about 12 kilometers at Northeast direction of Bandar Permaisuri. Pantai Penarik is one of the famous tourism area in Terengganu. Along the beaches, there are several restaurants, grocery store, and population nearby the coastline.

The nearest delta located about 8 kilometres at Northwest direction from the study area. At distance from 10 to 30 kilometres from the coastline, there are several islands at the sea which include Pulau Redang and Pulau Perhentian. This study cover 1 kilometre distance of the beaches. There are 11 checkpoints named as point 0 to point 10. Each point has distance of 100 meters apart. Cross section and sediments sampling are collected at each checkpoint.

Figure 1: The location of the study area

3. Research methodology

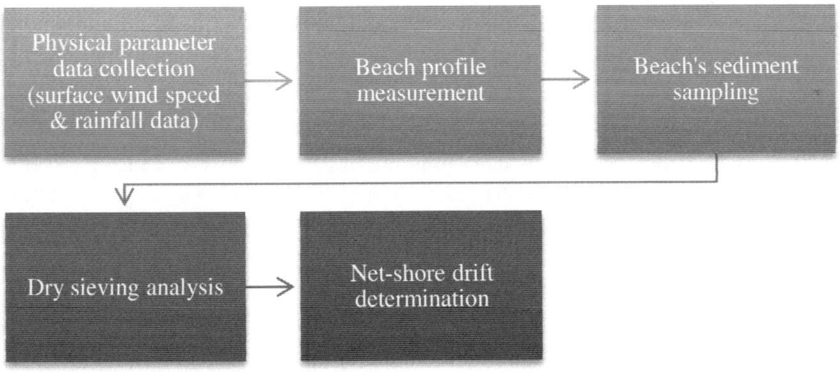

Figure 2: The flowchart of research methodology

4. Result and discussion

4.1. Physical parameter data collection

The monthly rainfall data and wind speed data of 2015 and 2016 were collected from Malaysia Meteorological Department to see the weather change based on each season. The result show that during November, the total rainfall in Terengganu is almost was the highest compared to another month, similar to the past study that have been done *[7]*. This show that the peak of Northeast monsoon happens in November. The surface wind speed data also show that wind speed on both years are high from December until February. The highest mean surface wind speed recorded was in February 2016 which is still within the northeast monsoon period. In short, the weather during northeast monsoon do become abrupt compared to the transition period and southwest monsoon *[4]*.

4.2. Beach profile measurement

The beach profile measurement for data before northeast monsoon was completed by using the total station equipment. The data consist of profile for 1 kilometre long beach with 11. Each checkpoint has 100 meter from each other. Based on the data collected, the beach width and maximum slope angle of each checkpoints are collected.

Two process either erosion or deposition of sediments happened along the beach. The erosion is identified when there is a decrease in profile from July compared to October and deposition is vice versa. The summary of the result are tabulated in Table 1.

Table 1: Summary of profiling result

Distance (m)	Beach width (meter)		Maximum Slope angle (°)		Activities
	July	October	July	October	
0	23.22	22.72	9.85	22.05	Erosion
100	36.12	43.24	10.55	9.05	Erosion
200	29.64	31.40	12.11	9.69	Erosion
300	33.92	30.67	9.63	17.95	Erosion
400	27.03	39.51	13.11	10.46	Deposition
500	25.74	30.00	15.60	14.74	Erosion
600	35.28	31.25	11.02	9.45	Deposition
700	29.86	32.65	13.53	9.89	Erosion
800	27.87	32.71	9.71	13.92	Deposition
900	25.04	28.10	10.47	8.59	Deposition
1000	27.18	28.98	9.47	7.76	Deposition

4.3. Sieve analysis

The results of mean size analysis and sediment sorting calculate using GRADISTAT method [6] from samples taken on July and October are used to generate a static model. The map from July data are shown in Figure 3(a) and 5(b). The sediments collected during July show that most of the sample in low tide area are coarse sand. In the middle tide area, the grain size are either fine sand or medium sand. The high tide area during July all has medium sand grains. The results of sorting show that the sediments at the low tide area on July are either poorly sorted or moderately sorted. At the middle tide area, the sorting are mostly either moderately well sorted or moderately sorted. In the high tide zone, most of the sediments are moderately sorted. The low tide area that has coarse sand and poorly sorted show the inconsistency in the wave energy. That area probably only experience low wave energy hence only transport the light fine grain.

During October, the low tide area has changed from coarse sand to medium sand in some area. Some area still has coarse sand. The middle tide area still has the medium or fine sand and the high tide area all show medium sand grains. During October, the sediments in the low tide zone are poorly sorted. Only three checkpoints are moderately sorted. The middle tide zone are all moderately sorted and the high tide zone mostly are moderately sorted. Except for point 0 where it is moderately well sorted. This show the consistency in the wave energy which probably quite strong and capable of transport coarse sand throughout the process.

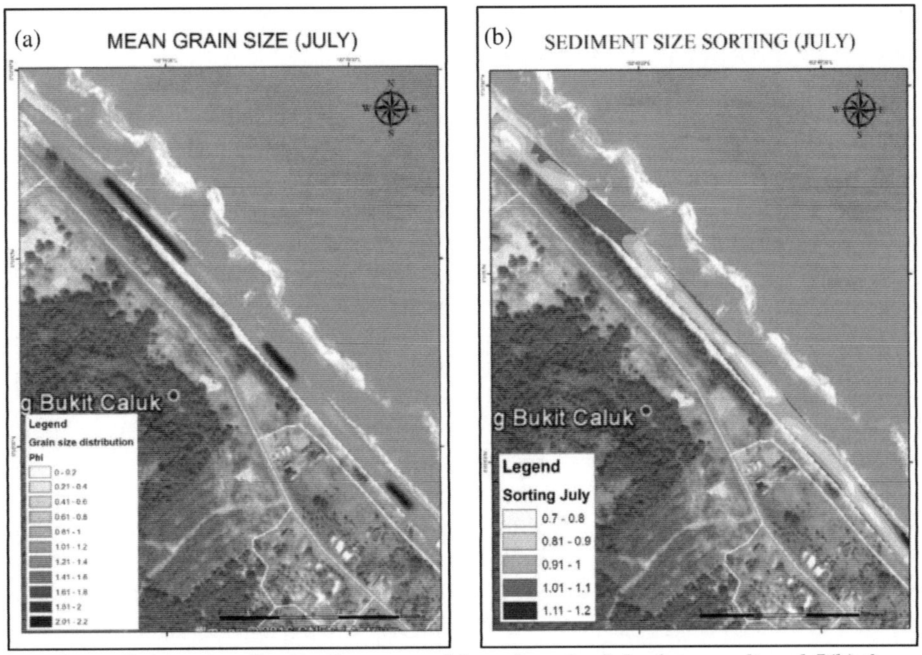

Figure 3(a) the sediment size distribution created based on the July data result and 5(b) the sediment sorting map of July data

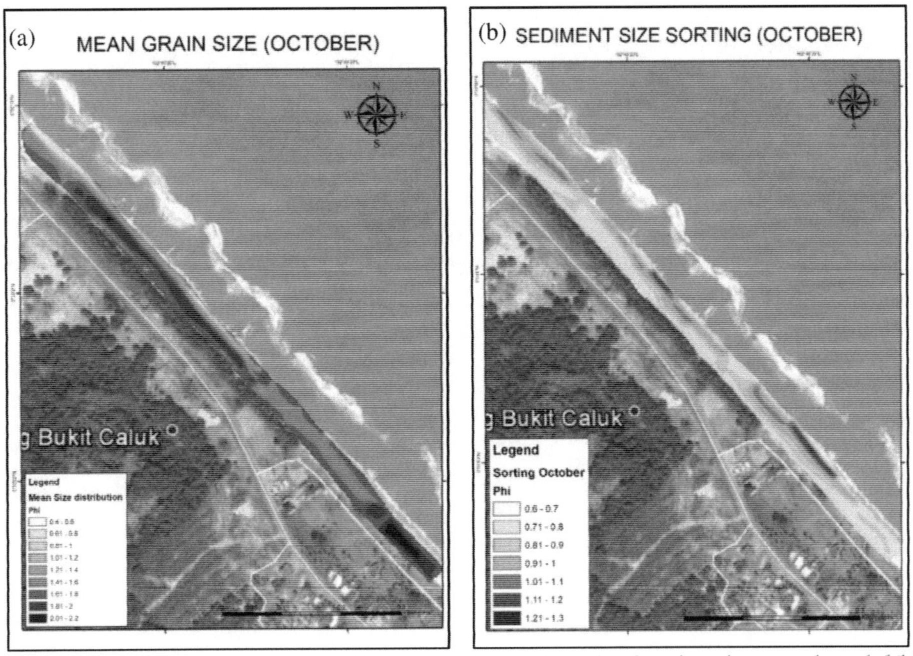

Figure 4(a) the sediment size distribution created based on the October data result and 6(b) the sediment sorting map of October data

4.4. Net-shore drift

Based on the profiling result and the sieve analysis result, the net shore drift can be determine based on the observation of the data. Based on profiling result in Table 1, morphological change show that the erosion happen in the northwest direction of the study area and the deposition happen in the southeast direction of the study area. This show that the sediments that eroded from the northwest area are transported and deposited at the southeast area.

The static model of the mean size distribution from July data (Figure 3) show coarser sediments accumulated at the northwest area. Although most of the sediments especially in the low tide areas are coarse sand, there is still more distribution of coarser sediment at the northwest area. This can be interpreted as the finer sediment have been transported toward southeast direction.

The mean size distribution map from October data (Figure 4) shown that there is change of distribution compared to July data. During this period, the sediments accumulated at the northwest area are mostly coarse sediments while the moderate sand sediments accumulated at the southeast area. This has strengthen the interpretation made from July data which stated that the fine sediments have been transported in southeast direction.

4.5. Economic value

Pantai Penarik is an attractive tourism spot in Terengganu. Generally, this area is a moderately populated area. On top of that, there are several shop and restaurant located along the road which located only 50 meters from the shoreline. The wave and coastal erosion can effect or destroy the houses and shops that located near the shoreline. During the second visit, a minor destruction on a small infrastructure was observed. This show that the wave have the capability to hit and destroy other infrastructure along the coastline. Other than that, the change could also causes the human activities to be affected which include the fisherman activities and tourism [8] [9].

5. Conclusion and recommendation

Based on the past studies such as Open Sandy Beach Morphology and Morphodynamic as Response to Seasonal Monsoon in Kuala Terengganu and Beach Erosion Variability during a Northeast Monsoon: The Kuala Setiu Coastline which stated about the sediment deposition disruption and possible coastal erosion due to the effect of Northeast monsoon, the conclusion that can be made is that more study need to be done not only in Terengganu, but also other location in East Coast. The study can be focusing on monitoring the beach change due to the wave action and effect caused by human activities. Other than that the reason the effect of the Northeast monsoon worsen also need to be study to understand the causes and learn the possible prevention action.

In term of this study, the comparison between the July data and end of October data show the difference especially in sediments analysis. The physical parameter data show that the month of October wind speed and rainfall data is high, but still in the average value. The maximum amount of wind speed and rainfall is on November. Because of this, the data is best collected during November so the major effect of Northeast monsoon can be study.

Other than that, the rules regarding the development, especially on the coastal line area need to be revise. The geological factors need to be emphasized in the regulations so the future development will have no effect to the environment or the effect can be controlled.

6. References

[1] A. Azid, C. N. Che Hasnam, H. Juahir, M. A. Amran, M. E. Toriman, M. K. A. Kamarudin, A. S. Mohd Saudi, M. B. Gasim and A. D. Mustafa, "Coastal Erosion Measurement Along Tanjung Lumpur to Cherok Paloh, Pahang During The Northeast Mosoon Season," *Jurnal Teknologi,* pp. 27-34, 2014.

[2] C. L. Wong, R. Venneker, S. Uhlenbrook, A. B. Jamil and Y. Zhou, "Variability of rainfall in Peninsular Malaysia," *Hydrology and Earth System Sciences,* pp. 5471-5503, 2009.

[3] K. Sopian, M. Y. Othman and A. Wirsat, "The Wind Energy Potential of Malaysia," *Renewable Energy,* pp. 1005 - 1016, 1995.

[4] E. H. Ariffin, M. Sedrati, M. F. Akhir, R. Yaacob and M. L. Husain, "Open Sandy Beach Morphology and Morphodynamic as Response to Seasonal Monsoon in Kuala Terengganu, Malaysia," *Journal of Coastal Research,* pp. 1032 - 1036, 2016.

[5] J. S. Pethick, An Introduction to Coastal Geomorphology, London: Edward Arnold, 1984.

[6] S. J. Blott and K. Pye, "GRADISTAT: A grains size distribution and statistics package for the analysis of unconsolidated sediments," *Earth Surface Processes and Landforms,* pp. 1237-1248, 2001.

[7] J. Suhaila, S. M. Deni, W. Z. Wan Zin and A. A. Jemain, "Trends in Peninsular Malaysia Rainfall Data During the Southwest Monsoon and Northeast Monsoon Seasons: 1975-2004," *Sains Malaysiana,* pp. 533-542, 2010.

[8] M. Muhammad , K. Idris, E. H. Ariffin, H. A. Mohamed Shaffril, B. Abu Samah and T. Suandi, "The Impact of Climate Change on Small-Scale Fisherman in Malaysia," *The Social Sciences,* pp. 3352-3356, 2016.

[9] G. H. Ching, T. W. Hin, R. Zainol, A. R. Roslan and M. Mohd Yusoff, "Significance of climate change at coastal tourism sites: A case study of East Coast, Peninsular Malaysia".

Numerical modelling for quantitative environmental risk assessment for the disposal of drill cuttings and mud

Mohd Amirul Faiz Abdul Wahab[1], Mohamad Shaufi Sokiman[1], Kim Parsberg Jakobsen [2]

[1] Geoscience Department, Universiti Teknologi PETRONAS, Bandar Seri Iskandar, 32610 Perak, Malaysia.

[2] Head of Department, Coastal & Marine, DHI Water & Environment (M) Sdn Bhd, Wisma Perindustrian, 11/F, Jalan Istiadat, Likas, Kota Kinabalu, 88400, Sabah, Malaysia

Abstract. To investigate the fate of drilling waste and their impacts towards surrounding environment, numerical models were generated using an environmental software; MIKE by DHI. These numerical models were used to study the transportation of suspended drill waste plumes in the water column and its deposition on seabed in South China Sea (SCS). A random disposal site with the model area of 50 km x 25 km was selected near the Madalene Shoal in SCS and the ambient currents as well as other meteorological conditions were simulated in details at the proposed location. This paper was focusing on sensitivity study of different drill waste particle characteristics on impacts towards marine receiving environment. The drilling scenarios were obtained and adapted from the oil producer well at offshore Sabah (Case 1) and data from actual exploration drilling case at Pumbaa location (PL 469) in the Norwegian Sea (Case 2). The two cases were compared to study the effect of different drilling particle characteristics and their behavior in marine receiving environment after discharged. Using the Hydrodynamic and Sediment Transport models simulated in MIKE by DHI, the variation of currents and the behavior of the drilling waste particles can be analyzed and evaluated in terms of multiple degree zones of impacts.

1. Introduction

Conservation of environment was part of consideration that needed to be taken into account in oil and gas businesses. Nowadays, many countries which operating the oil and gas activities encouraged the service companies to do the assessment on environmental impacts to conserve and protect the ecological habitants of surrounding environments. The author in collaboration with DHI Water & Environment (M) SDN BHD undertaken a study to determine the fate of drilling waste disposal (drill cuttings and mud) for two (2) different cases in terms of suspended drill waste particles in water column and its deposition on seabed based on degree zone of impacts using numerical modelling. Impact of the drill waste particles can be observed within the framework of their effect towards benthic communities such as corals, either through burial or chemical toxicity [2][3]. The deposition of the drill waste on seabed might also resulted in smothering of benthic organism regardless of the type of drilling mud used [1]. Uncontrolled discharge of the drill waste will affect the marine habitants in terms of elevated suspended sediment concentration in water column that can reduced the light penetration from the sea surface in which subsequently affects the photosynthesis process for the surrounding corals. In addition, excess sedimentation of the drill waste on seabed can also clog the filter mechanism of the benthic organism.

The proposed disposal site was located near Madalene Shoal, 8.3 km; 7.57° NW of Bintulu, Sarawak. The project duration will be categorized into three (3) different monsoon scenarios: a) North-East (NE) monsoon periods. b) Inter-monsoon (IM) periods. c) South-West (SW) monsoon periods.

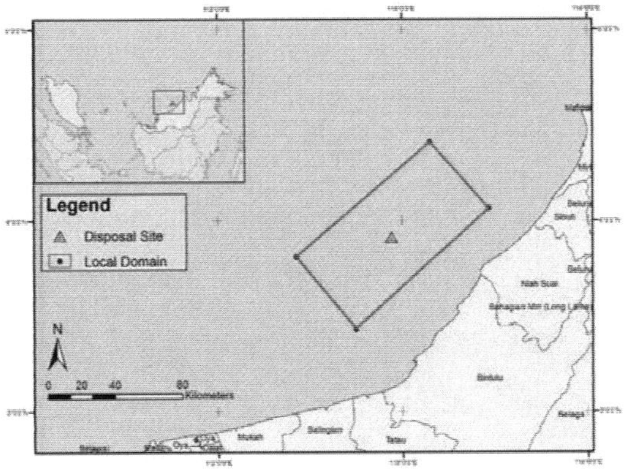

Figure 1. Location of the disposal site and the local domain within the study area.

There were also few limitations in this project such as the impact assessment criteria used were based on the general tolerance of the corals against the suspended sediment and its deposition. No actual coral positioning data at the study area were used due to limitation and confidentiality of the data. However, considerable amount of past researches had been done on studying the environmental impact from the effects of the operational drilling waste, particularly in United States and in Europe [2]. The drilling scenarios and the drill waste particle characteristics were also obtained from past researches and adapted from the actual data mostly from Norwegian Sea. Thus, different particle characteristics may affect the simulated results of the numerical models generated later on. Only two different cases (Case 1 and Case 2) were chosen to study the sensitivity of different particle characteristics towards its behaviour after discharge to open sea due to limitation of data.

2. Research methodology

2.1 Modelling Strategy

This project will focus on generation of hydrodynamic (HD) model and sediment transport model using environmental software, MIKE by DHI. The generation of numerical models were followed according to standard DHI modelling techniques. The process of generating the numerical models started with the generation of regional HD model for South China Sea (SCS) to see the regional variations of the ambient conditions as well as other meteorological conditions in SCS.

2.2 *Disposal Method*

For this project, the disposal method chosen would be the offshore discharge method. The drill cutting and mud were brought up to the platform and undergo the cleaning processes. The drilling fluid (mud) recovered from the cleaning processes will be re-used. However, the remaining drill cuttings and mud stain which known as drill waste were discharge to the marine receiving environment [1].

2.3 Drilling Scenario and Drill Waste Characteristics

This project was a sensitivity study to investigate the effect of different drill waste characteristics towards the behaviour of the drilling discharge in the marine receiving environment. Two different cases representing the different drill waste characteristics (Case 1 and Case 2) were selected with the same amount of total volume discharge. The differences in term of the drill waste particle sizes, drilling scenario, rate of discharges as well as other drill waste characteristic were determined for this study.

The drilling scenario and drill waste characteristics were obtained and adapted from past researches regarding waste disposal mostly from the Norwegian Sea for both Case 1 and Case 2. The drilling scenarios and particle size characteristics for both cases were described in the tables below:

Table 1. Drilling scenario for Case 1 and Case 2.

Case 1

Well Section	Hole Size (Inch)	Section Length (ft)	Drilling Time (hr)	Volume (bbls)	Weight (MT)	Drill Cuttings and Mud Discharge (kg/s)
1	13.38	227	19.2	240.1	91.5	1.584
2	12.25	3083	60.0	449.4	171.2	0.948
3	8.50	1487	40.8	104.4	39.8	0.324
Total		4797	120.0	793.9	302.5	2.868

Case 2

Well Section	Hole Size (Inch)	Section Length (m)	Drilling Time (hr)	Weight (MT)	Drill Cuttings and Mud Discharge (kg/s)
1	36	83 + 2 Batch	14	2103	41.726
2	8.5	339 + 1 Batch	19.75	2899.45	40.78
3	17.5	335 + 2 Batch	37	5389.63	40.463
Total			70.75	10392.08	245.938

Table 2. Drill waste characteristics for Case 1 and Case 2 used for simulating the sediment transport model.

Case 1

Fraction	Average Size (mm)	Contribution (%)	Settling Velocity (mm/s)	Density (g/cm^3)
1	2.073	5	1333.22413	2.65
2	1.225	15	528.12161	2.65
3	0.855	24	291.068519	2.65
4	0.345	12	71.9839585	2.65
5	0.094	8	8.51686484	2.65
6	0.061	8	3.7781812	2.65
7	0.031	20	1.0066431	2.65
8	0.002	8	0.0042481	2.65

Case 2

Fraction	Average Size (mm)	Contribution (%)	Settling Velocity (mm/s)	Density (g/cm^3)
1	2.203	11	1790.1831	3.0
2	1.625	10	1031.8218	3.0
3	1.330	2	725.5673	3.0
4	0.995	23	442.9635	3.0
5	0.462	8	131.5793	3.0
6	0.069	21	4.3542	2.5
7	0.043	10	1.7424	2.5
8	0.031	15	0.9146	2.5

2.4 Impact Assessment Criteria

The following tolerance limits and impact thresholds for corals habitats (Table 3 and Table 4) have been developed by DHI through a combination of literature reviews and previous experience from environmental mitigation and monitoring plans. The impact assessment criteria for both suspended sediments in water column and sediment deposition on seabed were tabulated as follows:

Table 3. Threshold values for suspended particles in water column

Zone of Impact	Criteria	Percentage
Zone of High Impact	> 25 mg/l	> 7 %
EPO: Total mortality allowed	> 10 mg/l	> 19 %
	> 5 mg/l	> 40 %
Zone of Moderate Impact	> 25 mg/l	2.5 -7 %
EPO: <30% mortality	> 10 mg/l	10 - 19 %
	> 5 mg/l	25 - 40 %
Zone of Influence (Low Impact)	> 25 mg/l	0.5 % - 2.5 %
EPO: 0% mortality	> 10 mg/l	0.5 % - 10 %
	> 5 mg/l	2.5 % - 25 %
No Impact	> 25 mg/l	< 0.5 %
	> 10 mg/l	< 0.5 %
	> 5 mg/l	< 2.5 %

Table 4. Threshold values for sediment deposition on seabed.

Zone of Impact	Sedimentation (mg/cm^2) or (mm/14 days)
Zone of High Impact EPO: Total mortality allowed	> 14 or > 4.9
Zone of Moderate Impact EPO: <30% mortality	5 - 14 or 1.7 - 4.9
Zone of Influence (Low Impact) EPO: 0% mortality	1 - 5 or 0.3 - 1.7
No Impact	< 1 or <0.3

3. Result and discussion

3.1 Impact Assessment

3.1.1 During NE monsoon period

3.1.1.1 Total suspended sediment concentration in water column.

Based on Figure 2, the dark brown color represented the high impact zone while the orange color indicated the area covered by moderate impact zone. The low impact zone were shown by the electron gold color. Major concentration of the suspended particles for low impact zone during NE monsoon in Case 1 were pointing towards SW.

Case 1 **Case 2**

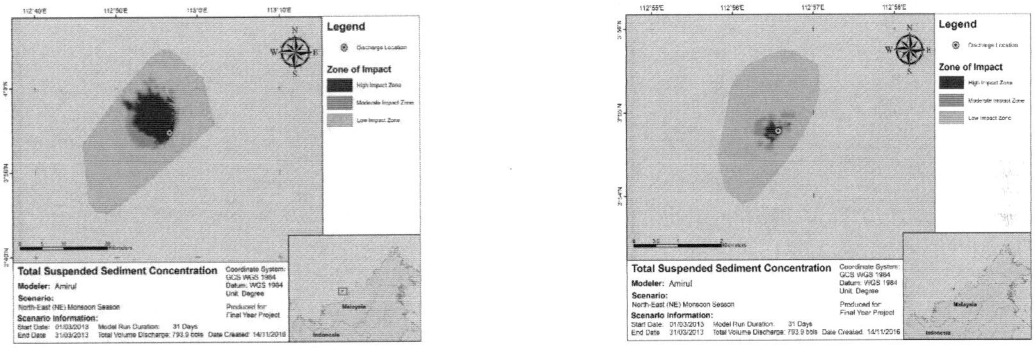

Figure 2. Impact assessment on TSSC model results for Case 1 and Case 2 during NE monsoon.

3.1.1.2 Sediment deposition on seabed

Case 1 **Case 2**

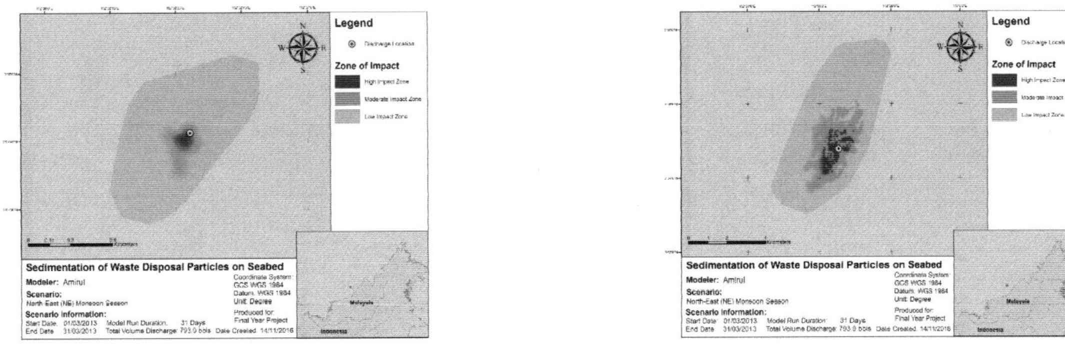

Figure 3. Impact assessment of waste disposal sedimentation on seabed for Case 1 and Case 2 during NE monsoon.

Based on Figure 3, the sedimentation of the particles was likely deposited at the SW direction from the disposal site during NE monsoon. This can be seen by the particles which possessed the high, moderate and low impact zone were mostly accumulated and deposited at the SW direction from the disposal site. For Case 2, the sediment also showed major deposition in SW direction from the disposal site. Minor deposition towards other directions were mainly due to the minor currents movement throughout NE monsoon period.

3.1.2 During SW monsoon period.

3.1.2.1 Total suspended sediment concentration in water column

Case 1 **Case 2**

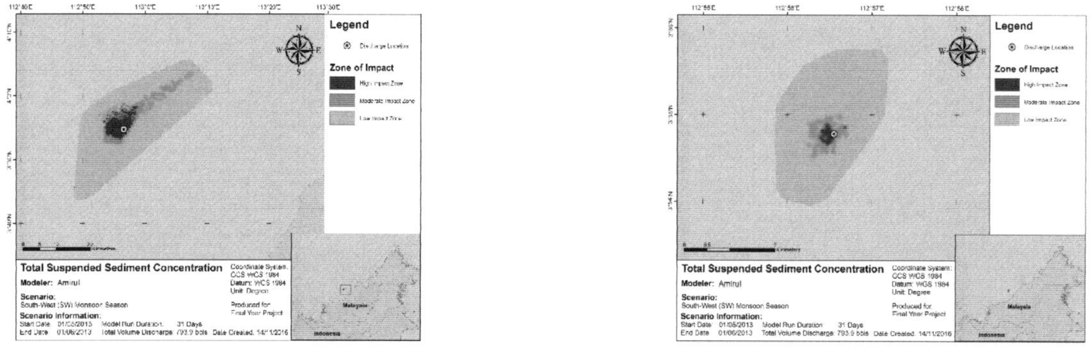

Figure 4. Impact assessment on TSSC model results for Case 1 and Case 2 during SW monsoon.

In terms of SW monsoon scenario, Case 1 showed major concentrations in NE direction for all impact zones. This supported the evidence that during SW monsoon, the major current movements were pointing towards NE direction hence, the concentration of the suspended sediments were likely pointing towards NE direction. Meanwhile, for Case 2; the covered area for low impact zone mostly pointing towards NE direction although there was few covered area in SW direction.

3.1.2.2 Sediment deposition on seabed

Case 1 **Case 2**

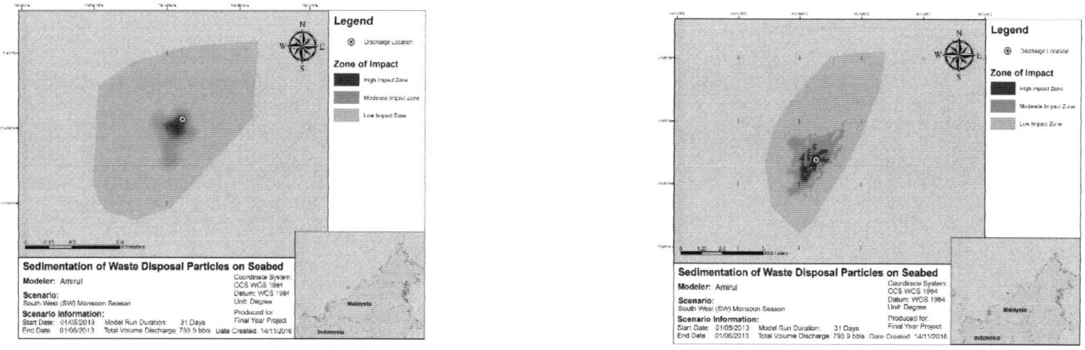

Figure 5. Impact assessment of waste disposal sedimentation on seabed for Case 1 and Case 2 during SW monsoon.

In terms of sediment deposition, Case 1 showed minor deposition towards NE direction for high and moderate impact zone during SW monsoon period. However, deposition of the particles which possessed low impact zone had major deposition towards NE direction compared from previous Inter-monsoon period. Other covered areas for all degree of impact came from previous NE and Inter-monsoon period. However, Case 2 showed major deposition towards NE direction during the SW monsoon period. The low impact zone showed the highest deposition towards NE direction compared to high and moderate impact zone.

IOP Conf. Series: Earth and Environmental Science **88** (2017) 012013 doi:10.1088/1755-1315/88/1/012013

4. Conclusion & Recommendation

Based on the hydrodynamic (HD) modelling for the proposed location, it was proven that during NE monsoon, the currents variations were most likely moved towards South-West direction meanwhile during SW monsoon periods, the currents variations were most likely showed dominant movement towards North-East direction. However, during Inter-monsoon period, there were no major direction which dominated the currents movement. Using the drilling scenarios for two (2) different cases (Case 1 and Case 2) as well as the estimated distributions and settling velocities for different particle sizes, transportation and deposition model were generated. Case 1 had the largest area covered for all degree of impacts in terms of suspended sediment concentrations compared to Case 2 scenario. However, in terms of sediment deposition on seabed, Case 2 proved to have largest covered area compared to Case 1. This was due to differences in drill waste characteristics such as settling velocities, density and particle sizes of the drilling discharge.

As for the recommendations on the impacts of drill waste discharge towards surrounding corals, a real time coral data should be applied to validate the degree zone of impact. The detailed modelling for studying the behaviour of each drill waste particle sizes in terms of suspended sediment in water column and also sediment deposition on seabed should be showing the impacts towards surrounding coral. Other factors such as flocculation effects, critical sheer stress for drill waste particles and topography of the seafloor should be included in the future modelling works.

Full Environmental Impact Assessment (EIA) should also be carried out in the near future for the proposed location to determine whether the proposed disposal site could be used as offshore disposal location for drilling discharge or not. If the impacts from the drilling discharges were to be prevented, it was necessary to reduce the volume discharged and discharge rate to allow the suspended sediment from the discharged materials to settle down on seabed. This could prevented the harm towards marine habitants due to abrading of gills of the marine habitants as well as giving the nearby coral time for self-cleaning.

References

[1] Bernier, R., Galland, E., Glickman, A., Jones, F., Mairs, H., Melton, R., Campbell, J. (2003). Environmental Aspects of the Use and Disposal of Non Aqueous Drilling Fluids Associated with Offshore Oil & Gas Operations. OGP Publications.

[2] Bryden, L., & Carles, I. (1997). Establishing a Simulation Strategy for the Dispersion of the Drill-Cuttings. International Offshore and Polar Engineering Conference. Vol. I, p. 521. Honolulu, USA: The International Society of Offshore and Polar Engineers.

[3] Bryden, L., & Carles, I. (1998). Characteristics of Offshore Drill-Cuttings for Simulation Purposes. International Offshore and Polar Engineering Conference. Vol. II, p. 1-6. Montreal, Canada: The International Society of Offshore and Polar Engineers.

IOP Conf. Series: Earth and Environmental Science **88** (2017) 012014 doi:10.1088/1755-1315/88/1/012014

Relationship of oil seep in Kudat Peninsula with surrounding rocks based on geochemical analysis

Nurul Izzati Azman[1]Siti Nur Fathiyah Jamaludin[1]

[1]Department of Geosciences, Faculty of Geosciences and Petroleum Engineering, Universiti Teknologi PETRONAS, 32610 Bandar Seri Iskandar, Perak.

fathiyah.jamaludin@utp.edu.my

Abstract: This study aims to investigate the relation of oil seepages on the Sikuati area with the surrounding rocks on Kudat Peninsula through geochemical analysis. Outcrops with highly carbonaceous laminated rocks belonging to the Sikuati Member at Kudat Peninsula indicate the presence of oil seepage in this area. A detailed geochemical analysis of the source rock samples and oil seepages from the Sikuati area was carried out for their characterization and correlation. Hydrocarbon prospective of Sikuati Member source rock is poor to good with Total Organic Carbon (TOC) value of 0.11% to 1.48%. The rocks are categorized as immature to early mature with Vitrinite Reflectance (VRo) values of 0.43% to 0.50 %Ro. Based on biomarker the distribution, from Gas Chromatography (GC) and Gas Chromatography-Mass Spectrometry (GC-MS) analysis, the source rock sample shows Pr/Ph, CPI and WI values of 2.22 to 2.68, 2.17 to 2.19 and 2.46 to 2.74 respectively indicating the source rock is immature and coming from terrestrial environment. The source rock might be rich in carbonaceous organic matter resulting from planktonic/bacterial activity which occurred in a fluvial to fluvio-deltaic environment. Overall, the source rock from outcrop level of Kudat Peninsula is moderately prolific in term of prospective and maturity. However, deeper beneath the surface, we can expect the pressure of mature source rocks that generate and expulse of hydrocarbon from the subsurface, migrating through deep-seated faults to form oil seepages in the Sikuati area.

Keywords: Kudat Peninsula, Oil Seepage, Kampung Minyak Sikuati, Geochemical analysis

1. Introduction

This study is conducted at the Sikuati area (Figure 1), within the central part of Kudat Peninsula, Sabah, Malaysia. Sikuati is a coastal village located on the west coast of the northern tip of Borneo which is about 23 km to the south of Kudat Town. The study involves an area of approximately 12.7km x 7.2km and in total of 92km^2 of area.

Oil leakages had been reported since Year 1881 until present day in Kampung Minyak, Sikuati. The area has become one of the tourist attraction in Kudat. Muda, 2010 indicates that the oil seepage is in a tidal mangrove swamp surrounded by an adjacent green area of primary and secondary vegetation. The geology of the area has been mapped as being underlain by the Sikuati Member, one of the members in the Kudat Formation and dated as Early Miocene (Foo & McDonald, 1983; Hall, 2013; Hutchison, 2005; Leong, 1999; Liechti et al., 1960; Sanudin & Baba, 2007; Tongkul, 1990; Van Hattum et al., 2013).

The occurrence of oil seepage in Kampung Minyak, Sikuati has long been known but it remains unclear what had caused the seepage to leak to the surface and where is the source of the migrating hydrocarbons due to complex tectonic regime and structural framework at the northern tip of Borneo. In general, this study focuses on finding the relationship of surface oil seepages with surrounding rock formations. The main aim of this paper is find a relationship of the oil seepage in Kampung Minyak with the surrounding rocks in the area, based on the conducted geochemical analysis.

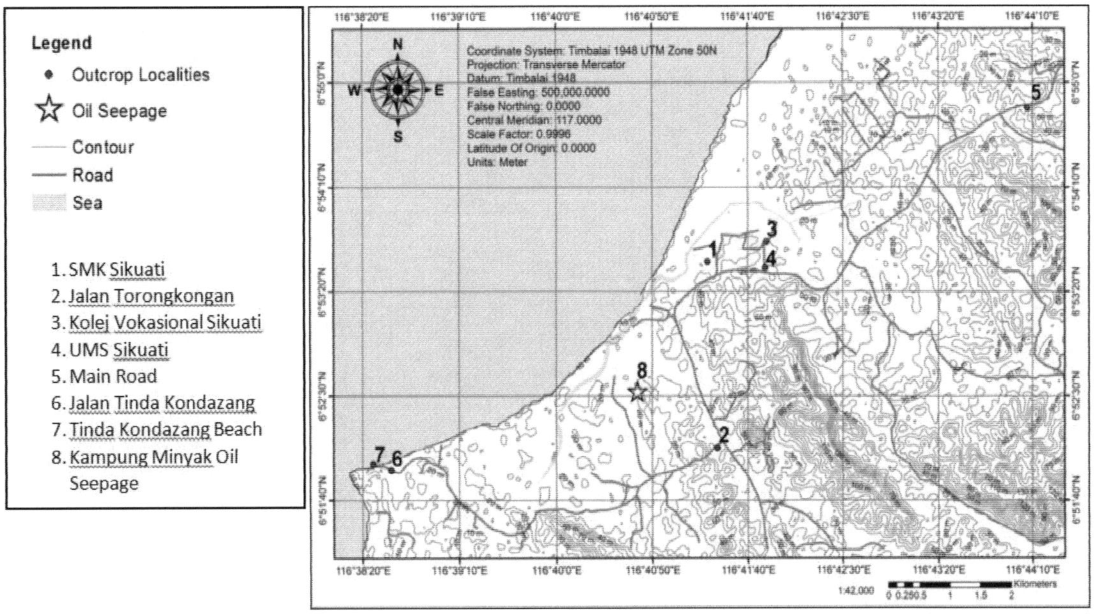

Figure 1: Map showing the location of the study area, topography and outcrop localities (with numbers) in Sikuati, Kudat.

2. Sampling and Methodology

Fieldwork was carried out on outcrops of the Sikuati Member of the Kudat Formation and West Crocker Formation. Eight (8) localities as shown in Figure 1 were visited. Thirteen (13) rock samples from the outcrops and one oil seepage from Pool 1 were collected for further analysis.

2.1 Geochemical Analysis

The geochemistry of the source rock and oil presence is a crucial element in the interpretation of the oil seepage at Kampung Minyak. The aim of this geochemistry study is to provide geochemical analyses on samples from the Sikuati region and also to enhance the result as to establish a good correlation between the Sikuati Member source rocks with the oil seepage present at Kampung Minyak. Figure 2 gives an overview of the workflow and fractionation of the samples and fluid collected during the fieldwork and analyses that have been done to achieve the objective of the study.

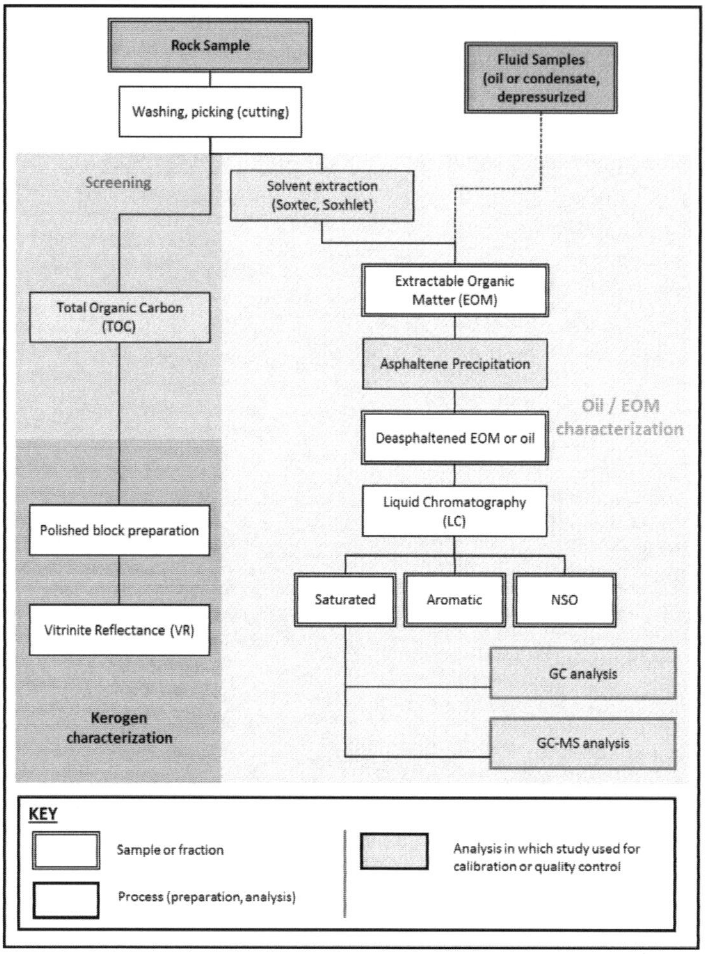

Figure 2: The flowchart showing laboratory analysis done in this study which include TOC, VR, GC and GCMS Analysis.

The samples were analysed for Total Carbon (TC) and Total Organic Carbon (TOC). The source rock samples such as shale and carbonaceous shale were crushed to analyse the carbon content using the Analitik Jena Analyzer multi N/C® 3100 instrument. Most of the samples tested for TC and TOC present some challenges due to uneven distribution of inorganic and organic carbon content within the samples.

Vitrinite Reflectance (VR) Analysis was also performed to indicate the level of maturity of the samples. Vitrinite Reflectance (VR) was performed using calibration numbers ranging from 40 to 100 due to variable amount of vitrinite clasts in the samples. Gas Chromatography (GC) and Gas Chromatography-Mass Spectrometer (GCMS) were also conducted on the samples. GC Analysis allows the calculation of mass hydrocarbon chain in the source rock and fluid from the sample was collected for geochemical characterization and correlation of the biomarker. On the other hand, GCMS Analysis helps in determining the environment of deposition and the maturity-parameter which is dependent on the degree of microbial degradation (Waples and Machiharia, 1991).

3. Results and Discussion

3.1 Total Organic Carbon (TOC) Analysis

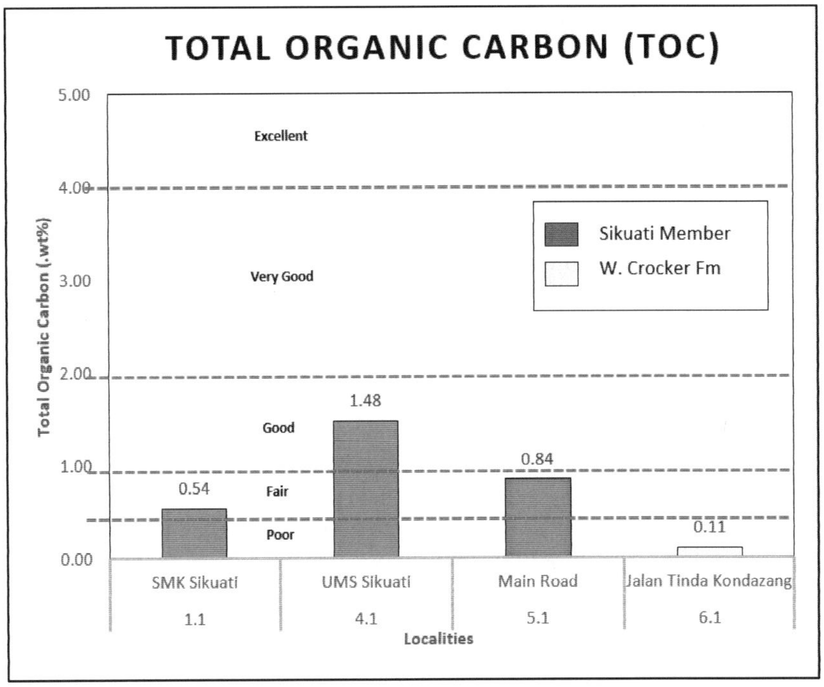

Figure 3: Total organic carbon (TOC) distribution according to localities

TOC analysis were conducted on samples collected from four localities, which are, one from the West Crocker Formation and the other three from the Sikuati Member. These samples were selected as they are shale, mudstone and carbonaceous rocks. The analysis shows that TOC for Sikuati Member varies from 0.54 to 1.48%wt and TOC from the Crocker Formation mudstone is 0.11%wt (Figure 3). The highest TOC content (1.48 %wt.) was recorded from the UMS Sikuati mudstone (Sample 4.1). All these samples can be characterized as good source rock based on Peters and Cassa (1994). In contrast, the SMK Sikuati mudstone (Sample 1.1) and Main Road (Sample 5.1) mudstone are classified as fair source rock with a TOC of 0.54%wt and 0.84%wt respectively. On the other hand, the sample from Crocker Formation (Sample 6.1) located at Jalan Tinda Kondazang is labeled as poor source rock potential.

3.2 Vitrinite Relfectance (VRo) Analysis

Mudstone samples from the Sikuati Member which are mainly shale with carbonaceous laminae give VRo values in between 0.43% to 0.50% Ro which indicate an immature to marginally early mature oil window. These results reflect that outcrops at the Kudat Peninsula had been buried with a short intense of thermal burial history.

IOP Conf. Series: Earth and Environmental Science **88** (2017) 012014 doi:10.1088/1755-1315/88/1/012014

3.3 Gas Chromatography (GC) Analysis

Pristine to Phytane Ratio (Pr/Ph) are used to determine the depositional environment through GC analysis. Source rock sample 5.1 had recorded a Pr/Ph ratio of 2.22, 2.68 for GC and GC-MS m/z 85 indicating that rocks from the Sikuati Member was deposited in a terrigenous environment. Based on Figure 4, the cross-plot of Pr/C17 vs Ph/C18 for Main Road (Sample 5.1) shows that the Sikuati Member deposited in a mixed environment with slight or no influence of marine environment. Meanwhile, biomarker distribution of the oil seepage collected at Kampung Minyak, Sikuati (Pool 1) recorded Pr/Ph ratio of 5.63 indicating that the oil seepage deposited in oxidizing environment. In addition, the Pr/C17 vs Ph/C18 cross plot showing the oil coming from peat-coal environment (Figure 4). These results showed that the rock originated from a swampy environment. Waxiness index was used to determine the amount of land-derived organic matter in sediments, based on the assumption that terrigenous material contributes high molecular weight normal alkane components (Peters et al., 2005). Both sample from Main Road (Sample 5.1) and Pool 1 showed relatively high waxiness index varying from 2.18 to 2.74, indicating high contribution from land plants for both source rock and oil seepage

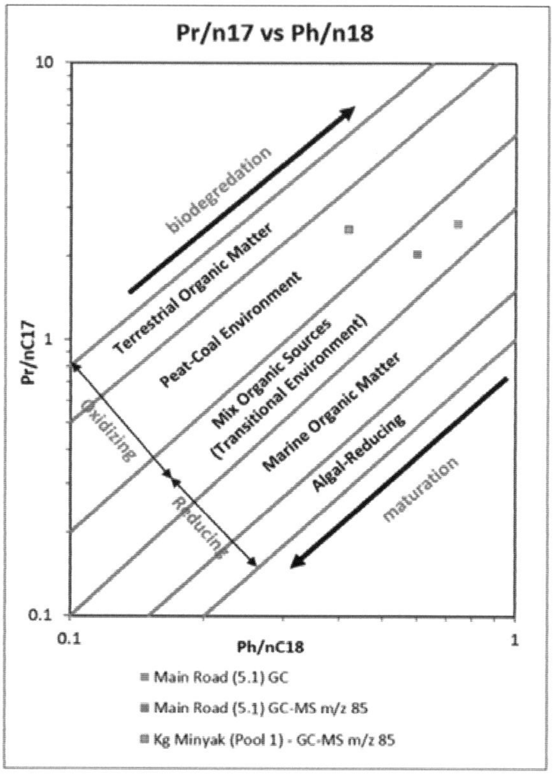

Figure 4: Cross-plot of Pr/C17 vs Ph/C18 showing depositional environment interpretation based on GC result.

3.4 Gas Chromatography- Mass Spectomertry (GCMS) Analysis

The maturity of the source rock and oil-seepage sample were tested based on GCMS Analysis. The Ts/ (Ts+Tm) values can be used to assess the maturity of a sample but could be influenced by maturation, organic matter typing and lithology (Moldowan et al., 1985). The ratio between Ts and Tm based on Ts/ (Ts+Tm) values for both samples recorded 0.08 and 0.42 respectively. The sample from Main Road (Sample 5.1) recorded a low C32-H reading of 0.39 which indicated the sample is immature. C30 M/C30 H and C23 Tri/ C30 H readings from Main Road (Sample 5.1) recorded 0.74 and 0.32 respectively. This indicates the immature condition which suggests the sample have yet to enter the oil window generation.

On the other hand, the oil seepage at Kg.Minyak recorded a C32-H reading of 0.62 indicating the sample is mature and had reached the oil window. This is suggested by C30M/C30H and C23 Tri/C30H readings of 0.15 and 0.03 respectively. Based on the source dependent parameter from the GCMS Analysis, both samples indicate that the source rocks originated from distributive fluvial system rich in organic matter.

4. Conclusion

Results gathered from geochemical analysis in this study showed that rocks from the outcrops are immature and originated from a mixed terrestrial/marine setting, but with a preponderance of terrestrial organic matter. On the other hand, oil seepages had indicated that the oil is coming from a peat-coal environment. Thus, the source rock might be rich in carbonaceous material organic matter originating from planktonic/bacterial activity which occurred in a fluvial-deltaic environment. From this study, the aim to investigate the relationship between the oil seepage in Kampung Minyak, Sikuati with the surrounding geology around the Sikuati area has been achieved. Overall, the source rock from outcrops in Kudat Peninsula are moderately prolific in terms of prospectivity and maturity. However, deeper beneath the surface, we can expect the existence of matured source rock generating and expulsing hydrocarbon from the subsurface and migrating through deep-seated faults to form oil seepages in this study area.

Acknowledgement
Our team would like express our gratitude to Universiti Teknologi PETRONAS for providing funding to conduct the fieldwork and geochemical analysis through STIRF Grant 0153AA-D67 as well as YUTP-FRG Grant 0153AA-E71 to present and publish this paper. Our special thanks to Dr Khairul Azlan Mustapha and Prof. Dr W.Hasiah Abdullah from Department of Geology, Universiti Malaya for providing the facilities and conducting the laboratory analysis at reasonable price at their Geochemistry Laboratory.

References

[1] Muda, J. (2010). Oil seepages at Kampung Minyak, Kudat Peninsula, Northern Sabah: Potential for geotourism development. Bulletin of the Geological Society of Malaysia, 56, 49-52.
[2] Tongkul, F. (2006). The structural style of Lower Miocene sedimentary rocks, Kudat Peninsula, Sabah. Geological Society of Malaysia Bulletin, 49, 119-124.
[3] Peters, K. E., & Cassa, M. R. (1994). Applied source rock geochemistry. Memoirs-American Association of Petroleum Geologists, 93-93.
[4] Waples, D. W., & Machiharia, T. (1991). Biomarkers for geologists.
[5] Foo, W. Y., & McDonald. (1983). *Sedimentology of Block 8T*. Offshore North Sabah Internal Report.
[6] Hall, R. (2013). Contraction and extension in northern Borneo driven by subduction rollback. *Journal of Asian Earth Sciences, 76*, 399-411.
[7] Hutchison, C. S. (2005). Geology of North-West Borneo: Sarawak, Brunei and Sabah: Elsevier.
[8] Leong, K. (1999). Geological Setting of Sabah *The Petroleum Geology and Resources of Malaysia* (pp. 473-497). Kuala Lumpur: PETRONAS.
[9] Liechti, P., Roe, F. W., Haile, N. S., & Kirk, H. (1960). The geology of Sarawak, Brunei and the western part of North Borneo.
[10] Tongkul, F. (1990). *Structural Style and Tectonics of Western and Northern Sabah*. Paper presented at the GSM Petroleum Geology Seminar 1989.
[11] Sanudin, H. T., & Baba, M. (2007). *Pengenalan kepada stratigrafi*: Unit Penerbitan, UMS.

Investigation of Thermophysical Parameters Properties as A New Work Flow for Enhancing Overpressure Mechanism Estimation. Case Study: Miri Area, West Baram Delta.

Kurniawan Adha[1], Wan Ismail Wan Yusoff[2], Luluan Almanna Lubis[3]

[1,2,3]Petroleum Geoscience Department, Universiti Teknologi PETRONAS
kurniawanadhaa@gmail.com[1], wanismail_wanyusoff@utp.edu.my[2],
luluan.lubis@utp.edu.my[3]

Abstract. Determining the pore pressure data and overpressure zone is a compulsory part of oil and gas exploration in which the data can enhance the safety with profit and preventing the drilling hazards. Investigation of thermophysical parameters such as temperature and thermal conductivity can enhance the pore pressure estimation for overpressure mechanism determination. Since those parameters are dependent on rock properties, it may reflect the changes on the column of thermophysical parameters when there is abnormally in pore pressure. The study was conducted in "MRI 1" well offshore Sarawak, where a new approach method designed to determine the overpressure generation. The study was insisted the contribution of thermophysical parameters for supporting the velocity analysis method, petrophysical analysis were done in these studies. Four thermal facies were identified along the well. The overpressure developed below the thermal facies 4, where the pressure reached 38 MPa and temperature was increasing significantly. The velocity and the thermal conductivity cross plots shows a linear relationship since the both parameters mainly are the function of the rock compaction. When the rock more compact, the particles were brought closer into contact and making the sound wave going faster while the thermal conductivity were increasing. In addition, the increment of temperature and high heat flow indicated the presence of fluid expansion mechanism. Since the shale sonic velocity and density analysis were the common methods in overpressure mechanism and pore pressure estimation. As the addition parameters for determining overpressure zone, the presence of thermophysical analysis was enhancing the current method, where the current method was the single function of velocity analysis. The presence of thermophysical analysis will improve the understanding in overpressure mechanism determination as the new input parameters. Thus, integrated of thermophysical technique and velocity analysis are important parameters in investigating the overpressure mechanisms and pore pressure estimation during oil and gas exploitation in the future.

Keywords – Overpressure, Thermal Conductivity, Temperature, Heat Flow

1. Introduction

Many methods have been suggested and used to predict the accurate pore pressure, even in shallow or in deep reservoir (i.e overpressure zone). These parameters are crucial for exploration, drilling and resource estimation since the accurate pore pressure can affect safety and profit. Eaton method (Eaton, 1975) has been frequently used for determining the pore pressure and overpressure zone. In this method, one of main parameters that should be constructed is normal compaction trend (NCT) of an attribute of slowness and the deviation which come from that parameters was used for developing the trend and constructing the exponent to predict the high pore pressure which named as Eaton-exponent (typically 3 or 4), which assumed that the higher pore pressure was only developed by undercompaction of sediments. The effective pressure or effective stress refers to the difference between the overburden pressure and pore pressure. Based on that concept, the velocity will decrease when the sound travels from normal pressure region to overpressure zone. This happened due to effective stress or effective pressure in overpressure zone is lower than normal zone (Adha et. all, 2017). In normal condition, the pore pressure is equal to hydrostatic pressure or water column from the formations depth to sea level. Formation pore pressure is constantly estimated or predicted based on well log analysis in combine with terzaghi hypothesis, which function of compaction the overburden stress and effective stress consist of matrix stress and fluid pressure. The difference between the *overburden pressure (Po)pore pressure (Pp)* is *effective stress (σ)*which can be written as following equation 1.

$$\sigma = Po - PP \quad (1)$$

If pressure is higher or less than the normal pressure, it is called as abnormal pressure. Overpressure generated by varies mechanism which has different characteristic with dissimilar treatment. Normally, abnormal pressure (above or below normal pressure) mechanism is caused by i.e *stress related mechanism, fluid volume increase mechanism and Fluid Movement and Buoyancy mechanism.* Those mechanisms were happened by the rapid change of formation due to different medium and fluid properties at particular depth. However, from the all mechanisms, two main mechanisms are highlighted, which are disqualibirium compaction and fluid expansion where that mechanism has significant contribution in overpressure generation (Swarbick and Hills, 1999) (Osborne Swarbick, 1997).

However, there are some weaknesses in pore pressure prediction through sounds method, due to lack of seismic or velocity data and lack of understanding on the effect of pressure on physical properties of medium (Huffman, 2002). The effective pressure tends to be overestimated resulting in underestimated of pore pressure (Zhang, 2011). In addition, on 1995 through noteworthy publication, Bowers suggested that in very high pore pressure the inclusion of thermal mechanism was essential for account that mechanism since the mechanical compaction alone could make underestimate in pore pressure determination.

In this paper, the relationship of thermophysical properties investigations including temperature gradient, thermal conductivity and heat flow were carried out to investigate and predict the pore pressure and overpressure mechanism along the well. The study was worthwhile for tackle the estimation in overpressure mechanism determination. Since the mechanism that reducing the velocity not only the function of overpressure zone, it is hope that this analysis will justify the existence of sonic velocity trend. This shall also investigate the relationship of overpressure occurrences with thermal and depth profile. With this observation, the anomalies of overpressure due to rock properties and temperature can be determined.

1.1 Regional Geology

The Baram delta province is elongated in triangular shape, with its apex occurring onshore and cantered in Brunei and north eastern coastal area of Sarawak. This province exists in onshore and expend into offshore with cover the whole width of Brunei water, West Baram delta (Sarawak Province), East Baram delta (Sabah Province) and covering an area of 7500 km^2, including 2500 km^2 on shore. In western area, the Baram delta is margin by the West Baram line, a large system, northeast heading, down-to-the-basin faults that separates the delta from the older balingan and central luconia province. Meanwhile, the eastern margin is defined by the Morris Fault-Jerodung. Miri formation itself represents the lower part of Sarawak basin (West Baram delta) which is late Eocene to recent in age (Tingay, 2009). Distribution of overpressure in Miri Field can be divided into outer shelf and inner shelf. Inner shelf and outer shelf have different petroleum play types and different stratigraphic units. Predominant mechanism which is disequilibrium compaction of pro delta shale located in outer shelf is gradually shallow towards Northwest, onset at top of pro delta shale, meanwhile fluid expansion mechanism of deltaic sequence in Inner shelf has suddenly varied in depth to onset and highly compartmentalized. Temperature plays important role in the development of overpressure in the Miri Field. In certain temperature, the mechanism of disequilibrium compaction will change to the fluid expansion. Thus, temperature may lead to fluid expansion and cementation as identified (Tingay 2011).

1.2 Thermophysical Properties Overview

Thermal conductivity is defined as the rate of energy which transfer across a unit area under potential of a unit temperature gradient which has been expressed in $W(m^oK)^{-1}$ which is the function of rock type, anistrophy, pressure and temperature. Thermal conductivity across a reservoir rock is the total thermal conductivity or bulk conductivity of all medium which consists of rock properties and fluids inside the medium.

The thermal conductivity can be assumed as a rock property based on two main reasons. Firstly, the mineral thermal is much higher than fluid conductivity. Secondly, the mass fraction of mineral is higher than total fluid per bulk volume. A study found that the relationship between thermal conductivity and pressure is nonlinear relationship regardless of the rock type (Illmutdon, 2006). Yet, in general the thermal conductivity will increase with overburden pressure, since the both parameters are the functions of rock properties and net vertical forces.

Heat transfer across the earth from the core to surface through conduction and convection, which can be expressed as the product of temperature gradient and thermal conductivity is given by Fourier's law. The heat flow (Q) is directly proportional to temperature gradient in express form in Equation 2.

$$Q=-KdT/dy \quad (2)$$

Where K is the function thermal conductivity and dt/dy is the temperature gradient. T is in unit of ^0C while Q is in mW/m^2.

2. Methodology

The analysis used the well data, RFT data and BHT data from offshore Miri, Baram delta. For answering the objective, the study was divided in several stages. It was started with simple petrophysical analysis of rock properties which was used to determine the thermophsyical parameters such as *thermal conductivity, temperature and heat flow*. After that stage, the analysis sin crossplot analysis was applied to determine and characterize the behavior of each thermal parameters with pressure and velocity. In the final stage, those parameters were tied together for analyzing the trends of overpressure distribution along the well (vertical scale) since this analysis was not involved in the seismic data (Figure 1).

The modified workflow was designed to achieve the objective of this study. In general, the stages of this workflow were divided into three main stages. First, petrophysical analysis as in the processes was started with data preparation and quality checked, counties with petrophysical analysis for determining the reservoir zone, volume of fine, porosity as the function of thermal conductivity and heat flow determination. In addition, the pore pressure estimation corrected with pressure data was also investigated in this stage. This section involved cross plotting of the field data.

For the next stage, the thermophysical parameters were conducted with the input from well data. Thermal facies, zoning, heat flow and compaction factor were determined in this stage. Thermal facies were determined using empirical equation which developed by previous study (Yusoff, 1993) presented in Table 1.

In the final stage, each zone interest and top of overpressure were investigated to determine and characterize the trend of thermophysical parameters which tied with velocity for overpressure determination.

IOP Conf. Series: Earth and Environmental Science **88** (2017) 012015 doi:10.1088/1755-1315/88/1/012015

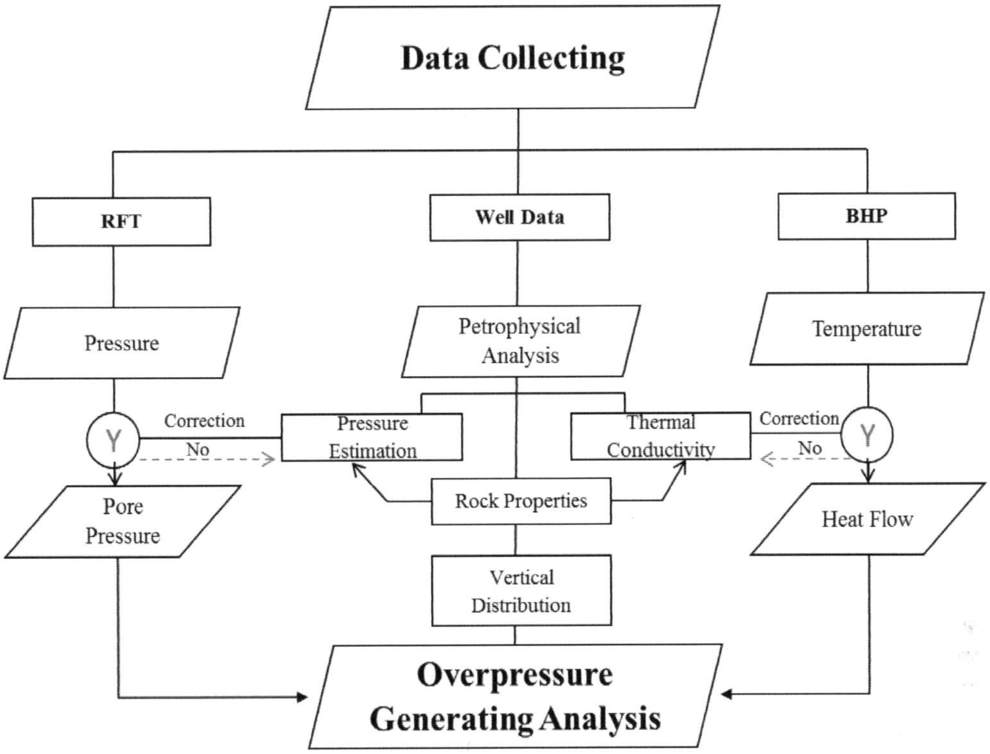

Figure 1. General work flow

Table 1. Empirical equation of thermal facies (Yusoff, 1993)

Thermal Facies. (Kf)	Limit of Log Parameters	Empirical Equation
Kf_1	Vf<20, Porosity < 36 %	$Kemp = 6.86(0.99)^{vf}(0.98)^{Porosity}$
Kf_2	20 ≤Vf< 40 , Porosity <35 %	$Kemp = 6.43(0.99)^{vf}(0.96)^{Porosity}$
Kf_3	40≤Vf<60 , 18% <Porosity <35 %	$Kemp = 10.59(1.00)^{vf}(0.98)^{Porosity}$
Kf_4	60≤Vf<80 , 18% <Porosity <39 %	$Kemp = 5.83(0.99)^{vf}(1.00)^{Porosity}$
Kf_5	13≤Vf<30 , 33% <Porosity <60 %	$Kemp = 13.41(1.09)^{vf}(0.91)^{Porosity}$
Kf_6	30≤Vf<40 , 35% <Porosity <49 %	$Kemp = 12.95(1.03)^{vf}(0.98)^{Porosity}$
Kf_7	40 ≤Vf>85 , 35% <Porosity <61 %	$Kemp = 15.24(1.01)^{vf}(0.94)^{Porosity}$

3. Result and Discussion

3.1 Petrophysical Analysis

Petrophysical analysis was conducted to carry out the rock properties of formation such as porosity and volume of fine or volume of shale. Volume of fine was determined by gamma ray while the neutron porosity (NPHI) was applied in this analysis to determine empirical thermal conductivity (Figure 4). In addition, the sandstone interval or zone was identified from several major traits which were from gamma ray (GR) log, resistivity log and also density-neutron log. In this well, the cut off for sand, silt and shale was determined. Clean sandstone (0-35 API) and siltstone were set at a range of 35-75 API while shale range is more than 75API (Figure 4). On the other hand, the interpretation of resistivity indicates the potential of fluid contents in the formation, while high resistivity indicates oil or gas (hydrocarbon). The gap between deep resistivity log (LLD) as well as shallow resistivity (LLS) log indicates the permeability of the layers identified. Coarsening upwards which is mostly stratigraphic sequence was found on this well. These facies indicate that the sediment was deposited in delta environment while the sedimentation was influenced by low energy current.

3.2 Thermophysical Parameters and Pressure Correlation

Thermal facies were used to differentiate the variance of thermal conductivity along the depth of well. It has been accepted that thermal conductivity of the formation are function of the lithology and porosity. In clastic sediments, the thermal facies were increasing due to the reducing of porosity as the function of burial and compaction. There were 4 thermal facies identified in "MRI 1" well. That was thermal facies 1 to thermal facies 4 from depth 1800 m ft-30480 ft.

According to the data, the thermal facies 3 had the highest values of thermal conductivity, followed by facies 1, facies 4 and facies 2. Not only the values of thermal conductivity, the distribution of thermal facies 3 was dominant along the wells (Table 2).

Table 2. Average thermal conductivity of Well MRI 1

Thermal Facies	Log Parameters		Average Thermal Conductivity
	Volume of Fine (Vf)	Porosity (Øn)	
Kf_1	17.4	33	3.91
Kf_2	21.3	29	1.56
Kf_3	57.20	24	5.97
Kf_4	63.24	22	2.73

The thermal conductivity was mandatory in heat flow determination where the heat flow is the function of thermal conductivity and temperature gradient. The heat traveled from the lowest thermal conductivity to high thermal conductivity. The movement of the heat transfer was influenced by the temperature, where the heat anomaly can be developed as the function of overpressure generating. In general, the high thermal conductivity tended to be developed high heat flow with the control by of formation temperature. In addition, the Figure 2 shows that similar behavior from velocity and thermal conductivity, where both parameters are the function of rock compaction and rock properties.

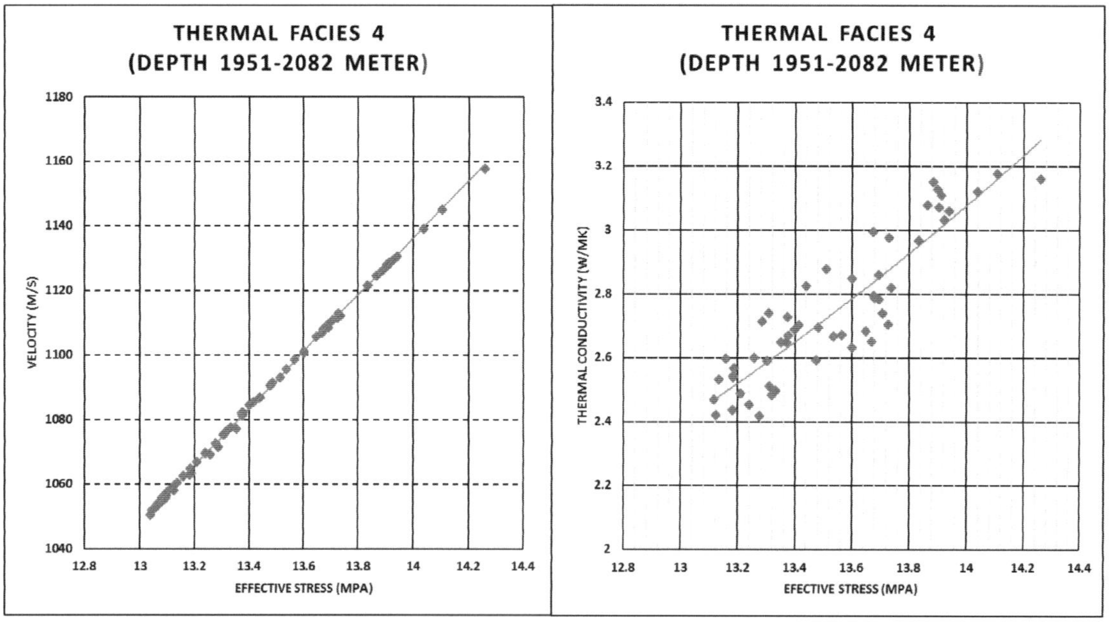

Figure 2. Velocity and thermal conductivity trends against effective stress for thermal facies 4, where both parameters were increased due to increasing of effective stress. That condition was proven by rock compaction theory.

3.3 Pressure and Thermal Depth Profiles

The pressure data was carried out from pore pressure estimation using Eaton method which was justified by the Repeat Formation Test (RFT) and Drill Stem Test (DST). The temperature and thermal conductivity were derived from estimated formation temperature which was justified by BHT data and thermal derived from petrophysical analysis. Meanwhile, the vertical heat flow profiles were determined by the multiple the average thermal and temperature gradient.

Based on this analysis, the top of overpressure was identified at 2872 meter depth (figure 3). On that particular depth, the pore pressure moved from the hydrostatic line to lithostatic line. In addition, the temperature and the heat flow show the similar trend where the pore pressure was increasing in 5.8 MPa followed by the increasing of 20^0C in the temperature.

The increasing of the heat flow and reducing of fluid flow indicator was interpreted as fluid expansion. That condition was interpreted as in fluid expansion the fluid was trapped and tried to escape and causing the increasing of heat flow and overpressure was developed. As mentioned in introduction, the fluid expansion is of mechanism that generates the overpressure. The movement of the fluids will be one of factors that increase the temperature. In addition, the convex form of the temperature indicates the movement of the fluid was horizontal. However, the horizontal movement of the fluid was tended to seal existence and hydrocarbon present (Figure 3).

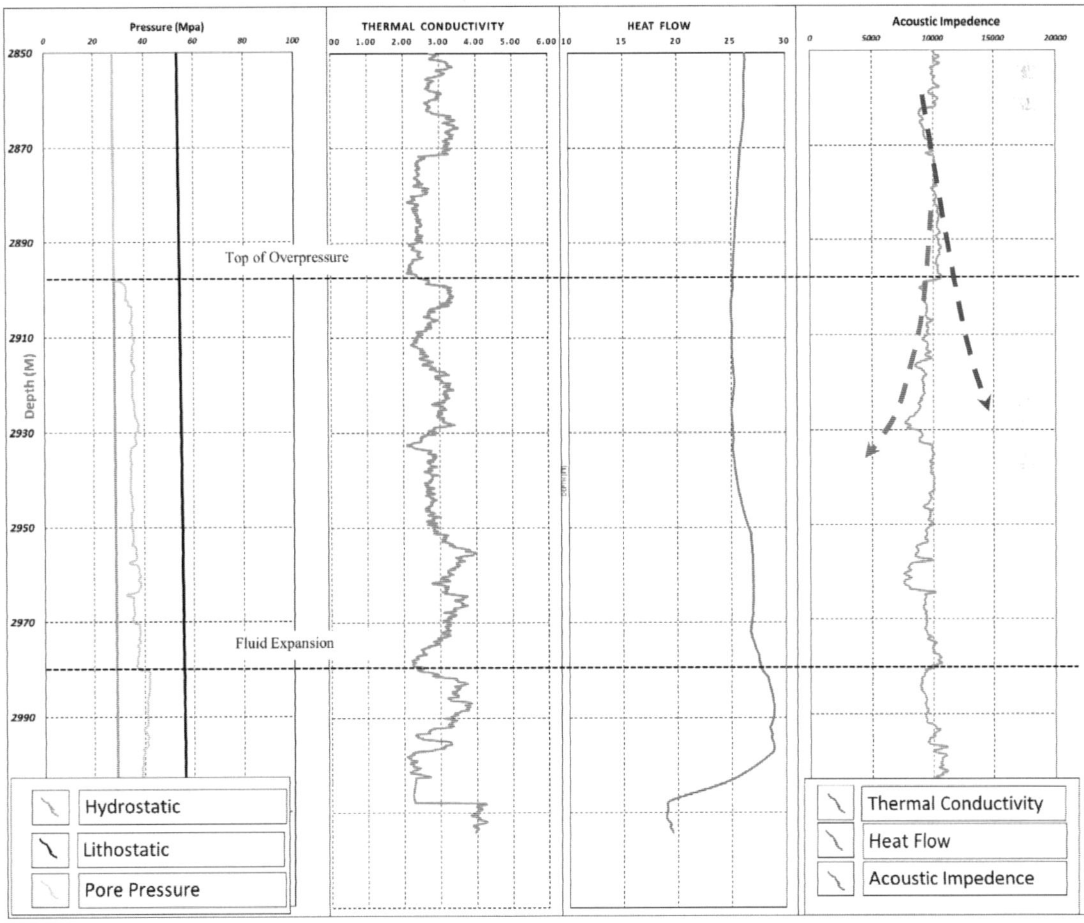

Figure 3. Overpressure mechanism respecting the thermophysical parameters, the top of overpressure (2872 m depth) was indicated when the pressure going high while the AI was decreasing. The trend shows that AI should increase due compression (blue line with arrow), but the AI going decreasing due to overpressure was developed.

4. Conclusion and Recommendation

Several conclusions can be drawn from this study. First, from petrophysical analysis of this well, the cut off for sand, silt and shale was determined. Clean sandstone (0-35 API) and siltstone were set at a range of 35-75 API while shale range is more than 75 API. On the other hand, the interpretation of resistivity indicates the potential of fluid contents in the formation, while high resistivity indicates oil or gas (hydrocarbon).

Through four thermal facies identified in "MRI 1" well (from the depth of 1800 m -30480 m), data shows the thermal facies 3 had the highest values of thermal conductivity, followed by facies 1, facies 4 and facies 2. In addition, this study found that velocity and thermal conductivity had similar behavior (trend) against effective stress. Along the well, the sandstone and sandy shale was dominated.

Lastly, based on this analysis, the top of overpressure was identified at 2872 m depth. On that particular depth, the pore pressure moved from the hydrostatic line to lithostatic line. In addition, the temperature and the heat flow show the similar trend where the pore pressure was increasing in 5.8 MPa followed by the increasing of 20^0C in the temperature. The increasing of the temperature was interpreted as fluid expansion. As mentioned in the introduction, the fluid expansion is of mechanism that generates the overpressure.

The investigation of thermophysical properties provides the enhancement for pore pressure estimation. In addition, by evaluating and considering the thermophysical parameters then correlating them with burial history or geochemical data, more accurate pore pressure estimation will be obtained.

5. Acknowledgements

The authors express their gratitude to Universiti Teknologi PETRONAS (UTP) and PETRONAS Sdn Bhd for providing data and financial assistance on this study.

6. References

1. Adha Kurniawan, Wan Yusoff Wan Ismail Lubis Luluan Almanna, Mhd Hanapiah Nurhajeerah. "Chapter 33 Characteristics of Pore Pressure and Effective Stress Changes in Sandstone Reservoir Through Velocity Analysis Approach Due to Hydrocarbon Production", Springer Nature, 2017

2. Bell, D.W., 2002, Velocity Estimation for Pore Pressure Prediction, in Huffman, A. L. and Bowers, G.L., eds, Pressures Regimes in Sedimentary Basin and Their Prediction, AAPG Memoir 76, p. 177-215

3. Bowers, G. (1995). Pore pressure estimation from velocity data: Accounting for overpressure mechanism besides undercompaction. *SPE Drilling & Completion*, 89-95.

4. Bowers, G. (2002). Detecting high overpressure. *The Leading Edge*, 174-177.

5. Huffman, A.R(2002). The future of pressure prediction using geophysical methods. *Pressure regimes and sedimentary basins and their prediction: AAPG Memoir 76, 217-233*

6. G.N Boitnott, et all (2009). Pore Pressure Depletion and Effective Stress Issues in Gulf of Mexico's Lower Tertiary Play. *SPE Annual Technical Conference.* New Orlans

7. Madon, M. (2006). Overpressure history of the Malay Basin, Ofshore Peninsular Malaysia. *Petroleum Geology Conference & Exhibition* (pp. 135-144). Kuala Lumpur: Geological Society of Malaysia.

8. Madon, M., Abolins, P., Hoesni, M., & Ahmad, M. (1999). The petroleum Geology and Resources of Malaysia. In PETRONAS, *Malay Basin* (pp. 173-185). Kuala Lumpur: Petroliam Nasional Berhad (PETRONAS).

9. Tingay, M., Hillis, R., Swarbrick, R., Morley, C., & Damit, R. (2011). Origin of Overpressure and Pore Pressure Prediction in the Baram Delata Province, Brunei. *AAPG Memoir*, 65-85.

10. Tingay, M., Hillis, R., Swardick, R., Morley, C., & Damit, A. R. (2009, January). Origin of overpressure and pore-pressure prediction in the Baram province, Brunei. *AAPG Bulletin*, pp. 51-74.

11. Wahab, M., Asraff, A., Ismail, J., & Ibrahim, A. (2013). Significant Hydrocarbon Accumulation in Deep Overpressured Play of Baram Delta, Malaysia. *EAGE Conference & Exhibition.* London, UK.

12. Yusoff_, Wan Ismail Wan (1993) Geothermics of the Malay basin, o_shore Malaysm, Durham theses, Durham University. Available at Durham E-Theses Online: http://etheses.dur.ac.uk/5537/

13. Zhang, J. "Pore pressure prediction from well logs: Methods, modifications, and new approaches", Earth-Science Reviews, 201109

Diagenetic history of late Oligocene-early Miocene carbonates in East Sabah, Malaysia

N S Zainal Abidin[1], R R Raymond[2] and N S I Bashah[3]

[1,2,3] Department of Geosciences, Universiti Teknologi PETRONAS, 32610 Bandar Seri Iskandar, Perak, Malaysia.

norsyazwani.z@utp.edu.my

Abstract. Limestones are particularly susceptible to drastic early diagenesis modifications, mainly cementation and dissolution. During the early Miocene, a major tectonic deformation has caused a widespread of uplift in Sabah. This has resulted change in depositional environment from deep to shallow marine, which favours the deposition of Gomantong Limestone. This study aims to investigate the diagenetic history of Gomantong Limestone in East Sabah. Thorough understanding of the diagenetic processes may provide data to unravel the tectonic activities which affected the reservoir quality of the carbonates. Combining the data from comprehensive petrographic analysis, and Scanning Electron Microscopy (SEM) of 30 samples, two main cements type were identified. These are microcrystalline cement and Mg-calcite cement of granular and blocky mosaics which are dominantly seen in all samples. The sequence of diagenesis events are determined as (1) micritization; (2) grain scale compaction; (3) cementation (pore-filling); (4) mechanical compaction and cementation infilling fractures and (5) chemical compaction. These diagenetic events are interpreted as reflection of changes in diagenetic environment from shallow marine to deep burial. The massive cementation in the Gomantong Limestone has resulted into a poor reservoir quality.

1. Introduction

Tertiary carbonates in Southeast Asia are extensive and diverse particularly on the east side of Borneo. These carbonate deposits are among the major hydrocarbon exploration targets, including drilling that has been carried out in the Sulu Sea in offshore eastern part of Sabah [1]. Through an unpublished seismic data, Gomantong Limestone may have an extension in the offshore where it forms isolated limestone mounds occurring along the boundary between Labang Formation and Tanjong Formation [2].

Diagenetic processes in carbonates can either create or destruct porosity and permeability. Hence, it is also important to note the influences of diagenetic process on the reservoir quality of Gomantong Limestone as diagenesis holds a major control in determining the economical values of the reservoir rocks. However, due to the lack of exposure in the past, little study has been done on the onshore carbonate deposits in Sukau, East Sabah. The outcrops are now being exposed due to rapid plantation development and quarrying activities (Figure 1).

This paper documents the diagenetic features and processes that have affected carbonates of the Gomantong Limestone in Sukau, east of Sabah, Malaysia. The occurrence and distribution of the diagenetic features may provide further information and understanding of limestone mounds that may have extended to the offshore Sabah Basin. This study was based on field observations and petrographic studies of ten outcrops and samples from identified locations across 50 km road of Sukau. Around 20 rock samples were collected from different sampling locations (Figure 1). Petrographic analyses are based on examination of 20 thin sections impregnated with blue-dyed epoxy. Petrographic examination of the samples was performed using a polarizing microscope and selected samples were examined with a scanning electron microscope for further information on the cements.

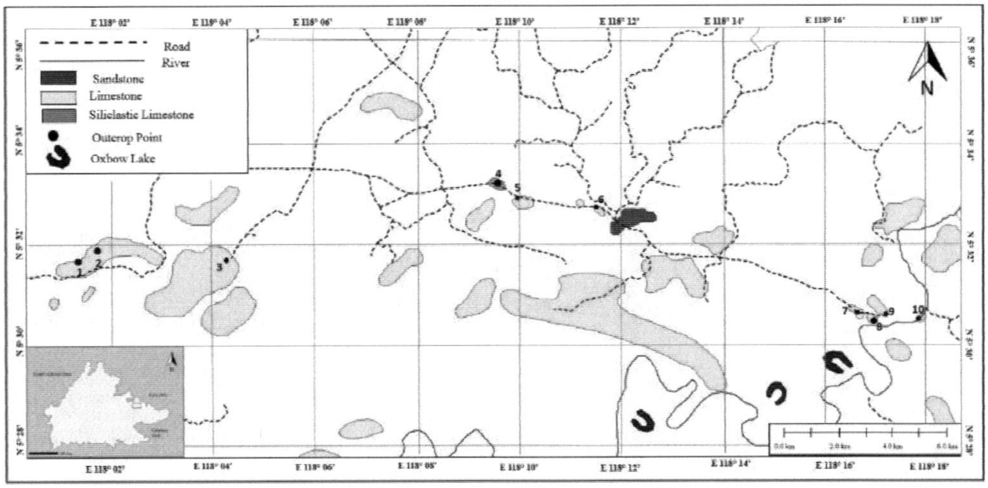

Figure 1. Geological map showing the overview of limestones distribution in Sukau.

2. Geological Setting

Gomantong Limestone is a shallow water coralline limestone formed in the late Oligocene-early Miocene period when there was a sudden change of shallowing of environment, due to uplift event. The outcrops are located at the northern part of the lower Kinabatangan River in Sukau, east of Sabah and can be dated back to early Miocene where there was period of progressive tectonic deformation in the fore-arc region during the subduction and widespread mélange development in Sabah. There are series of isolated exposures of limestone forming the East-West trending belt of steep-sided hill stretching from the Lahad Datu Road (close to the Kinabatangan River) to the east of Sukau. Bands of limestone outcrops can be found extending approximately 15km in a North-South direction. Some of these limestones can be accessed by the Sukau Road. It is argued that the limestone of Gomantong may extend to the offshore, forming isolated limestone mounds occurring along the boundary between Labang Formation and overlying Tanjong Formation (Figure 2) [2].

The formation of Gomantong Limestone occurred during the same period as the Ayer melanges [3]. This can be seen in the Gomantong Caves, where the Gomantong Limestone reaches approximately 300m in thickness. This cave suggested that the Gomantong Limestone apparently overlies the Oligocene Labang Formation unconformably. A detailed study on Gomantong Limestone suggested the possibility of simultaneous deposition (Figure 2) of both clastic (Labang and Kulapis Formations) [4] and carbonate rocks (Gomantong Formations) during Oligocene to Miocene period [2]. Evidence can be seen in one of the outcrops which contains well cemented Labang clast up to 25cm in diameter which in turn suggested an unconformable contact. On the contrary, the limestone is interbedded with several thin, grey-green fissile, clastic mudstone beds which also contain Labang Formation clast [4]. Several other outcrop were found to have small clastic clast or fine-grained clastic material within the limestone deposits. However, these assumptions still remained as assumption which required further study on the lithostratigraphy of the area.

Recent study proposed three hypotheses in explaining the formation of the Gomantong Limestone and mélanges as shown in Figure 3 [5]. The first hypothesis is that the limestone is deposited above the mélange. The second hypothesis is that the limestone was deposited together with the mélanges. The third hypothesis states that the limestone might deposited below the mélange. However, the three hypotheses on the formation of Gomantong Limestone may not be supported by sufficient numbers and variation of rock samples and thus this study will be undertake more precisely and detailed sampling in order to confirm the diagenesis history of Gomantong Limestone Formation.

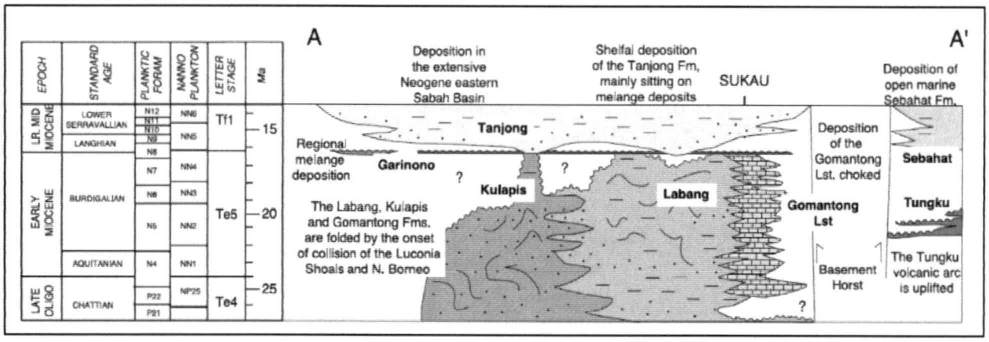

Figure 2. The chronostratigraphic section of the Oligocene-early Miocene of eastern Sabah [2].

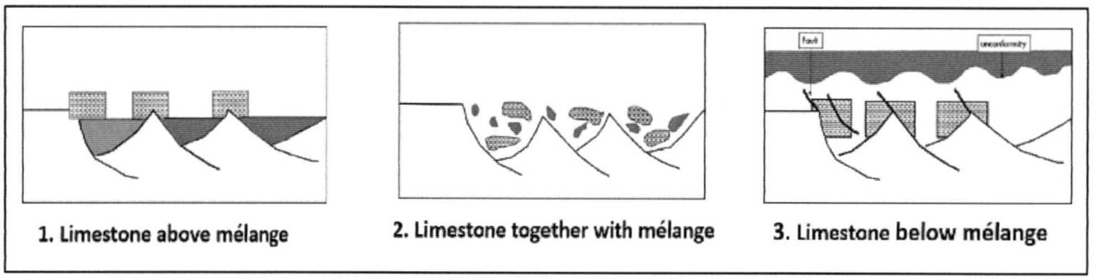

Figure 3. Hypothesis on deposition of limestone in relative to melanges [5].

3. Results and Discussions

The diagenesis of Gomantong Limestone is similar to other Tertiary platforms from neighbouring islands of Borneo such as Tonasa Platform (Sulawesi, Indonesia) [7] which showed limited early diagenesis (marine or meteoric) together with prevalent of dissolution, cementations and compaction that is related to shallow to deeper burial diagenesis [6-8]. The types and morphology of cement are discuss in the subsequent sections (Figure 4A-4F):

3.1 Types and morphology of cements

3.1.1 Microcrystalline cements

Microcrystalline cement formed by micritization that can be seen as light to dark brown rims on the outer part of the allochems. Grain micritization is almost seen in all thin sections but the area of the rim is said to be less than 2-5%. The size of micritic rim is observed dominantly between 20 and 30μm. Micritisation is the first alteration process that occurs at the early stage of diagenesis which affected almost all of the samples. Moreover, it is consider pre-date of other diagenetic features as those features cross-cut the micrite envelopes and also micritic rims surrounding the allochems. The micritic rim found is said to be due to infilling of microborings that formed by endolithic organisms as the micritisation mostly are encroached into the bioclasts. Micritizations are abundant in wackestonespackstones and also boundstones in platform top. It may due to the reason of high endolithic organism activity in shallow water which are influenced by low to moderate energy.

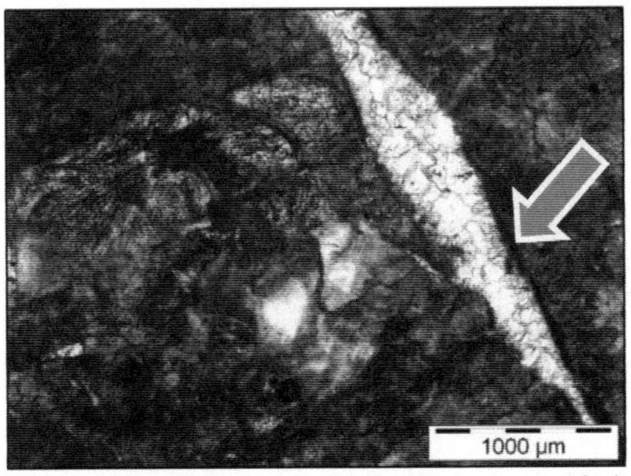

Figure 4A. Thin layer of micritic cement (dark brown, red arrow) line the former *Lepidocyclina* sp. Note crystals of blocky calcite cement occupy the secondary porosity of the forams. Plane polarized light.

Figure 4B. Packstone with many *Lepidocyclina* sp. and *Miogypsina* sp. forams which are partly broken. Mechanical burial diagenesis caused grain distortion and cross-cutting of fracture within the grains which then infilled by granular mosaic calcite cements. Plane polarized light.

Figure 4C. Grainstone in shallow water platform top with abundant of *Lepidocyclina* sp. forams and minor mud matrix. Note many of grain breakage and minor fractures between grains due to mechanical burial diagenesis of post-depositional alteration. Plane polarized light.

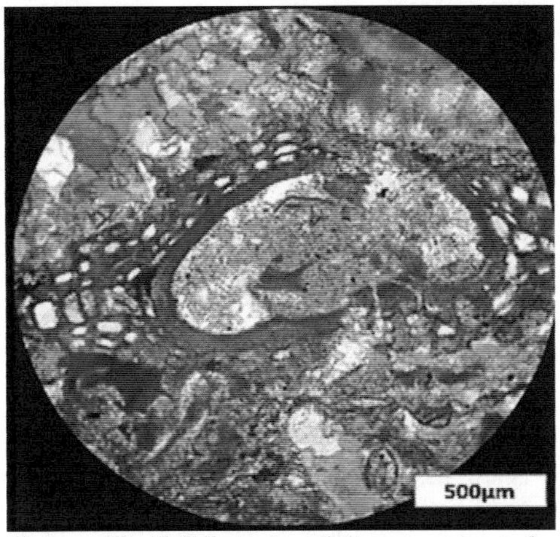

Figure 4D. Poikilotopic calcite cements at the left upper part of a thin section from Outcrop 2. Plane polarized light.

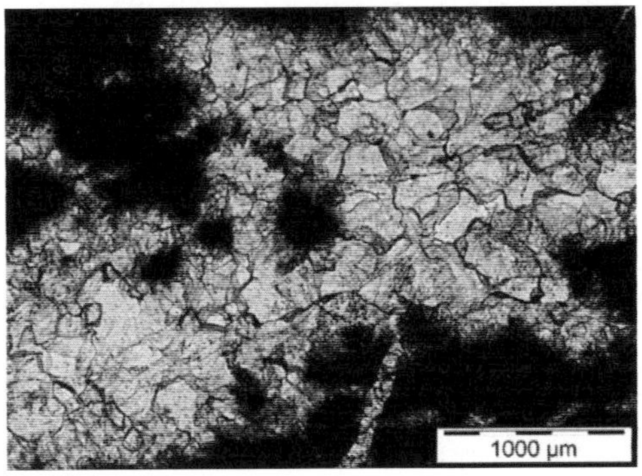

Figure 4E. Blocky cements infilled within the fractures and in most of the intergranular pores.

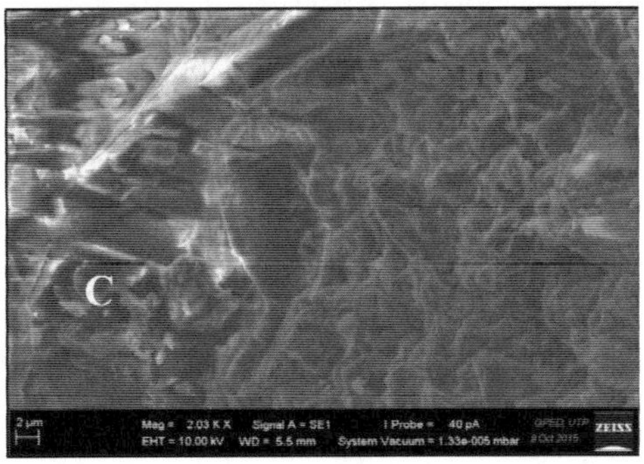

Figure 4F. SEM image of Figure 4C showing the presence of blocky calcite cements (C).

3.1.2 Crystalline cements

Detailed observations on most of the samples indicate the presence of Mg-calcite cement. The calcite cement are dominantly occupied both inter- and intragranular porosity and secondary fractures by the granular mosaic and blocky mosaic and predominantly by the poikilotopic cements. The blocky calcite cements can be seen as equidimensional crystal up to 200μm and euhedral. Poikilotopic crystals are larger than 400μm which enclose a number of allochems and this cements are only seen in one thin section from Outcrop 2 (Figure 4D). The cementations are seen to be after grains-related compaction and mostly are blocky mosaics. The diagenetic of shallow marine to shallow burial environment is deduced as the granular mosaic and blocky calcite mosaics postdates most of the grains breakage events.

3.2 Diagenetic process and environment

Through petrographic analysis, there are five diagenetic features that can be observed, which consist of micritization, grain scale compaction, cementation, mechanical compaction and chemical compaction (Table 1). These studies reveal that there are three main phases that reflecting the environment of diagenesis that have affected the Gomantong Limestone in Sukau; (1) the presence of micritic envelope shows that the sample has gone through pervasive shallow marine micritization (Figure 4A), followed by (2) moderate burial diagenesis shown by the grain distortion, mechanical grain breakage and closer grain packing due to mechanical compaction and pore-filling cementation by granular calcite cement (Figure 4B, 4C); and (3) moderate - deep burial diagenesis shown by cross-cutting fractures due to compaction followed by cementation of blocky calcite cement and poikilotopic cement (Figure 4D, 4E, 4F)., and chemical dissolution that resulted the formation of stylolites.

Micritic envelopes appear at an early stage of submarine cementation. The micrite cement is found interbedded with and surrounding calcite crystallites of the allochems, indicating the original structure of the allochems (e.g. forams). The original structure of the leached bioclasts (e.g forams) is preserved by the stable micritic envelope formed during stabilization in marine pore fluids (Figure 4A). The presence of micritic rims is thought to be the result of high carbonate concentrations during the commencement of cement precipitation [9] and possibly controlled by the distribution and metabolic activity of bacteria [10] or physicochemical process that caused rapid precipitation [11].

Diagenetic alteration continues with burial compactional features by the evidence of the grain distortion, mechanical grain breakage and closer grain packing which latter includes the tangential and concavo-convex grain contacts that is prevalent throughout the samples of packstone-grainstone facies (Figure 4B, 4C). However, the degree of closeness between grain packing, grain breakage and the sutured grain contacts are highly variable between samples from different outcrops which might indicate the increased burial compaction relative to lithification and depth. However, there might also influenced by the lithology-influenced. It can be seen that the closer grain contacts leading to pressure dissolution which occur over the shallow marine to moderate burial depth (Table 1) where the depth ranges are lower than those micrite dominated matrix samples. The grain-related compactional features are prevalent in samples which have (1) contain grains or lithologies which are prone to grain breakage (2) strongly affected by burial prior to cementation and lithifications.

Cross-cutting fractures (Figure 4B) were observed more than individual grains (including the grain breakages). The relative timing of fracturing varies with every samples and some samples may have undergone several phases of fracturing. Granular and blocky mosaics are most commonly observed as infilled in all the fractures with crystal size commonly <100μm. The fracturing is mostly related to mechanical compaction which is formed in both shallow and deep burial depth as the diagenetic environments. Fractures filled by blocky cements may have burial or tectonic structuration origins depending on their relative timing and later cements fills. In this case, the relative timing is said to be post-date of micritization, grain breakage and cementations processes as it crosscut all of the features. Moreover, the filling of the fractures are mostly by blocky cements induced that this mechanical compaction occurred further down the shallow burial depth and may have influenced of meteoric influence after the sediments underwent an uplift.

Stylolites with features of continuum of jagged like to 'anastomosing' dissolution seams [8] are observed in certain thin sections. The stylolites are mostly postdate all other diagenetic features. Stylolites observed are mostly bed parallel and also some occur as circum-clast stylolites. The stylolites and dissolution seams can be spotted as dark brown-black color. Stylolites, which is the chemical compaction features, post-date all the other diagenetic features based on the cross-cutting relationships. This features are mostly form in moderate to deep burial environments.

Table 1. Summary of diagenetic features and environment.

Diagenetic Features	Diagenetic Environment			
	Shallow Marine	Shallow Marine – Moderate Burial	Moderate – Deep Burial	Meteoric Influence
Micritization	■			
Grain Scale Compaction		■		
Cementation (Inter- & Intragranular Pores)		■		
Fracturing		■	■	
Fracture Filling Cements		■		■
Stylolites & Dissolution			■	

3.3 Reservoir Quality

Diagenesis comprises a wide spectrum of geochemical, physical and biological post deposition events where the original sedimentary mineral assemblages and their interstitial pore waters interact in attempt to reach the textural and thermodynamic equilibrium with their environment [12]. The Gomantong Limestone Formation has low porosity and permeability due to the pervasive cementation. The porosity of the Gomantong Limestone decreases with the increasing depth due to compaction process. However, burial diagenetic settings cause dissolution between bioclasts which can enhance the porosity (Figure 4).

4. Conclusion

In the Gomantong Limestone Formation is well preserved and variation of lithofacies can be observed throughout the ten outcrops identified in Sukau. The diagenetic processes and environment is determined through detailed petrographic analysis and scanning electron microscopy results. The diagenetic processes of the Gomantong Limestone Formation are mostly minimal due to the lack of sub-aerial exposure and restricted to shallow marine and burial diagenetic environment. It is found that there is little variation of cements with dominant granular and blocky cement types. Over 90% of the samples are fractured indicating shallow burial effects. The Gomantong Limestone Formation has very low reservoir potential due to the high cementations. However, a few samples indicating the burial diagenetic settings showed enhancement of reservoir quality where the dissolution processes created pores between bioclasts.

Acknowledgements
We would like to express our deepest gratitude to the parties involved in completing this paper especially STIRF (0153AA-D45) from Universiti Teknologi PETRONAS for funding this project.

References

[1] King, D. W., and Morton D.J. 1987. Sentry Bank Reef - a Case History. Proc. 13th ASEC Conference, *Australian Society of Exploration Geophysicists,* vol. 18, pp 114 – 120.

[2] Noad, J. 2001. The Gomantong Limestone of eastern Borneo: a sedimentological comparison with the near-contemporaneous Luconia Province. Elsevier, vol. 175(1), pp. 273-302.

[3] Lim, P. S. 1985. Geological Map of Sabah (3rd Edition). Geological Survey of Malaysia.

[4] Chuah, T. B. 1984. The General Geology of Suan Lamba - Lukau Area, Sandakan Sabah, East Malaysia. Universiti Malaya.

[5] Wong, Y. L. 2013. Stratigraphy and Structural Position of Carbonates Relative to the Melanges in East Sabah. Final Year Project, Universiti Teknologi PETRONAS.

[6] Saller, A. H., and Vijaya, S. 2002. Depositional and diagenetic history of the Kerendan carbonate platform, Oligocene, central Kalimantan, Indonesia. *Journal of Petroleum Geology,* vol. 25, pp. 123-150.

[7] Wilson, M. E., Bosence, D. W., and Limbong, A. 2000. Tertiary syn-tectonic carbonate platform development in Indonesia. Elsevier, vol. 47(2), pp. 395-419.

[8] Arosi, H.A. and Wilson, M. E. J. 2015. Diagenesis and fracturing of a large-scale, syn-tectonic carbonate platform. 109-134.

[9] Neumeier, U. 1999. Experimental modelling of beachrock cementation under microbial influence. *Sediment.Geol.,* vol. 126 (1–4), pp. 35–46.

[10] Beier, J.A. 1985. Diagenesis of Quaternary Bahamian beachrock: petrographic and isotopic evidence. *J. Sediment. Petrol.* vol. 55, pp. 755–761.

[11] Vieira, M. M. and Ros, L. F. D. 2007. Cementation patterns and genetic implications of Holocene beachrocks from northeastern Brazil. *Sediment.Geol.,* vol. 192 (3–4), pp. 207–230.

[12] Tucker, Maurice E., Wright, V., and Paul Dickson, J. A D. 2009. Carbonate Sedimentology, pp. 1-482.

IOP Conf. Series: Earth and Environmental Science **88** (2017) 012017 doi:10.1088/1755-1315/88/1/012017

Depth Imaging Enhancement Using Reverse Time Migration

S.Y. Moussavi Alashloo[1,*], D.P. Ghosh[1], Y. Bashir[1] and W.I. Wan Yusoff [1]

[1] Department of Petroleum Geoscience, Universiti Teknologi PETRONAS, Sri Iskandar, Malaysia

E-mail: y.alashloo@gmail.com

Abstract. Imaging the subsurface with complex structures and steeply dipping salt boundaries is a challenging task in seismic exploration. Between two main categories of seismic migration, wavefield-continuation methods have been more successful than ray-based ones. Wavefield-continuation migration constructs source and reflection wavefields as a function of depth or time by directly solving one-way or two-way wave equations. Consequently, this technique has two benefits. First, compared to Kirchhoff migration, it does not use high-frequency approximation and can more accurately propagate wavefields in shallow depths. Secondly, it can naturally handle multi-paths or multi-arrivals. Reverse time migration (RTM) is a wavefield-continuation method which is accepted as the best migration method currently available for imaging complicated geology. The main objective of this research is to improve imaging of complex structures by utilizing the RTM technique. Two models, involving anticlines, faults, etc., are employed to test the technique. The imaging results demonstrated that the RTM method succeeded to image the flanks, remove noises and improve the resolution.

1. Introduction

Migration is a seismic processing method aiming to move dipping reflections to their true subsurface positions and collapse diffractions, hence resulting in an image of the subsurface geology and increasing spatial resolution. For conventional Kirchhoff migration, the Green's function is extracted from asymptotic ray theory, which is a high frequency approximation to the true Green's function. Therefore, the asymptotic approximation of the Green's function involves errors. Another type of migration is known as wavefield-continuation migration which constructs source and reflection wavefields as a function of depth or time by directly solving one-way or two-way wave equation. Consequently, wavefield-continuation migration techniques have two benefits. First of all, these methods do not use high-frequency approximation and can more accurately propagate wavefields in shallow depths [1]. Secondly, they can naturally handle multi-paths or multi-arrivals [2, 3].

Wavefield-continuation migration includes two main categories: downward continuation and reverse time methods [4]. The main advantage of downward continuation techniques is that they can deal with lateral velocity variations because it extrapolates wavefields in the space domain. However, since it uses the one-way wave equation, it still has inaccuracy for steep dips [5]. The limitation of all mentioned imaging methods can be handled using the most powerful migration method which is called reverse time migration (RTM). RTM is based on two-way wave equation, and it can involve all frequencies and dip angles and adapts to any type of velocity variations [6]. RTM was first introduced by Whitmore [6], McMechan [7], and Levin [8]. Two main feature of RTM are: constructing wavefields along the time axis instead of depth, and extrapolating the wavefield with a two-way wave equation. These features make RTM able to be executed for any type of velocity variation, migrates all frequencies, deals with all dips, and handles all types of waves such as prismatic reflections and turning waves [3, 9, 10]. Despite its advantages listed above, RTM was extremely expensive in both computing resources and computational time. The increase of computer speed and decrease in the cost of computer facilities, especially pc-clusters, have allowed RTM to be utilized as a principal prestack seismic imaging technique.

This paper presents the application of RTM, and its comparison with Kirchhoff imaging. A simple layered model with an anticline structure, and the Marmousi model, as a complex model, are used to investigate the strengths and limitations of RTM. The RTM results demonstrate that flanks and steep dips are imaged properly, noises are remove and the resolution is improved.

2. Methodology

Similar to all wavefield-continuation migration methods, RTM includes two steps: wavefield extrapolation followed by application of an imaging condition. For poststack zero-offset RTM, wavefield extrapolation is applied as a back propagation of the stacked data. For prestack RTM, however, wavefield extrapolation is divided into two substeps. One is forward extrapolation to construct the source wavefield at each time sample, and the other is back propagation of surface recorded data to reconstruct the reflected wavefield at each time sample. Figure 1 illustrates the workflow of RTM which comprises three key steps:

1. The source wavefield u_s is extrapolated forward in time using the source location, the source wavelet and the velocity model.

2. The receiver wavefield u_r is propagated backward in time, from all receiver positions using the recorded data as well as the velocity model.

3. Applying an imaging condition to calculate the zero lag cross-correlation between the two simulated wavefields u_s and u_r at all model grid points to locate the subsurface reflectors.

For the acoustic wave equation, the forward extrapolation can be expressed as [11]:

$$\left[\nabla^2 - \frac{1}{v^2}\frac{\partial^2}{\partial t^2}\right] u_s(\boldsymbol{x},t) = \mathrm{f}(\boldsymbol{x},t) \quad , \tag{1}$$

where u_s is the source wavefield, v is the acoustic velocity of the medium, and for a point force of arbitrary orientation located at a point $\boldsymbol{x}_s$, $\mathrm{f}(\boldsymbol{x},t)$ is the body force distribution which is defined by

$$\mathrm{f}(\boldsymbol{x},t) = \mathbf{f}s(t)\delta(\boldsymbol{x}-\boldsymbol{x}_s) \quad , \tag{2}$$

in which s(t) is the source time function, the variation of the amplitude of the force as a function of time, and $\delta(\boldsymbol{x}-\boldsymbol{x}_s)$ is the Kronecker delta. **f** is a unit vector in the direction of the point force, and $\boldsymbol{x}_s$ is the source position vector.

Likewise, the backward extrapolation can be written as

$$\left[\nabla^2 - \frac{1}{v^2}\frac{\partial^2}{\partial t^2}\right] u_r(\boldsymbol{x},t) = 0 \quad , \qquad u_r(\boldsymbol{x}_0,t) = d(x_0,t) \tag{3}$$

where u_r is the reflected wavefield and d is recorded data on the surface which is used as a boundary condition. The boundary condition keeps the wavefront shape of the back-propagated wavefield the same as the record. However, the wavefront shape of the back-propagated wavefield is the same as the integral of the record. In equations 1 and 3, u_s and u_r represent the vertical component of particle velocity for geophone-recorded data or pressure for hydrophone-recorded data.

Besides solving the acoustic wavefield, it is also significant to eliminate artificial reflections from the model boundary. This is because seismic waves propagate into the whole earth, while in practice the computations are restricted to a limited area. This generates artificial boundaries which in turn cause artificial reflections. In order to suppress these unwanted reflections, a suitable boundary condition should be included in the numerical simulation; otherwise, these artifacts will contaminate the simulated wavefields. We employ a sponge absorbing boundary condition (ABC) for constructing a nonreflecting boundary condition. It is based on gradual reduction of the amplitudes in a strip of nodes along the boundaries of the mesh [12].

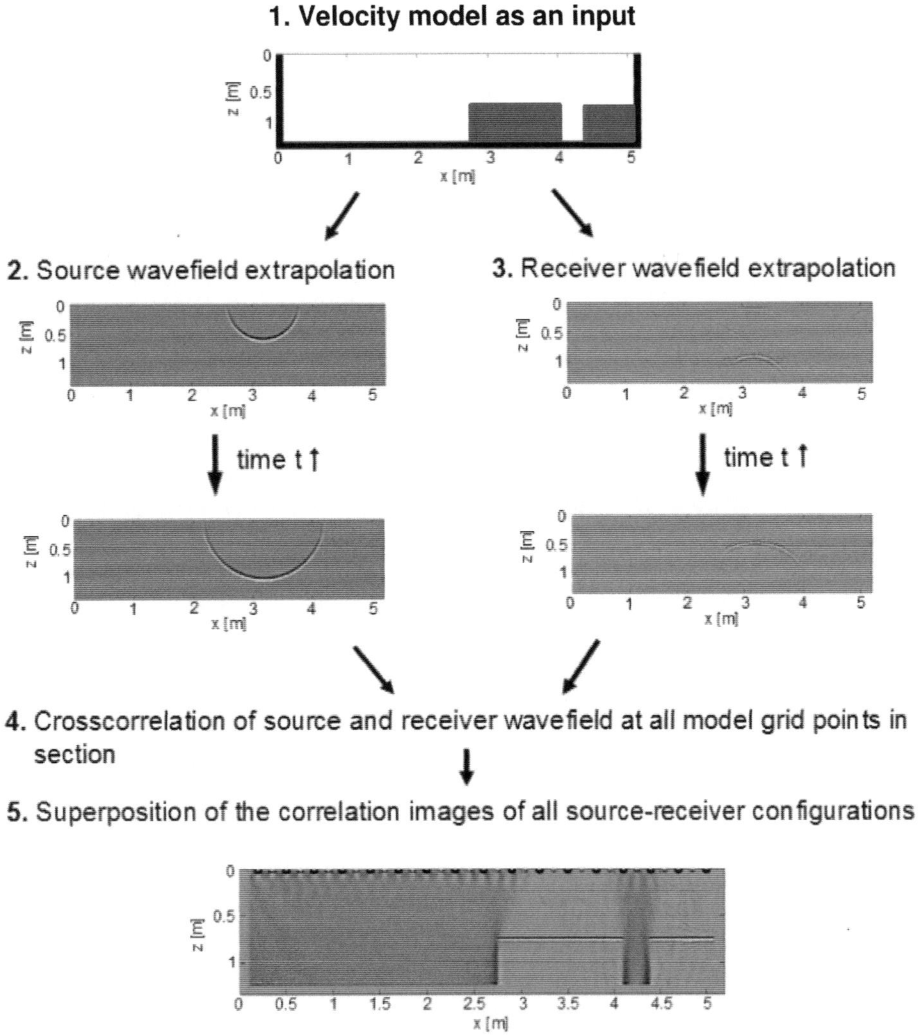

Figure 1: Flowchart of RTM [13].

The last key step of RTM is the imaging condition to extract the image from the constructed wavefields. In this research, the zero-lag cross-correlation imaging condition is used which is given by [14]:

$$I(\boldsymbol{x}) = \int u_s(\boldsymbol{x},t)u_r(\boldsymbol{x},t)dt. \quad (4)$$

where $I(\boldsymbol{x})$ is the migration image value at point $\boldsymbol{x}$. It can be carried out by cross-correlating the modelled source and backward-continued wavefields, and then extracting the value of the zero-lag (the sample at t = 0 in the cross-correlation signal). If the two wavefields are similar to each other, after cross-correlation, the zero-lag will be the biggest value in the cross-correlation. On the other hand, if the two wavefields are different, for example if the wavefields have different arrival times, the zero-lag cross-correlation value is normally small.

3. Numerical examples

The result of RTM on different models are investigated in this section. A simple layered model with an anticline structure is used to study the behavior of RTM to image a simple layering media. Next, the Marmousi model is chosen because it contains thin layers, steep faults, anticlines, pinch-out and other structures. This model is a complex model and more realistic, thus it can assist to find the strengths and limitations of RTM algorithm. Finally, we compare the RTM results with the Kirchhoff images.

3.1 Simple Layered Model with Anticline

The model contains three layers with velocities from 2 to 3 km/s. An anticline with steep dip flanks is embedded in the model. The depth of model is 2 km and the lateral distance is 8km. Forward modeling conducted by using a source signature corresponding to a 30 Hz Ricker wavelet with 800 receivers with 10 m spacing interval on the surface. An isotropic finite difference method is employed to generate the records. The direct arrivals are muted from the gathers since they do not comprise any reflector information (Figure 2).

Figure 3 illustrates the dataset of the simple model including an anticline with 82° dip for the left flank and 75° dip for the right flank, and its results of Kirchhoff depth imaging, and RTM. The data comprises diffractions in sharp points, and events in wrong positions. Although Kirchhoff migration has imaged the flat interfaces correctly, it is obvious that it is not able to image the steep dips, and its image contains distortions. However, the RTM method has succeeded to image the flanks, remove noises and improve the resolution. It can provide more detail about corners and wrinkles.

In RTM image (Figure 3c), red arrays show that flank imaging has executed much more better than the Kirchhoff result, and the yellow circles demonstrate how RTM is able to image more details. It is also obvious that the reflectors in RTM image are even, whereas reflectors in the Kirchhoff image are wavier specially those parts below the anticline. Overall, by conducting different depth imaging algorithms on the anticline model, it is proved that the RTM outcome, regarding to image steep dips, is better than the Kirchhoff migration, and it can produce excellent results.

3.2 Marmousi Model

Marmousi model is utilized to show how RTM works for a model with several challenging features. A 20 Hz Ricker wavelet is used as the source signature. Figure 4 displays Marmousi model and two shot gathers. The result of RTM on Marmousi model is showed in Figure 5. The RTM result provides an accurate image of dip reflectors and the pinch-out. Comparing the RTM and Kirchhoff images demonstrates a considerable improvement. RTM adds more detail to the image, and the resolution is enhanced. By considering two pinch-outs at the bottom of the images, it can be clearly find out that RTM has boosted their amplitude, and they can be detected better than in the Kirchhoff image. Furthermore, the anticline, which is located at the middle and bottom of the model, are imaged perfectly in RTM, while in the Kirchhoff image, only some features of anticline can be seen.

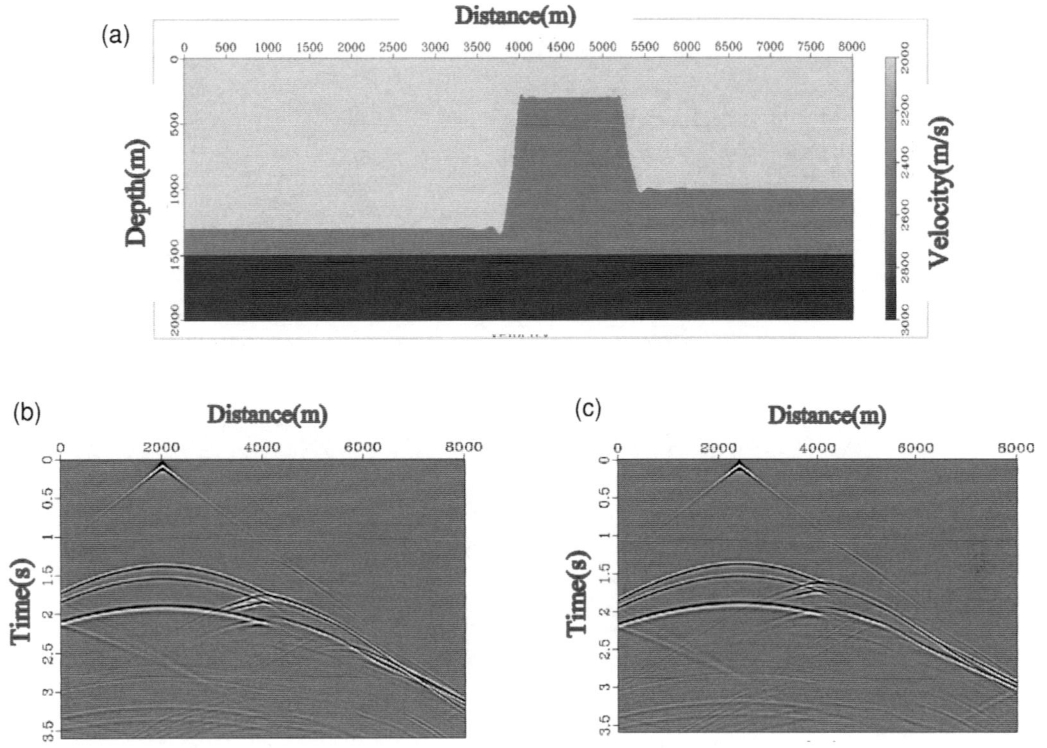

Figure 2: (a) Simple three-layer model with an anticline. (b) Shot gather of a source at 2 km and (c) shot gather of a source at 2.2 km.

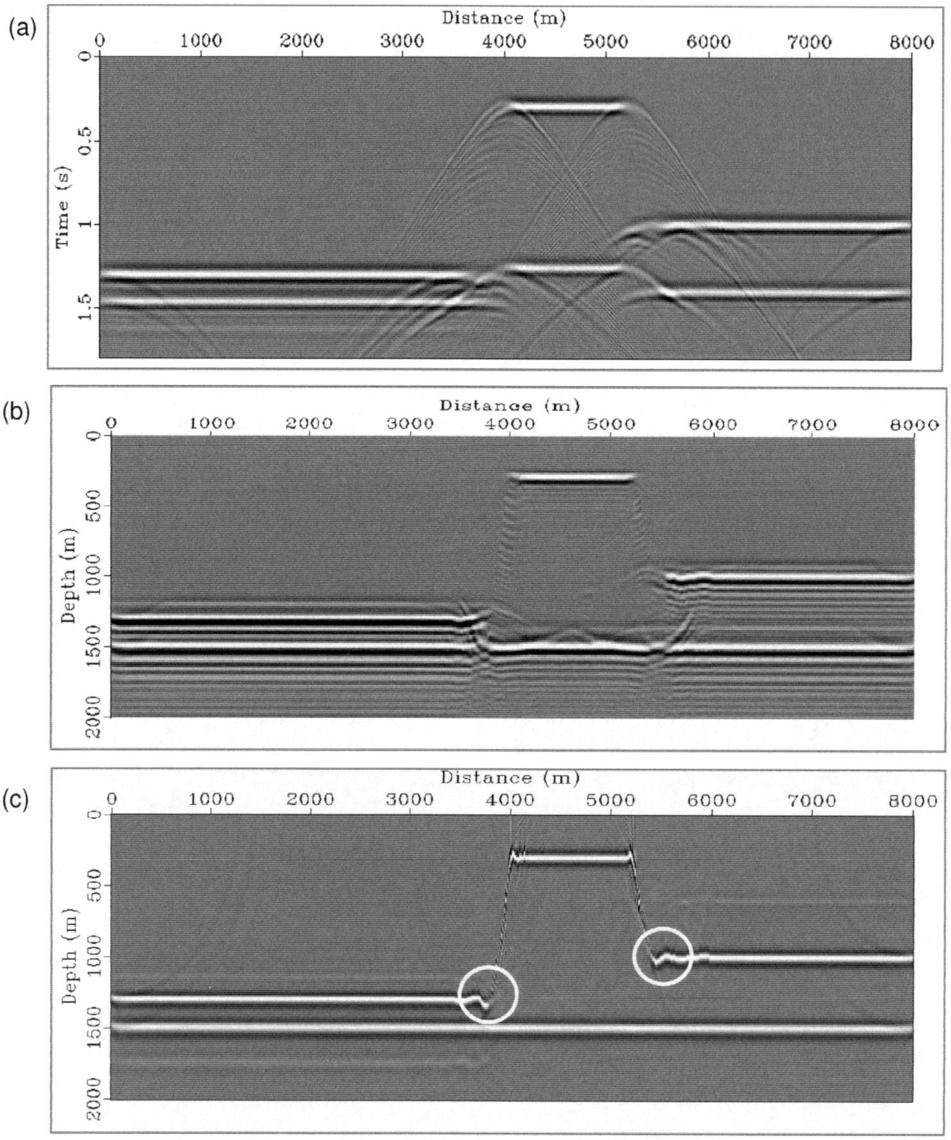

Figure 3: (a) Stack dataset of simple anticline model, and comparison of its images obtained by (b) Kirchhoff migration and (c) RTM.

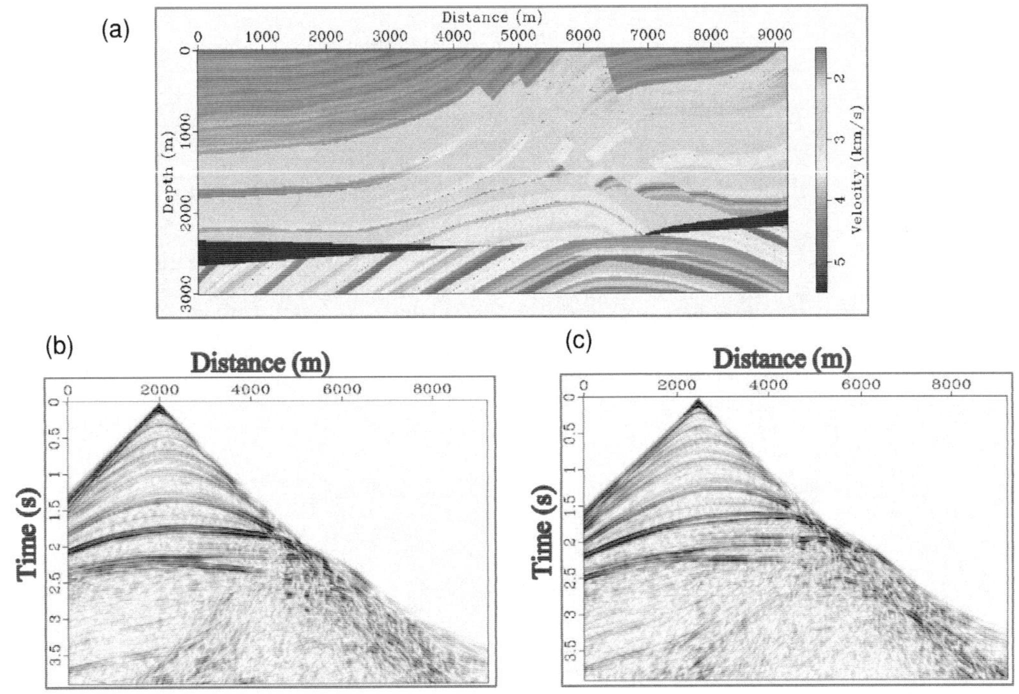

Figure 4: (a) Marmousi model involving complex structures. (b) Shot gather for a source at 2 km, and (c) shot gather for a source at 2.2 km.

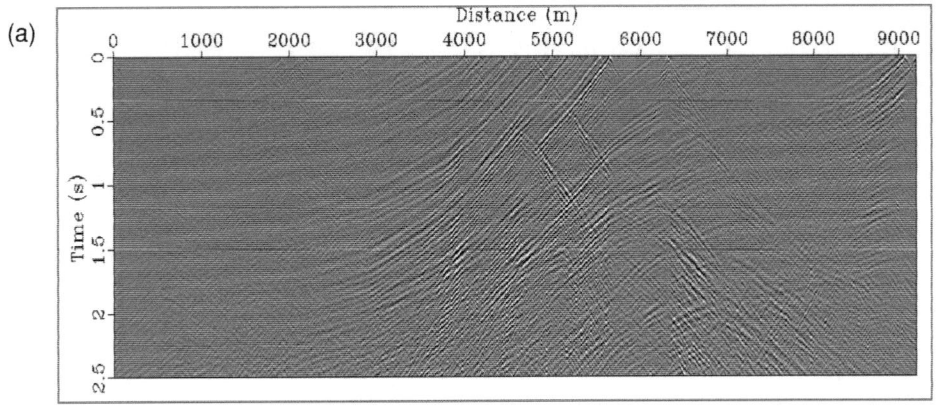

(b)

(c)

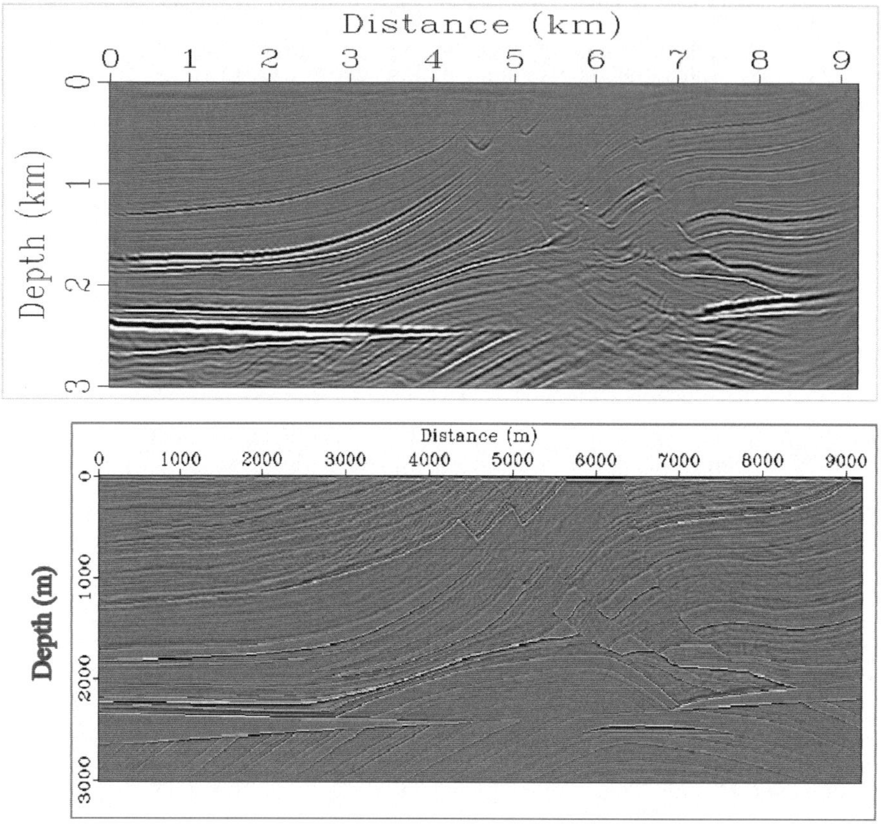

Figure 5: (a) Marmousi stack data and its images acquired using (b) Kirchhoff, and (c) RTM.

4. Conclusions

In this research, an RTM approach was studied. Although Kirchhoff migration has imaged the flat interfaces correctly, it is obvious that it is not able to image the steep dips, and its image contains distortions. However, the RTM method has succeeded to image the flanks, remove noises and improve the resolution. RTM could provide more details about corners and wrinkles. Overall, despite that RTM is an expensive method for imaging subsurface, it is crucial to employ when the structure is complex. The proposed RTM is appropriate for isotropic media. It can be modified for anisotropic imaging by using anisotropic wave propagator.

Acknowledgements

The authors acknowledge the contribution of CSI members, and CSI/UTP for providing the resources.

References

[1] P. Stoffa, *et al.*, "Split-step Fourier migration," *Geophysics,* vol. 55, pp. 410-421, 1990.

[2] B. Biondi, *3D seismic imaging*: Society of Exploration Geophysicists, 2006.

[3] W. Wang, *et al.*, "Analysis of up/down decomposed acoustic reverse time migration images," *Geophysics,* vol. 81, pp. S253-S259, 2016.

[4] N. Whitmore, "Iterative depth migration by backward time propagation," in *1983 SEG Annual Meeting*, 1983.

[5] G. McMECHAN, "Migration by extrapolation of time-dependent boundary values," *Geophysical Prospecting,* vol. 31, pp. 413-420, 1983.

[6] E. Baysal, *et al.*, "Reverse time migration," *Geophysics,* vol. 48, pp. 1514-1524, 1983.

[7] R. L. De Araújo and R. d. C. Pestana, "REVERSE TIME MIGRATION BY INTERPOLATION AND PSEUDO-ANALYTICAL METHODS," *Revista Brasileira de Geofísica,* vol. 32, pp. 753-764, 2014.

[8] S. A. Levin, "Principle of reverse-time migration," *Geophysics,* vol. 49, pp. 581-583, 1984.

[9] B. Biondi and G. Shan, "Prestack imaging of overturned reflections by reverse time migration," in *SEG Technical Program Expanded Abstracts*, 2002, pp. 1284-1287.

[10] C. Stork, "Eliminating nearly all dispersion error from FD modeling and RTM with minimal cost increase," in *75th EAGE Conference & Exhibition incorporating SPE EUROPEC 2013*, 2013.

[11] G. Yao and H. Jakubowicz, "Least-squares reverse-time migration in a matrix-based formulation," *Geophysical Prospecting,* 2015.

[12] C. Cerjan, *et al.*, "A nonreflecting boundary condition for discrete acoustic and elastic wave equations," *Geophysics,* vol. 50, pp. 705-708, 1985.

[13] M. Grohmann, "Reverse time migration: Introduction of a new imaging technique for ultrasonic measurements in civil engineering," 2015.

[14] J. F. Claerbout, "Toward a unified theory of reflector mapping," *Geophysics,* vol. 36, pp. 467-481, 1971.

Analysis on the overpressure characterization with respect to depositional environment facies: Case Study in Miri Area, Baram Delta.

N. Mhd Hanapiah[1], W.I. Wan Yusoff[2], M.N.A.Zakariah[3]

[1,2,3] Petroleum Geoscience Department, Universiti Teknologi PETRONAS, Malaysia

Email: hana.nurhaj@gmail.com, wanismail_wanyusoff@utp.edu.com

Abstract. Overpressure studies in oil and gas exploration and production are carried out in order to mitigate any losses that could happen while drilling. These concerns can be addressed by enhancing the understanding of overpressure characterization in the fields. This research emphasizes in determining the pore pressure trend in Miri area to assist pore pressure prediction for future hydrocarbon exploration and production. Generally, pore pressure trends are related to mechanisms that contribute to the overpressure generation. In the region predominant overpressure are disequilibrium compaction within the prodelta shales meanwhile in outer shelf overpressure generation controlled by fluid expansion in deltaic sequence of inner shelf area. The objective of this research is to analyze the pore pressure profile of wells for determining vertical trends of pore pressure for various depositional environment facies of Miri area. Integration of rock physics and pore pressure analysis and relating the trends to environment depositional environment facies within shale underlying sand interval. Analysis done shows that overpressure top is characterize by depositional environment facies within shale underlying sand interval.

Keywords: Overpressure, Overpressure mechanism, Depositional environment facies

1. INTRODUCTION

The Miri area located in the South China Sea in between Kalimantan and Brunei. Miri area is known to be one field of the Sarawak Basin and petroleum in Sarawak is produced from Neogene rocks that occur mostly in the offshore areas of Sarawak Basin [6]. Some authors claimed that the Sarawak Basin continue westwards into Indonesia as the East Natuna Basin and is separated from the Sabah Basin to the east by a major tectonic discontinuity called the West Baram Line as a major transform fault [2]. Evidence shows that this fault has significantly influenced the sedimentary evolution of the NW Borneo margin, as separating evolution of the carbonate sedimentation (Central Luconia Province) from deltaic siliclastic sedimentation (Baram Delta Province) [8].

In this study, field-scale distribution of overpressure trend is conducted in Miri area and basic analysis of pore pressure characteristic is demonstrated as a first step in pore-pressure estimation. The research choose Miri area as the study area due to some concern which is uncertain distribution overpressure mechanism in Miri area. The research suggest to investigate the origin of overpressure mechanism generated in Miri area. So, the study comply the understanding on relationship between the trends of pore pressure profile with the environment of deposition. Other concern of the study is due to under prediction of high overpressure buildup by seismic velocities in Miri area is one of the concern in this study. Thus, the study try to provide the essential foundation for subsequent vertical and lateral overpressure trend useful for pore pressure estimation.

2. LITERATURE REVIEW

2.1 Geological Setting

The location of these study area is in the Northeastern part of Sarawak, Malaysia. The project site are at two wells known as MA-3 and MA-5. MA-3 is located proximal to the shelf area of Baram Delta meanwhile MA-5 is distal from the shelf area with the distance of two well is around 50km apart as shown in Figure 1.

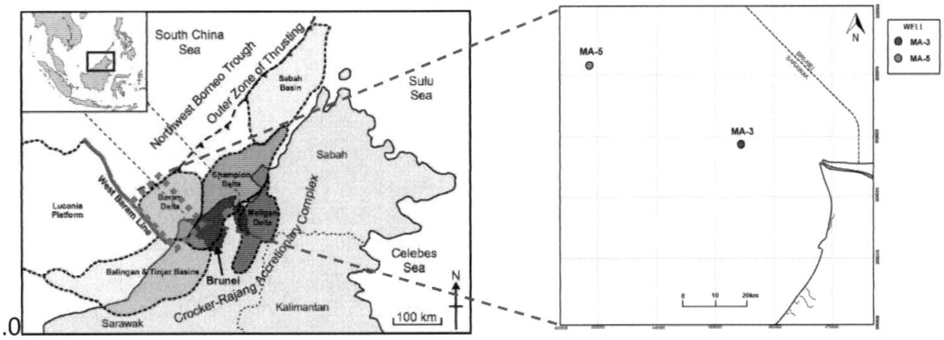

Figure 1:A map of location of Baram Delta and the well location of MA-3 (purple circle) and MA-5 (green circle) in the Miri area. Modified from (Peter, Derek & Robert, 2015).

2.2 Pore pressure Mechanism

The overpressure mechanism in Miri area is caused by two different mechanisms which are disequilibrium compaction and fluid expansion. The behaviour of widespread occurrence of overpressure have strong correlation between the disequilibrium compaction and fluid expansion. Disequilibrium compaction or under compaction is commonly described as the main contribution to overpressure generation in Baram Delta, it was driven by the rapid sedimentation underneath the prograding delta [14]. Other factors that control the high overpressure accumulation is in inner shelf of deltaic sequence overpressure being generated by any conventional fluid expansion mechanism (kerogen-to-gas-maturation) and was proposed that these overpressures have been vertically transfer into reservoir unit via faults, from prodelta shales [13].

3. METHODOLOGY

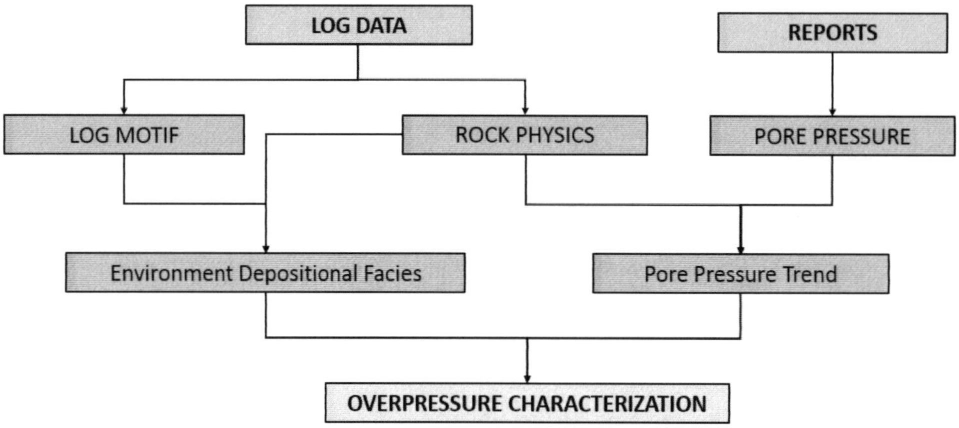

Figure 2: The methodology of this research study.

The study focuses on the offshore of Miri area to emphasize on understanding and predicting of overpressure zone using the well log motif approach. This study stressed on the rock physics analysis resulting on the trends of overpressure in the study area.The research began with overview structural, sequence and rock physics analysis on well data from the Miri area. Subsequently, well log analysis done for environment depositional facies studies. Significant cross plots were created in order to analyze the pore pressure trend. Further analysis was conducted on the pore pressure analysis for the well located in the Miri area. Pore pressure analysis was conducted to see the trend of porosity and density log from its normal compaction trend in Miri area. The report on the pore pressure trend that comprise of all the significant analysis produced at the end of the research. Thus, the resulted to the trend of pore pressure profile respected to overburden pressure as well as environment of deposition.

4. RESULT AND ANALYSIS

4.1 Pore pressure profile

Pore pressure is the essential parameters for physical properties and for geomechanical and geological analyses. Pore pressure is defined as fluid pressure in the pore spaces in the porous formation. Hydrostatic (normal) pressure is associated with a column of water from surface to depth of interest. Lithostatic pressure occurs when the pore pressure exceeds the hydrostatic values in a confined pore volume at depth.

Right profile in Figure 3 shows the variation of pore pressure from MA-3 with depth in Miri area. Top overpressure begin at depth 2960m. The transition overpressure zone present at 2960-3040m. At the depth of 3040m the hard overpressure shows a sharp change in pressure with depth until reach 3110m. Deep overpressure is recognise at depth of 3110m until it reach the highest pore pressure at depth 3212m. Left profile in Figure 3 shows the variation of pore pressure from MA-5 with depth in Miri area. Top of overpressure begin at depth of 2090m. Transition overpressure zone present at 2090-2140m is increase gradually with depth. Hard overpressure present at depth 2140m shows sharp change in pressure with depth until it reach 2468m. Deep overpressure is recognise at depth of 2468m until it reach the highest pore pressure at depth 2786m.

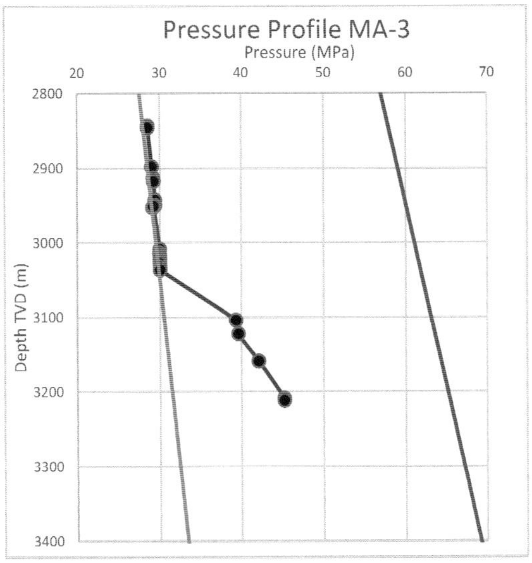

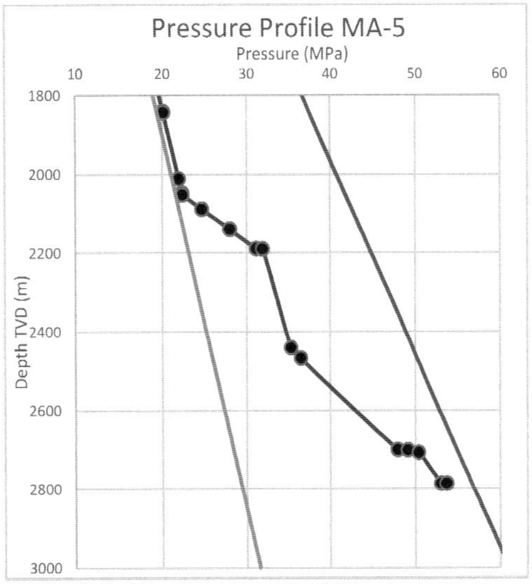

Figure 3: Pore pressure profile of MA-3 and MA-5. Hydrostatic gradient represent in green line (9.84 MPa/km) and lithostatic pressure represent in red line (20.36MPa/km). Pore pressure profile of MA-5. Hydrostatic gradient represent in green line (10.52MPa/km) and lithostatic pressure represent in red line (20.36MPa/km).

Focusing on the location of the well, MA-3 is located in inner shelf and MA-5 is located in outer shelf. Based on previous studies, researchers believe that both well are controlled by different overpressure mechanism either disequilibrium compaction or fluid expansion. Even though, both well are located 50km apart but they actually shows the same trend of pore pressure which is pore pressure shows the gradual changes in depth. Identical of pore pressure trend between two well means there are some similarities present in the zone that lead to deviation of the overpressure generation in the well at certain depth. This might be due to the matching lithology and structural style of the well. By having integration of lithology and structural style with the trend of pore pressure studies can improve the overpressure prediction of the wells for other oil and gas fields.

4.2 Facies characterization

Sequence stratigraphic shows the geometry of the sediment change by depth and each well experience different sequence stratigraphic. So, identifying log motif is important for recognizing the depositional environment of the well. Log motif is useful for the depositional environment analysis of the well located in Miri area. Facies characterization is necessary in order to distinguish the trend of the overpressure in the different sand packages. It give us better correlation that show the depositional environment could respond to the pore pressure trend in Miri area. Besides, it enhances the visualization of pore pressure variation on that particular depth interval. Firstly, facies characterization done in this studies by identify the cut off of Gamma ray log for different lithology such as sand, silt and shale. Facies then further divided based on the log motif as shown in Table 1. Here, log motif is categorize into different trend such as funnel, bell, boxcar, irregular and symmetrical shape.

Table 1: Table of log motif that been used as reference for depositional environment analysis for the sand package of MA-3 and MA-5. Log motif adapted from (Rider,2002).

Facies	Log motif	Sequence Stratigraphy	Depositional Environment	Gamma Ray GR ->
Coarsening Upward Section (CUS)	Funnel 	Prograding	• Crevasse splay • River mouth bar • Delta front • Shore faces	
Fining Upward Section (FUS)	Bell 	Retrograding	• Fluvial point bar • Tidal point bar • Tidal flat • Marine channel	
Cylindrical	Boxcar 	Aggrading	• Braided fluvial • Distributary channel-fill • Eolian sand	
Serrated	Irregular 	Aggrading	• Fluvial floodplain • Storm-dominated shelf • Distal deep-marine slope	
Hour glass	Symmetrical 	Prograding & Retrograding	• Regressive to transgressive shore face delta • Offshore shoal • Marsh	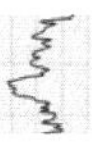

The result from well log motif from well MA-3 and MA-5 shows that most of the package shows coarsening upward section (CUS). It can be expressed that MA-3 and MA-5 both are dominated by delta and mouth bar. Based on the location of wells shown in Figure 1, even though MA-3 and MA-5 are located far apart but it share the same important feature in the environment of depositional facies. Since than assumption can be done as MA-3 and MA-5 experienced the same depositional environment facies, despite influence of location of the well that is far apart. Thus, the change of the trend of pore pressure has the relation to the environement of depositional facies. So, depositional environment analysis is important in order to enhance the understanding of the pore pressure for future hydrocarbon discovery.

4.3 Porosity

Porosity give the meaning of void volume against the total volume. Porosity plays an important role in geology as it can control the fluid storage. Besides, porosity has the relation to the fluid transport and rock properties. Thus, it can be used for pore pressure analysis in the reservoir. Porosity analysis begin with correction on the Neutron porosity with the Density log and Sonic log for MA-3 and MA-5. The correction was done in the clean wet sand package since the Neutron log give the best result in this sand package as it measure the hydrogen ions that present in the formation. Thus, it will set as reference to the correction of porosity in shale package. The observation done by compare the corrected porosity with the porosity equation that being used previous research such as Gassman, Wyllie and Gardener equation .

Observations based on porosity density and porosity sonic of MA-3 and MA-5 shows that deviation occur at the depths which are top of overpressure. At the depth below the porosity trend show decrease with depth, meanwhile when go deeper depth the porosity trend deviate at the certain depth then remains constant up until certain depth. Even though absent of sonic data in MA-5 but porosity density shows the significant changes. Porosity constant happens due to pressure retaining at that depth which present within the overpressure zone. So, because of the porosity is one of the element that important in the study of overpressure characterization. Thus, the trend of porosity can be an indicator of the overpressure top for the well. Thus, the properties of the shale itself will be used to further analyse the normal compaction trend of the well.

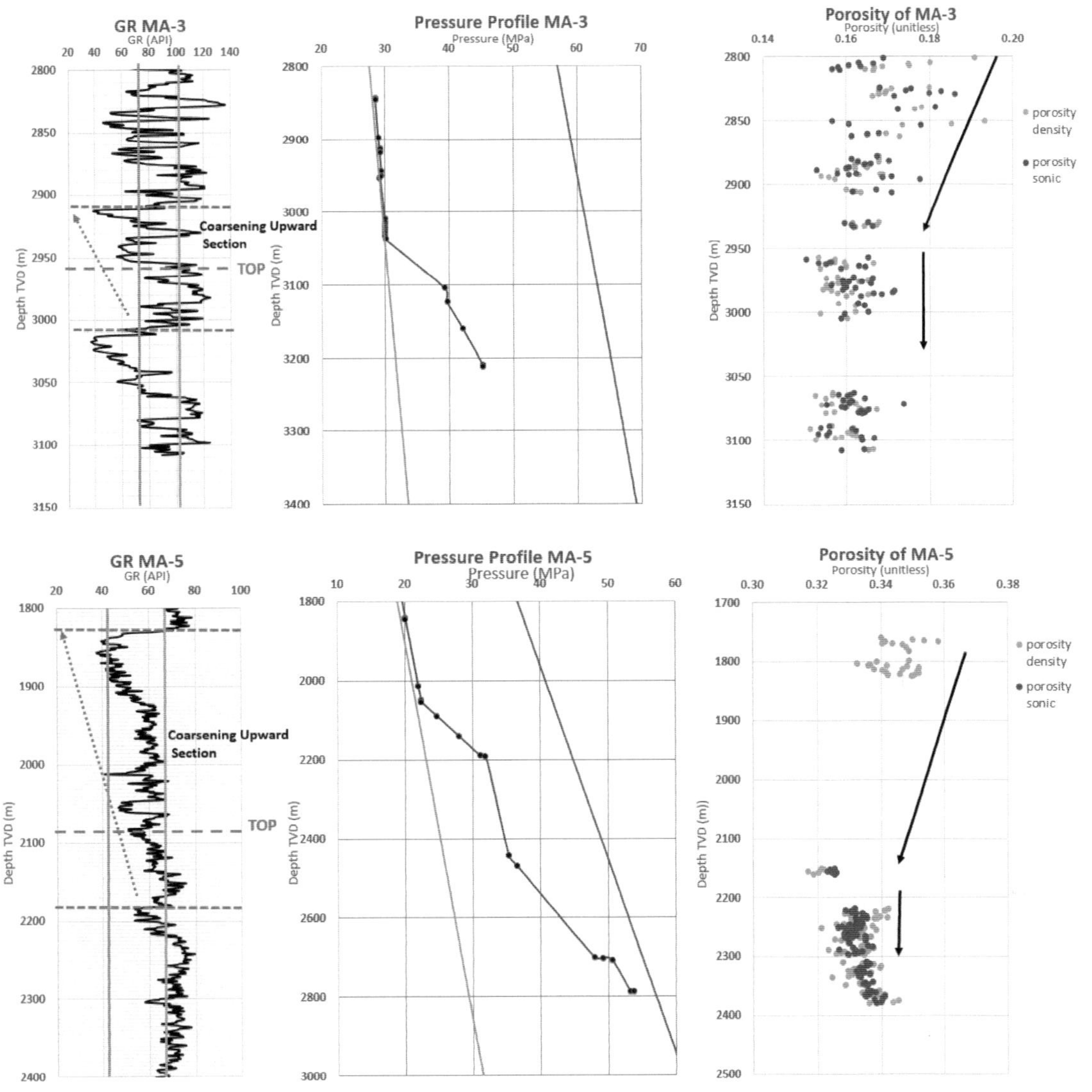

Figure 4 : The porosity trend of Sonic log ang Density log from MA-3 and MA-5.

Analysis on the pore pressure trend can be used for detection of fluid pressure and porosity evaluation from the well logs. In this study normal compaction trend was detected using the sonic and porosity data from the well logs. Since the porosity analysis shown that the shale experienced more compaction compare to sandstone, so the analysis on the normal compaction trend is more significant in the shale package. Sonic log and porosity log respond to the changes of pore pressure with depth. Thus, analysis on the normal compaction trend will enhance the understanding on the pore pressure distribution in Miri area. As refered to Figure 4, it shows how the change in sonic log and porosity log to pore pressure in the well. So, overpressure can be defined and identified based on the deviations of the sonic and porosity log from the normal compaction trend.

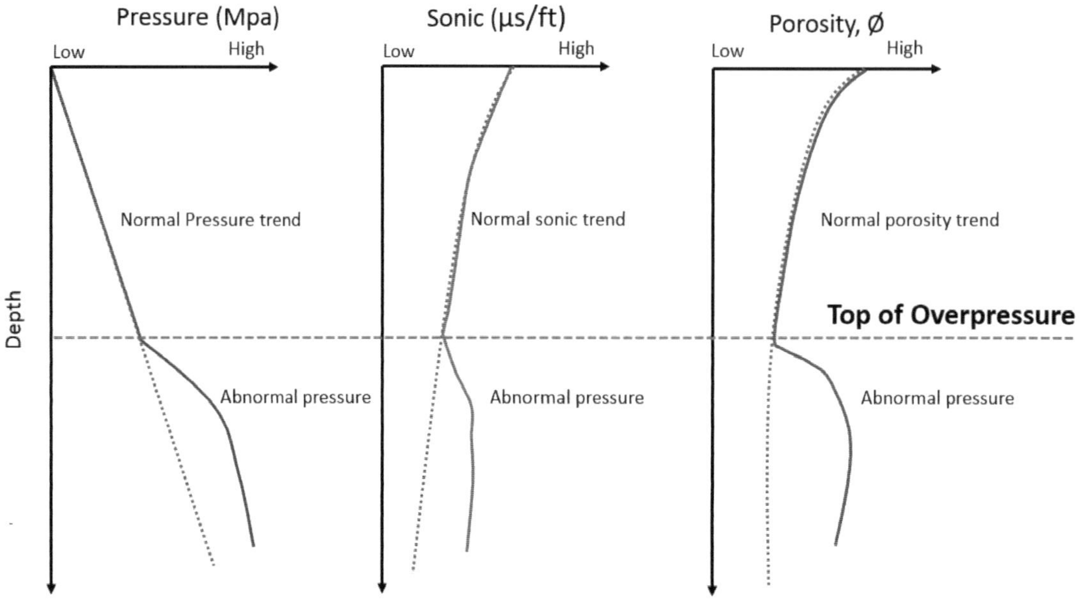

Figure 5: Schematic response of the Sonic and Porosity log to the normal compaction and overpressure zone. Modified from (Hermanrud et.al,1998).

5. CONCLUSION

Based on the observations of the porosity analysis, the results show that the corrected porosity of Density log as well as Sonic log is more close and realiable to the neutron porosity data from the well compare to the porosity that is obtained from the existing porosity formula. This happens because the formula is empirically and only suitable to be used in the area that has being tested. Thus, this research will use the corrected porosity to further analysis on the pore pressure trend of Miri area. The result from porosity of density and sonic logs shows the variations based on different facies characterization. Porosity analysis was further analyze to identify the compaction factor of shale package for MA-3 and MA-5.

The trend of pore pressure is related to depositional environment facies. Thus, both have similar depositional environment facies and pore pressure characteristic by overpressure top located within the shale underlying sand interval. As for MA-3 and MA-5, overpressure zone occur in the prograding sand package that shows in coarsening upward motif. Shale experience compaction higher than sand. The deviation of the data from its normal compaction trend shows in MA-3 (2960m) and MA-5 (2090m) indicate the overpressure interval. Thus, shale is significant in order to study the compaction trend of the formation.

Acknowledgment

I would like to express my deepest gratitude to the lectures of Petroleum Geoscience Department of Universiti Teknologi PETRONAS for giving me an opportunity to further the research study and to PETRONAS for the williness to share the data for my research study. I would like to thank my father, mother and family as well as colleagues for their continuous support.

References

[1] Bowers, G. (2002). Detecting high overpressure. *The Leading Edge*, 174-177.

[2] Hafizan, A. M., & Jennifer, C. (2011). Deep overpressure play: Second Lifeline for West Baram delta, East Malaysia. *Petroleum Geology Conference and Exhibition* (p. 172). Kuala Lumpur: Geological Society of Malaysia.

[3]Lawrence M.A., & David R.C., (2015). Characterization and Analysis of Porosity and Pore Structures. Mineralogical Soceity of America.

[4]Madon, M. (2006). Overpressure history of the Malay Basin, Ofshore Peninsular Malaysia. *Petroleum Geology Conference & Exhibition* (pp. 135-144). Kuala Lumpur: Geological Society of Malaysia.

[5]Mansor, M. Y., Rahman, A. A., Menier, D., & Pubellier, M. (2014). Structural Evolution of Malay Basin, its link to Sunda Block tectonics.

[6]Mazlan, H., Abolins, P., Hoesni, M., & Ahmad, M. (1999). The petroleum Geology and Resources of Malaysia. In PETRONAS, *Malay Basin* (pp. 173-185). Kuala Lumpur: Petroliam Nasional Berhad (PETRONAS).

[7]Miller, T., Luk, C., & Olgaard, D. (2002). Interrelationships between overpressure mechanism and in situ stress. *AAPG Memoir*, 145-161.

[8]Morley, C., Tingay, M., Richard, H., & Rosalind, K. (2008). Relationship between structural style, overpressure and modern stress, Baram Delta Province, northwest Borneo. *Journal of Geophysical Research* , 1-23.

[9]Tingay, M., Hillis, P., Swarbrick, R., Morley, C., Damit, & Razak, A. (2007). 'Vertically Transferred' overpressure in Brunei: Evidence for new mechanism for the formation of high-magnitude overpressure. *Geology*, 1023.

[10]Wahab, M., Asraff, A., Ismail, J., & Ibrahim, A. (2013). Significant Hydrocarbon Accumulation in Deep Overpressured Play of Baram Delta, Malaysia. *EAGE Conference & Exhibition.* London, UK.

[11]Yussof, W. l. W. (1993). Geothermics of Malay Basin, offshore Malaysm.

[12] Zhang, J. (2011). Pore pressure prediction from well logs: Methods, modification and new approaches. *Earth-Science Reviews (1008)*, 50-63

[13] Tingay, M., Hillis, P., Swarbrick, R., Morley, C., Damit, & Razak, A. (2009). Origin of overpressure and pore pressure prediction

[14] Ibrahim, C. A., Light, L., Kong, J. N.-C., & Mennie, J. (2012). Foresee the Unforseen: Modeling West Baram Delta Overpressure, Offshore Sarawak. *AAPG International Convention and Exhibition.* Singapore.

[15] Hermanrud, C., L. Wensaas, G.M.G. Teige, E. Vik, H.M. Nordgard Bolas & S. Hansen. (1998). Shale porosities from well logs on Haltenbanken (Offshore Mid-Norway) show no influence of overpressuring. Abnormal pressure in hydrocarbon environment: *AAPG Memoir 70*, 65-85.

Seismic input optimisation for Joint Migration Inversion on complex structures, Malay Basin.

H.H Hairi[1], A.G Md Rafek[2], A.H Abdul Latiff[3]

[1,] Research and Development Building, Geoscience Department, PETRONAS University of Technology
[2, 3] Block 16, Geoscience Department, PETRONAS University of Technology

Email: eezreeq@gmail.com

Abstract. The field of seismic imaging has made a recent breakthrough in migration technique with the introduction of Joint Migration Inversion to aid the imaging process by incorporating multiple scattering into the algorithm. Initially, seismic multiples have been regarded as noise and should be removed prior to migration, but today these multiples could be used to enhance the final seismic images since they carry a vast percentage of energy that was generated from the source during acquisition phase. This new technique could be used to improve the final image of complex structures such as the steeply fractured basement reservoir in Anding Utara Field in Malay Basin. As of now, the efforts of improving the steeply-fractured basement image have been done without the incorporation of multiple scattering. Thus, the application of JMI might shed some light on the potential of this technique. Joint Migration Inversion make use of a Closed-Loop imaging process which means that less wave energy is wasted during processing making the final image more consistent with the input data. In conventional migration process, multiple scattering is removed thus vast amount of energy is wasted that leads to a poor final subsurface image. The two main elements in Joint Migration Inversion are Full-Wavefield Migration and Full-Wavefield Inversion. Apart from including seismic multiples in the imaging process, Joint Migration Inversion also allows the user to automatically update the velocity model of the area under investigation while maintaining very little user intervention.

Keywords: JMI, migration, inversion, seismic, imaging, multiples, Full-Wavefield,

1. Introduction

Recent developments in research in the field of imaging geophysics has brought forth a new insight on how multiple reflections could be used to enhance the final images of the subsurface. The concept of constructing primary reflection data out of multiples goes back to the work of Claerbout (1968) on seismic interferometry. Based on Claerbout (1968), if at two locations passive measurement of random sources that are distributed in the earth are being made, the Green's function between these two locations is obtained by cross-correlation of these two measurements. In simpler words, any wavefield that reaches the surface which gets reflected back down as multiples and gets recorded at larger offsets, can also be seen as reflections from different source.

Recently, research has shown that multiple reflections could in fact assist in improving the final image of subsurface through the implementation of a method called Joint Migration Inversion or JMI. JMI is a combination of closed-loop full-wavefield imaging and closed-loop full-wavefield tomography (Staal and Verschuur, 2012). The implementation of a closed-loop algorithm instead of an open-loop algorithm would eventually be better because the final image of the subsurface is more consistent with the input data since no part of the initial data is removed such as the multiples. In another words, less energy is removed during processing and better energy preservation for the final image.

The next-generation seismic migration and inversion technology considers multiple scattering as vital information, allowing the industry to derive significantly better reservoir models with more detail and less uncertainty while requiring a minimum user intervention (Berkhout, 2012). There are two main aspects that constitute to this revolutionary JMI technique and they are Full-Wavefield Migration (FWM) and Full-Wavefield Inversion (FWI). Both processes have their own key roles to play in the process producing final seismic images by incorporating multiple scattering.

2. Joint Migration Inversion

2.1. Multiple Scattering

Multiples scattering has evolved from just being treated as noise to its incorporation in imaging process to improve the final seismic image of the subsurface. In order to utilise multiple scattering, we must first understand what are multiples, its types are and why it is important to include them in imaging process. Essentially, seismic multiples are reverberation or repetitions of primary seismic waves travelling within the earth. According to Verschuur et. al. (2013), multiples can be categorised by considering the interface where they have their shallowest downward bounce. In general, internal multiples as well as surface-related multiples are the subdivision of multiples which uses the basis of the shallowest downward bounce to categorise multiple reflections.

Apart from classifying multiple reflections based on the shallowest interfaces on which the bounces take place, a distinction could also be made between short-period multiples and long-period multiples. Essentially, long-period multiples are multiple reflections that could be decomposed into primary ray paths. Short-period multiples, on the other hand, are those multiple reflections that cannot be distinguished as separate events in the seismic section from the primaries that generate them. Multiples, in general, not only contributes to a large part to the finally transmitted energy, but they also generate dispersion effects, and an overall delay of the acoustic energy.

2.2. Joint Migration Inversion

The classic seismic processing method make use of open-loop method where only primary reflections are used as input and as a result, multiple reflections were removed prior to migration. This has been the setback in imaging geophysics since multiple reflections that were removed carry a vast amount of wave energy that was generated from a source that could have been used to improve the final image of the subsurface. In general, JMI utilises seismic reflection data in a closed-loop process in a way of matching forward-modelled responses to measured field data in a full waveform manner, with minimum user intervention. The inclusion of primary and multiple reflections means the final image is more consistent to the input data and less energy is wasted for imaging process. JMI aims to combine the best of elements of Full Waveform Migration (FWM), migration velocity analysis as well as Full Waveform Inversion (FWI).

Traditionally, the standard migration practice will offer limited information about the inconsistency between the input and the output seismic images. Well known migration algorithms such as the Reverse Time Migration and Wavefield Extrapolation Migration apply an open-loop algorithm that means multiple reflections are ignored. If multiple scattering is to be included in the migration processes, an open-loop approach will no longer be feasible.

In essence, FWM and FWI utilise multiple scattering in their own ways, and both are closed-loop type processes between the input and the output data. The difference between these two is that, in FWM the subsurface is described in terms of grid point reflectivity, while in FWI, the subsurface is described in terms of grid point contrast. In addition, FWM up and down wavefields are computed by recursive wavefield extrapolation while in FWI wavefields are computed by non-recursive, iterative wavefield modelling process (Berkhout and Verschuur, 2016).

IOP Conf. Series: Earth and Environmental Science **88** (2017) 012019 doi:10.1088/1755-1315/88/1/012019

3. Methodology

Mainly, in the first phase of these project, the original seismic data of Anding Field will be reprocessed to obtain the input necessary of JMI process. The reprocessing steps aim to remove unwanted noise, such as linear and swell noise from the raw data. The raw seismic data will be reprocessed using conventional seismic processing module and the steps are illustrated in part 4. Next, with the incorporation of multiple scattering, JMI algorithm will be applied to the reprocessed data in MATLAB to see whether the seismic image of the basement reservoir in Anding Field could be generated and to see whether it has been improved from the previous image or otherwise. Based on the final seismic image obtained by using JMI algorithm, the improvements made to the data will be analysed and compared to the previous image generated without the incorporation of multiple scattering. The effects that the new algorithm has on seismic data will also be analysed to further understand the potential and limits of JMI. The general research workflow is illustrated in Figure 1.

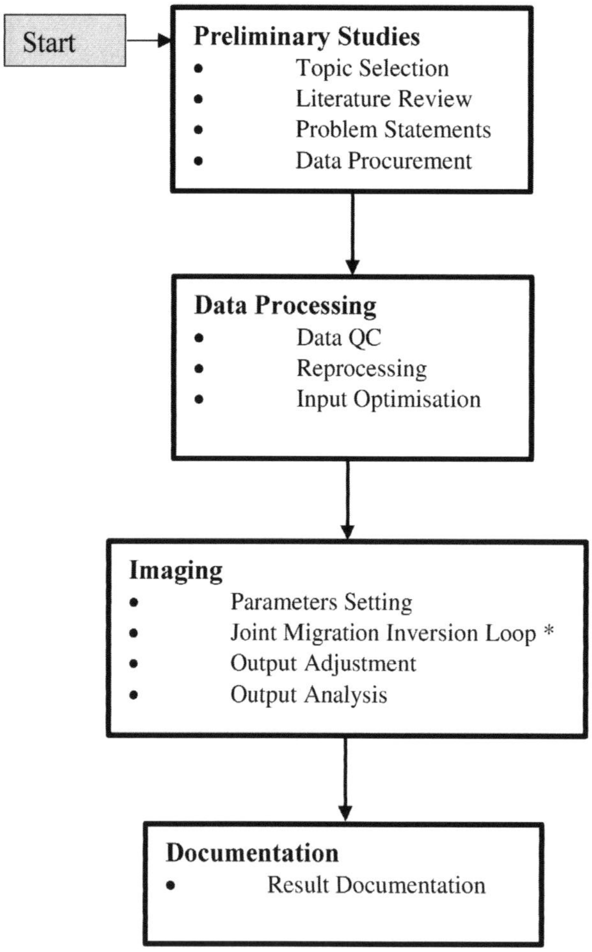

Figure 1. General research workflow.

IOP Conf. Series: Earth and Environmental Science **88** (2017) 012019 doi:10.1088/1755-1315/88/1/012019

4. Results and Analysis

The main objective of the reprocessing stage is to eliminate unnecessary noise prior to migration. Apart from that, multiple reflections are the essential elements in JMI technique. Thus, the reprocessing steps of the seismic data should not include multiple reflection elimination. Multiple reflections should be preserved in an effort to test the robustness of JMI and to enhance the final image using JMI.

Figure 2 shows the initial stacked section of the raw seismic data of the Anding Field. As expected, the seismic data is contaminated with much noise originated from the acquisition process as well as some instrumental noises. After each processing step, the data is QCed by analysing its stacked section to see whether the particular process has eliminated the targeted noise or whether the parameter was too harsh that some of the much needed data is also removed. The main key is to find the suitable processing parameters such as the frequency filter and linear noise attenuation parameters to remove unnecessary noise yet still maintain the primaries. Apart from presence of noise in the section, the issue of unbalanced amplitudes can also be observed from the stacked section. Thus, amplitude balancing is also included to the optimisation process. When observed in the shots record domain, a few bad traces as well as dead traces could also be observed and to mitigate this problem, trace editing and interpolation steps are also applied.

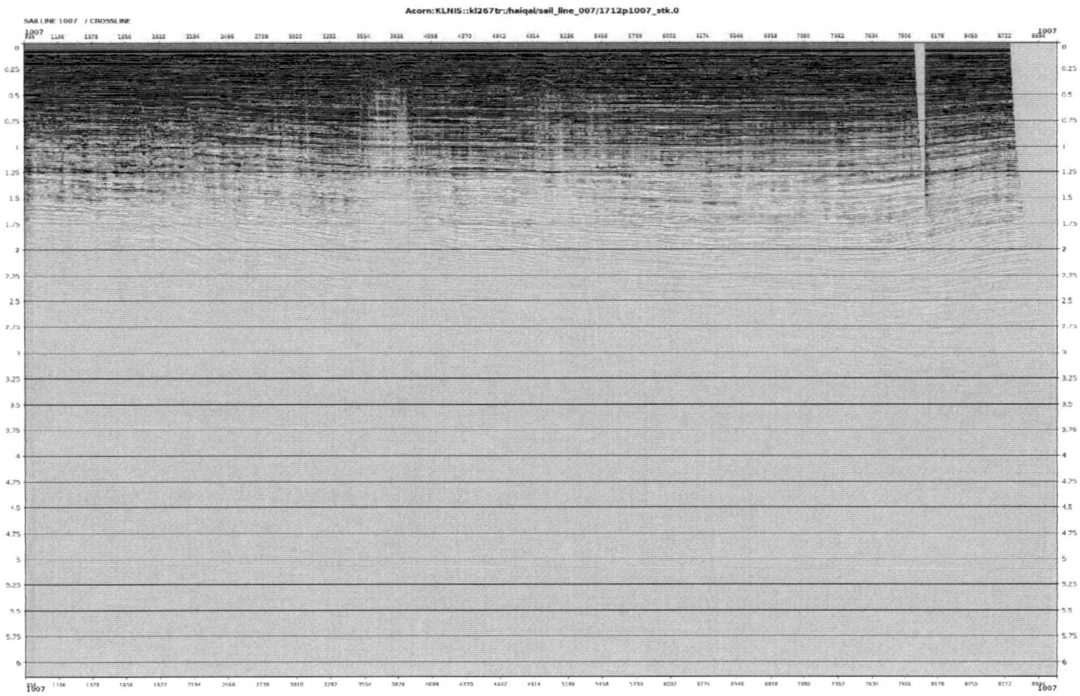

Figure 2. Shows stacked section of raw seismic data.

Figure 3 shows the stacked section of the same line after de-noise process. Generally, the processing workflow was done by utilising resources from CGG Asia Pacific located in Kuala Lumpur. The entire algorithm used was develop by CGG. Initially, the author considered utilising 3D data for JMI migration. Unfortunately, time constraint limited the project to use a single source and cable configuration, to save computing time.

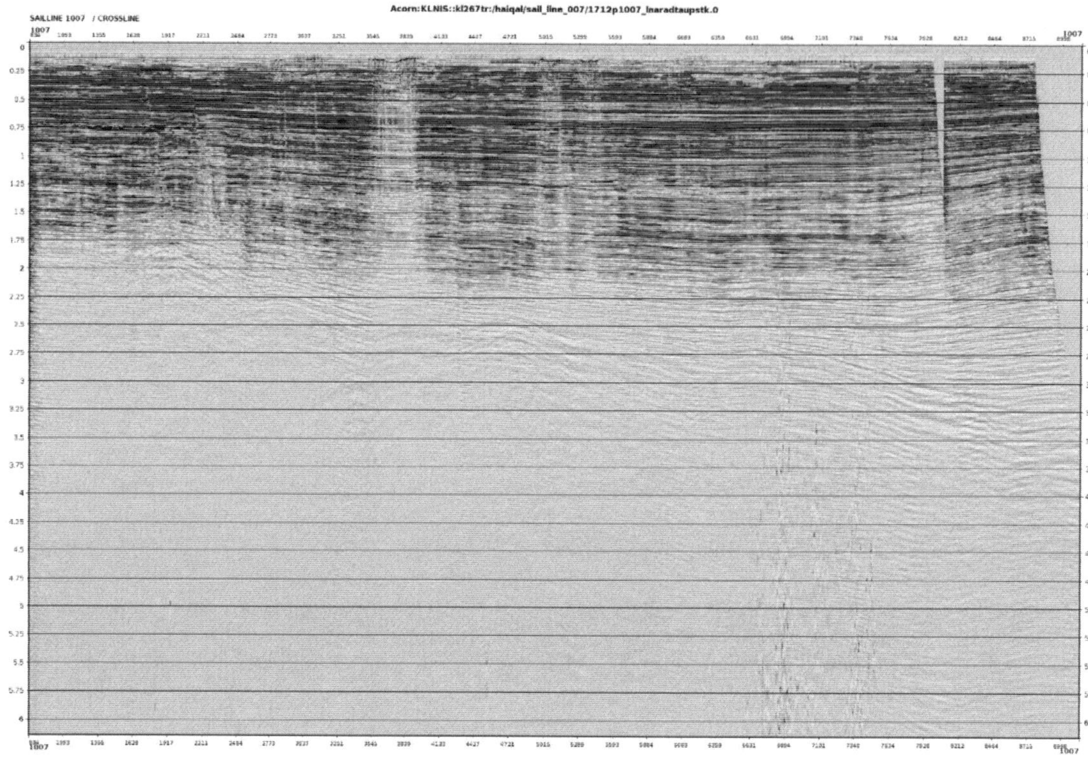

Figure 3. Shows seismic stacked section after de-noise. Low frequency noise is significantly reduced but remnant noise is still present after linear noise attenuation due to de-noising module was set too harshly. Further refining is needed. Amplitude balancing is still under par.

The de-noised stack section was produced by conducting noise removal modules. The workflow is illustrated by Figure 4. Subsequently, after noise attenuation process is done, the data will be tested if it is compatible with JMI input and adjustment will be made if necessary.

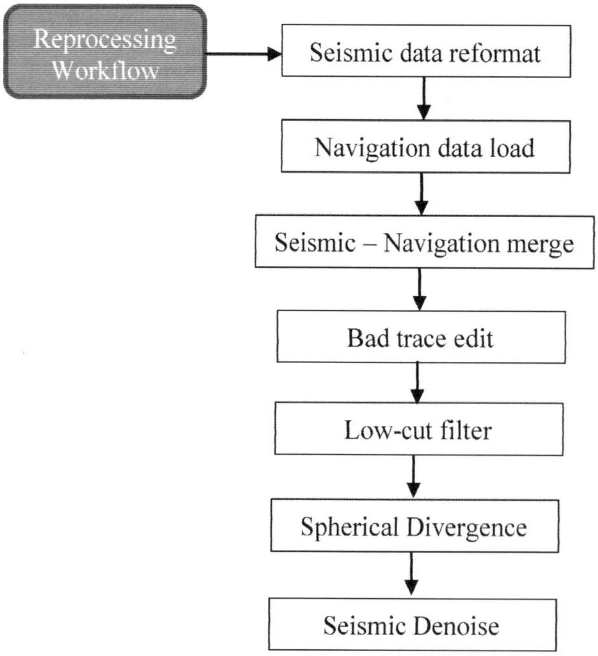

Figure 4. Processing steps in removing noise from the seismic section.

Noise removal stages start with the elimination of bad traces which were recorded during seismic data acquisition due to receivers malfunction. Bad traces could damage the overall data quality if they were kept, thus, they need to be removed. The noise removal processes continued with the Low-cut Filter process where the low frequency noise is removed. Low frequency waves that were < 2 Hz was removed by designing a filter from the filter module so that only waves of frequency of 2 Hz and above were kept. Next, seismic de-noise process aims to remove seismic noise such as linear noise, swell noise and random noise. By utilising specific module, linear noise and swell noise were reduced. Unfortunately, as depicted in Figure 3, some remnant noises are still present in the lower section. This is due to a few parameters used in the module such as the linear noise velocity and tau-p parameters that were set incorrectly thus, it did not successfully remove some of the noise. Therefore, in the future, some refining should be done to address this issue.

5. Conclusion

The primary objective of this work is to reprocess the raw seismic data for JMI application. A few issues have been identified concerning the size of the data and its quality and require some adjustments and modifications such as reprocessing and reducing the size of the data by only selecting certain shot points instead of using the overall data to be migrated with JMI algorithm. If the seismic data could not be used with the JMI, a set of synthetic data could be considered instead. This is due to the fact that synthetic data is much smaller in size, with no noise and in return, shorter computing time. Currently, the applicability of this new migration technique is still limited especially to a real dataset. Research on JMI has mostly revolved around synthetic data. Thus, for future work, this research will focus on making JMI technique more feasible to be used with real and larger dataset. Producing workflow for JMI could be important if this technique to be widely used in the future.

Acknowledgement

My utmost appreciation and gratitude to my supervisor, Prof. Abdul Ghani Md Rafek who is very motivative, inspiring and helpful in supporting progress of this research. Highest gratitude also goes to my Co-supervisor, Abdul Halim Abdul Latiff for his support and idea for this research. Not to forget, to Todd Wordford, manager of CGG Kuala Lumpur for allowing me to conduct the research at CGG. Last but not least, to my parents who were very helpful, encouraging and endlessly giving their support and ideas. Thank you.

References

[1] Berkhout, A.J. (2012). Combining Full-Wavefield Migration and Full-Wavefield Inversion: A Glance in the Future of Seismic Imaging. Geophysics Vol. 77. S43-S50.

[2] Berkhout, A.J., Verschuur, D.J. (2016). Enriched Seismic Imaging by Using Multiple Scattering. The Leading Edge, 128-133.

[3] Claerbout, J.F,. 1968. Synthesis of a layered medium from its acoustic transmission response, *Geophysics, **33**, 264-269*

[4] Staal, XR & Verschuur, DJ (2012). Velocity Model estimation by full waveform tomography of seismic reflection data. In Panos Kalemis (Ed.), Extended Abstracts of the 74th EAGE conference and exhibition incorporating SPE EUROPEC 2012 (pp. 232-235). Houten: EAGE.

[5] Verschuur, DJ & Staal, XR (2013). Part II, Structural Imaging , Chapter 12: Joint Migration Inversion: Integrating Full Waefield Migration and velocity updating. In D.J. Verschuur (Ed.), Delphi: The Multiple Removal and Structural Imaging Project (2012) (From Seismic Measurements to Rock and Pore Parameters, XXIV) (pp. 181-200). Delft, The Netherlands: Delft University of Technology.

Quantification of Organic richness through wireline logs: a case study of Roseneath shale formation, Cooper basin, Australia.

Maqsood Ahmad[1] Omer Iqbal[2], , [3]Askury Abd Kadir
[1]Petroleum Engineering, University teknologi Petronas (UTP), Perak Malaysia, maqsood.ahmad@utp.edu.my
[2]Petroleum Engineering, University Teknologi Petronas (UTP), Perak, Malaysia
[3]Petroleum Geoscience, University Teknologi Petronas (UTP), Perak, Malaysia

Abstract. The late Carboniferous-Middle Triassic, intracratonic Cooper basin in northeastern South Australia and southwestern Queensland is Australia's foremost onshore hydrocarbon producing region. The basin compromises Permian carbonaceous shale like lacustrine Roseneath and Murteree shale formation which is acting as source and reservoir rock.
The source rock can be distinguished from non-source intervals by lower density, higher transit time, higher gamma ray values, higher porosity and resistivity with increasing organic content. In current dissertation we have attempted to compare the different empirical approaches based on density relation and Δ LogR method through three overlays of sonic/resistivity, neutron/resistivity and density/resistivity to quantify Total organic content (TOC) of Permian lacustrine Roseneath shale formation using open hole wireline log data (DEN, GR, CNL, LLD) of Encounter 1 well. The TOC calculated from fourteen density relations at depth interval between 3174.5-3369 meters is averaged 0.56% while TOC from sonic/resistivity, neutron/resistivity and density/resistivity yielded an average value of 3.84%, 3.68%, 4.40%. The TOC from average of three overlay method is yielded to 3.98%.
According to geochemical report in PIRSA the Roseneath shale formation has TOC from 1 – 5 wt %.There is unpromising correlations observed for calculated TOC from fourteen density relations and measured TOC on samples. The TOC from average value of three overlays using Δ LogR method showed good correlation with measured TOC on samples.

1. Introduction

The common source rocks are shales and micritic limestones (Tissot, and Welte, 1984). The organic matter is important constituent in source rock. The rock considered source rock with commercial amounts of hydrocarbons with more than 1 wt% total organic content and the commerciality also depend upon other factors like type of organic matter, expulsion/drainage history (Hunt, 1996). The non-source rock also contain organic matter but less than 1 wt% (Passey et al.1990). The methods for evaluation of organic richness and maturity of source rock through laboratory methods are total organic carbon (TOC) analyses, pyrolysis, elemental analysis, vitrinite reflectance, thermal alteration index, gas chromatography and visual kerogen description). The interpretation of organic matter from wireline logs is based on physical properties i.e lower density, higher compressional slowness, high gamma rays, higher resistivity and high hydrogen and carbon concentration. Therefore the logs preferred for source rock evaluations include density, sonic, gamma ray, neutron and resistivity (Serra, 1986; Herron, 1988). In this paper comparative analyses has been carried out to compute total organic richness through different proposed empirical relations and their correlations with geochemical results performed on samples).

1.1 Geological setting

The studied area lies in Nappameri trough of Cooper basin, South Australia. The Cooper basin is predominantly gas producing with light liquid component (Heath, 1989; Boreham and Summons, 1999. The Complete geochemical analyses has been carried out in selected well (Encounter1) with complete set of wireline logs for Roseneath Formation (Beach). This formation comprises mostly shale with streaks of sandstone. In Encounter 1 according to geochemical report the studied formation is most prospective source rock interval within gas window between 3174.5-3369m. Thus the studied area only restricted to this column.

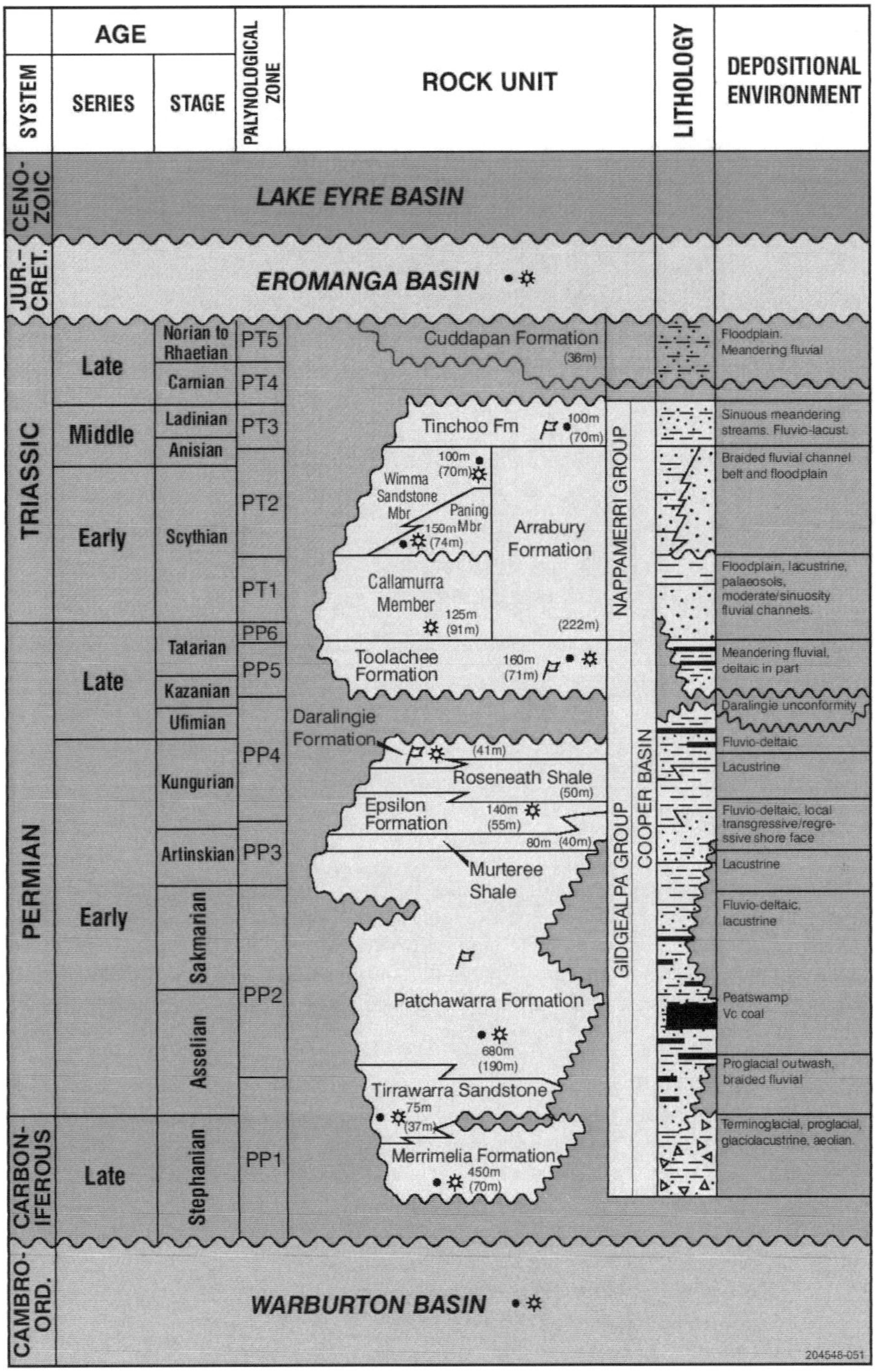

Figure 1: Stratigraphic column for Cooper basin (BEACH ENERGY).

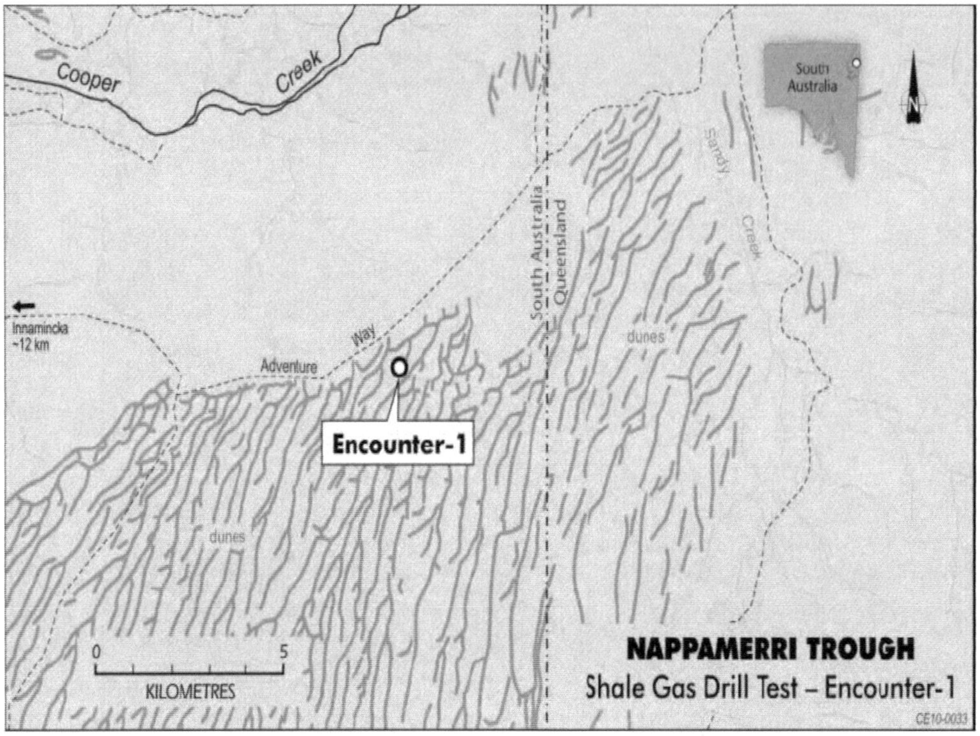

Figure 2: Location map for studied well (BEACG ENERGY).

1.2 Log response to organic –rich rocks

The responses of available logs (GR, Rt, φNΔt,ρb) to organic content can be summarized as follows:

1.2.1 Gamma ray Logs

It has been common to identify and quantify organic richness using well logs. Schmoker (1981) proposed the relationship between total gamma ray intensity and organic shales in the Appalachian Basin. The Organic rich source rocks can have high radioactivity and attributed to high gamma ray reading. This natural radioactivity is attributed to uranium, thorium and potassium contents. It can be concluded that high concentration of uranium in source rocks is due to planktons because planktons absorbs uranium ions from sea water and trace elements that contributed for Uranium enrichment in source rocks.

1.2.2 Formation Density logs

The difference in densities between rock matrix and OM can be used as a tool in identification and quantification of organic richness in source rock. The common matric densities varies from 2.6-2.9 gm/cc. While OM has specific gravity in the range of 0.95-1.05 gm/cc. This low specific gravity of OM is in line with specific gravity of fresh water. The organic matter is part of the matric in source rock and reduction in bulk density can be expected with OM (Autric,2002).

1.2.3 Resistivity and sonic logs

The increase in maturity and hydrocarbon generation replace the conductive pore water with nonconductive hydrocarbons that drastically increase resistivity (Meissner (1978)). The resistivity increases in mature source rocks due to generation of no conducting hydrocarbons (Nixon (1973), Meissner(1978), Schmoker and Hester (1989)). This response makes it possible tool for maturity indicator in organic rich source rock. The OM has low velocity while high transit time 180μs/ft (Mendelson and Toksoz 1985). The log response to OM depend upon distribution of OM in matrix.

The OM has low velocity while high transit time 180μs/ft (Mendelson and Toksoz 1985). The log response to OM depend upon distribution of OM in matrix.

1.2.4 Neutron logs

The neuron log porosity is relatively high in source rocks than non-source rocks because neutron tool measure hydrogen concentration which is present in water and hydrocarbons in rocks. The neutron response to OM is average 67 porosity units (Mendelson and Toksoz 1985) while matrix response are relatively close to zero.

2. Literature review

The premise studies have revealed the potential importance of wireline logs for source rocks evaluation. Beers, 1945, Swanson,1966, Schmoker (1981), Supernaw et al., 1978: fertl and Rieke , 1980 used gamma-ray spectral log for identification and quantification of organic richness with concept that gamma ray reading increase with increase in organic matter and organic matter is attributed to uranium associated with organic matter Schmoker (1981) proposed the relationship between total gamma ray intensity and organic richness for Devonian shales but he noted that extreme need to be taken when applying this technique to evaluate source rock in areas outside of specific Appalachians region. Schmoker and Hester (1983) used density log for estimating total organic content with observation that density log can be effected with heavy minerals such as pyrite that must be corrected base on linera relationship between pyrite and organic matter. Dellenbach et al. (1983) ,Autric and Dumesnil (1984,1985) developed method using gamma ray and sonic log to provide parameters that linearly relate to organic richness. Murray (1968), Nixon (1973), Meissner (1978) and Schmoker and hester (1989), used resistivity logs for estimation of organic content. Meyer and Nederlof (1984) introduced the method involving combination of resistivity, density and sonic logs by crossplot of density/resistivity or a transit time /resistivity but their method discriminates source and non-source rocks without attempting to quantify the organic richness (Passey et al,1990). This limitation has been fulfilled by Pessey et al,1990 by introducing the DlogR overlay methodA subsection

3 Methodology

The comparative analysis has been carried out for quantification of total organic contents. The TOC analyses were carried out in interactive petrophysics software using fourteen empirical relationships based on density log and Δ log method proposed by Passey et al,1990. The software also give TOC in Wt% by doing regression analyses based on density log. While Δ log method was used with an overlay of the resistivity log with sonic, density, or neutron logs and then their average TOC wt% were estimated. The flowchart of methodlogy is given below:

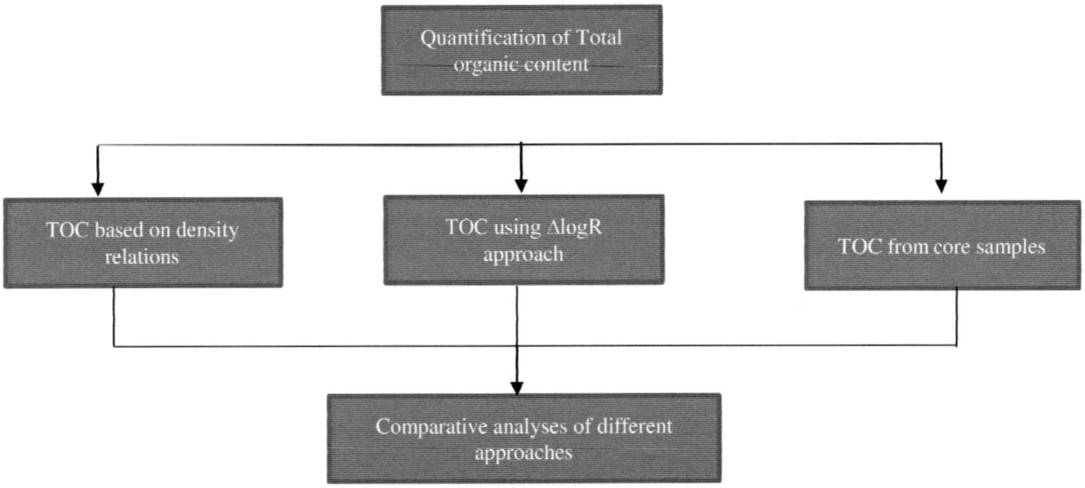

The Δ log R method used the resistivity log with properly scaled on porosity logs. The technique used the concept that two curves parallel to each other and can be overlain in organic clean rocks while separation between two curves occur in organic rich rocks. This separation is termed as Δ log R which is used to compute TOC in wt%.
Therefore, density (ρb) , sonic (Δt), neutron (φN) and resistivity (LLD) were properly scaled (-100 μsec/ft, 0.50 fraction porosity per two logarithmic resistivity cycles, i.e. a ratio of -50 μsec/ft to one resistivity cycle). The baseline conditions were calculated in fine-grained non-source rocks when two curves overlain each other. The separation of two curves were established as organic rich intervals while non-source rock intervals were established on overlain conditions of two curves. The expressions used to calculate Δ log R are::

$$\Delta \log R_{sonic} = \log 10(\frac{R}{R_{baseline}}) + 0.02(\Delta t - \Delta t_{basline}) \;.............(1)$$

$$\Delta \log R_{Neutron} = \log 10(\frac{R}{R_{baseline}}) + 4.00(\phi N - \phi N_{basline}) \;.......(2)$$

$$\Delta \log R_{Density} = \log 10(\frac{R}{R_{baseline}}) - 2.50(\rho b - -\rho b_{basline}) \;........(3)$$

Where Δ log R is separation, R, Δt, ρb and φN are resistivity in ohm-m, compressional slowness in μsec/ft and neutron porosity in fraction respectively and Rbaseline, are resistivity corresponding to Δtbaseline , φNbaseline and ρb baseline. The constant values 0.02, 4.00 and 2.50 are the ratios between scales of resistivity and each of sonic, neutron and density log. The empirical relations used to compute TOC is ::

$$TOC = (\Delta \log R) * 10^{(2.297 - 0.1688 * LOM} \;.........(4)$$

Where TOC is total organic content (wt%) and LOM is measured level of maturity which is obtained from vitrinite reflectance or thermal alteration index using maturation indicator (Hood et al(1975).

3.1 Use of Sonic/resistivity overlay

The resistivity log with range 0.1-1000ohm-m in logarithmic scale and sonic log with 200-0 μs/ft in linear scale showed good overlay. The expression used to compute Δ log R is :

$$\Delta \log R_{sonic} = \log 10(\frac{R}{R_{baseline}}) + 0.02(\Delta t - \Delta t_{basline})$$

Where Δ log R is separation between curves, R is resistivity in Ohm-m , Rbaseline is resistivity corresponding to Δtbaseline , 0.02 is based on ratio of compressional slowness cycle amount per one resistivity cycle. For Enconter 1 well, Rbaseline =10ohm-m and Δtbaseline= 100 μs/ft. The TOC were computed using empercial relations given below:

$$TOC = (\Delta \log R) * 10^{(2.297 - 0.1688 * LOM}$$

The TOC calculated using this relation is compared with TOC measured from geochemical analyses of samples represented by black dots.

4 Results and Discussion

The correlation were established between TOC from logs and TOC from samples which clearly indicated accuracy of each method. The TOC from density relation is 0.56 wt% (averaged) which is inconsistent with TOC from Δ log R method (3.98 wt%) and core samples as shown in Figure 3-7 with volume of shale and lithology determined using gamma ray log. The comparative analyses were carried out between TOC from density relations, Δ log R and TOC from samples. There is promising correlations were observed between Δ log R averaged and TOC from samples. While the induvial response of the sonic/resistivity, neutron/resistivity and density/resistivity overlay did not indicate the appreciable correlation with organic richness from samples. Similarly the unpromising correlation observed between organic richness by density relation and organic richness from samples accuracy of the standard deviation of difference (absolute error in wt%) between TOC from samples and TOC from log were calculated . The standard deviation of TOC from averaged Δ log R method and samples is ±1.7 which indicate very promising correlation between them while standard deviation of difference for individual sonic/resistivity, neutron/resistivity and density/resistivity overlay are ±2.3, ±2.10 and ±2.01 respectively. The Standard deviation for TOC based on density regression and density average are 2.07 and 2.48 respectively as shown in Table 1-5 below.

Table.1: Comparison of accuracy between TOC averaged Δ log R and Sample.

Sample No.	TOC _sample (wt%)	TOC_Lo g (wt%)	Absolute Error (wt%)
1	9.02	4.4017	4.6183
2	4.75	3.9791	0.7709
3	1.95	2.05	-0.1
4	6.12	4.3711	1.7489
5	2.35	3.9708	-1.6208
6	6.43	3.6706	2.7594
7	6.82	6.7	0.12
8	3.72	3.6903	0.0297
9	2.03	2.1	-0.07
10	2	4.2	-2.2
11	2.11	2.11	0
12	2.48	3.93	-1.45
13	2.31	3.9492	-1.6392
14	1.85	1.9	-0.05
15	2.8	2.8	0
16	1.58	1.58	0
17	1.8	3.333	-1.533
18	2.17	3.7003	-1.5303
Standard deviation of difference = ±1.7			

±1.7

Table 2: Comparison of accuracy between TOC from sonic/resistivity overlay and Sample

Sample No.	TOC _sample (wt%)	TOC_Log (wt%)	Absolute Error (wt%)
1	9.02	0	9.02
2	4.75	0	4.75
3	1.95	0	1.95
4	6.12	0.2276	5.8924
5	2.35	0	2.35
6	6.43	0	6.43
7	6.82	0	6.82
8	3.72	0	3.72
9	2.03	0	2.03
10	2	0	2
11	2.11	2.4184	-0.3084
12	2.48	0	2.48
13	2.31	0	2.31
14	1.85	0	1.85
15	2.8	0	2.8
16	1.58	0	1.58
17	1.8	0.0146	1.7854
18	2.17	0.6856	1.4844
Standard deviation of difference = = ±2.3			

±2.3

Table 3: Comparison of accuracy between TOC from neutron/resistivity overlay and samples.

Sample No.	TOC _sample (wt%)	TOC_Log (wt%)	Absolute Error (wt%)
1	9.02	4.0635	4.9565
2	4.75	3.7269	1.0231
3	1.95	3.7819	-1.8319
4	6.12	4.1051	2.0149
5	2.35	3.7861	-1.4361
6	6.43	3.6253	2.8047
7	6.82	4.2676	2.5524
8	3.72	3.5184	0.2016
9	2.03	4.1555	-2.1255
10	2	4.0198	-2.0198
11	2.11	3.8055	-1.6955
12	2.48	3.8793	-1.3993
13	2.31	3.8792	-1.5692
14	1.85	3.6564	-1.8064
15	2.8	3.887	-1.087
16	1.58	3.5459	-1.9659
17	1.8	3.1681	-1.3681
18	2.17	0.6856	1.4844
Standard deviation of difference = = ±2.10			

±2.10

Table 4: Comparison of accuracy between TOC from density/resistivity overlay and samples

Sample No.	TOC _sample (wt%)	TOC_Log (wt%)	Absolute Error (wt%)
1	9.02	4.1321	4.8879
2	4.75	3.7856	0.9644
3	1.95	3.8345	-1.8845
4	6.12	3.944	2.176
5	2.35	3.613	-1.263
6	6.43	3.5975	2.8325
7	6.82	4.1397	2.6803
8	3.72	3.4197	0.3003
9	2.03	3.6838	-1.6538
10	2	3.7894	-1.7894
11	2.11	3.5164	-1.4064
12	2.48	3.8306	-1.3506
13	2.31	3.5022	-1.1922
14	1.85	3.5125	-1.6625
15	2.8	3.5705	-0.7705
16	1.58	3.3708	-1.7908
17	1.8	2.8132	-1.0132
18	2.17	2.7838	-0.6138
Standard deviation of difference = = ±2.01			

±2.01

Table 5: Comparison of accuracy between TOC from density regression and samples.

Sample No.	TOC _sample (wt%)	TOC_Log (wt%)	Absolute Error (wt%)
1	9.02	4.4017	4.6183
2	4.75	3.9791	0.7709
3	1.95	4.0754	-2.1254
4	6.12	4.3711	1.7489
5	2.35	3.9708	-1.6208
6	6.43	3.6706	2.7594
7	6.82	4.3119	2.5081
8	3.72	3.6903	0.0297
9	2.03	4.1038	-2.0738
10	2	4.0793	-2.0793
11	2.11	4.1292	-2.0192
12	2.48	3.9521	-1.4721
13	2.31	3.9492	-1.6392
14	1.85	3.8227	-1.9727
15	2.8	3.9609	-1.1609
16	1.58	3.6756	-2.0956
17	1.8	3.333	-1.533
18	2.17	3.7003	-1.5303
Standard deviation of difference = = ±2.07			

±2.07

Table.6: Comparison of accuracy between TOC from averaged density relations and Samples

Sample No.	TOC _sample (wt%)	TOC_Log (wt%)	Absolute Error (wt%)
1	9.02	0.6122	8.4078
2	4.75	0.2852	4.4648
3	1.95	0.8058	1.1442
4	6.12	1.2296	4.8904
5	2.35	0.7699	1.5801
6	6.43	0	6.43
7	6.82	0	6.82
8	3.72	0.9223	2.7977
9	2.03	0.6077	1.4223
10	2	0.7204	1.2796
11	2.11	2.878	-0.768
12	2.48	0	2.48
13	2.31	0.9647	1.3453
14	1.85	0.4037	1.4463
15	2.8	0.4287	2.3713
16	1.58	0.3422	1.2378
17	1.8	1.0673	0.7327
18	2.17	1.5766	0.5934
Standard deviation of difference = = ±2.48			

±2.48

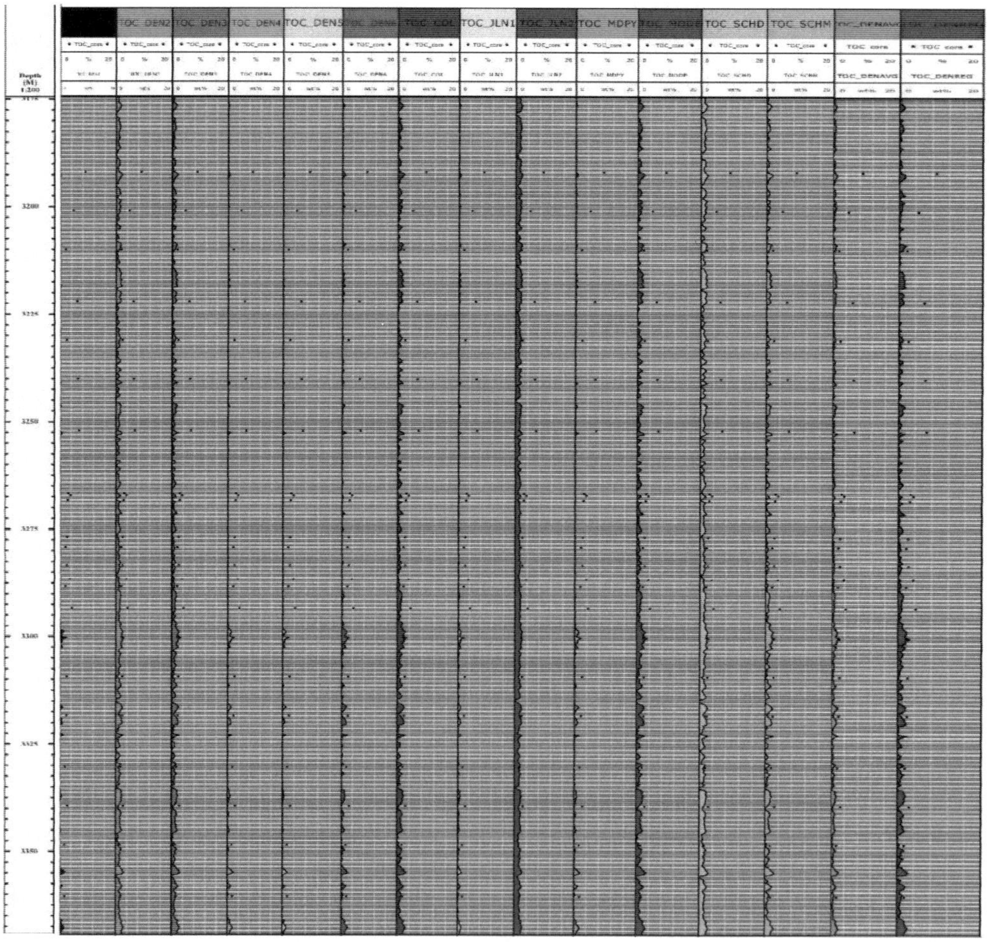

Figure3: Showing comparison between TOC calculated from density relations and TOC measured on samples for Roseneath Formation.

 doi:10.1088/1755-1315/88/1/012020

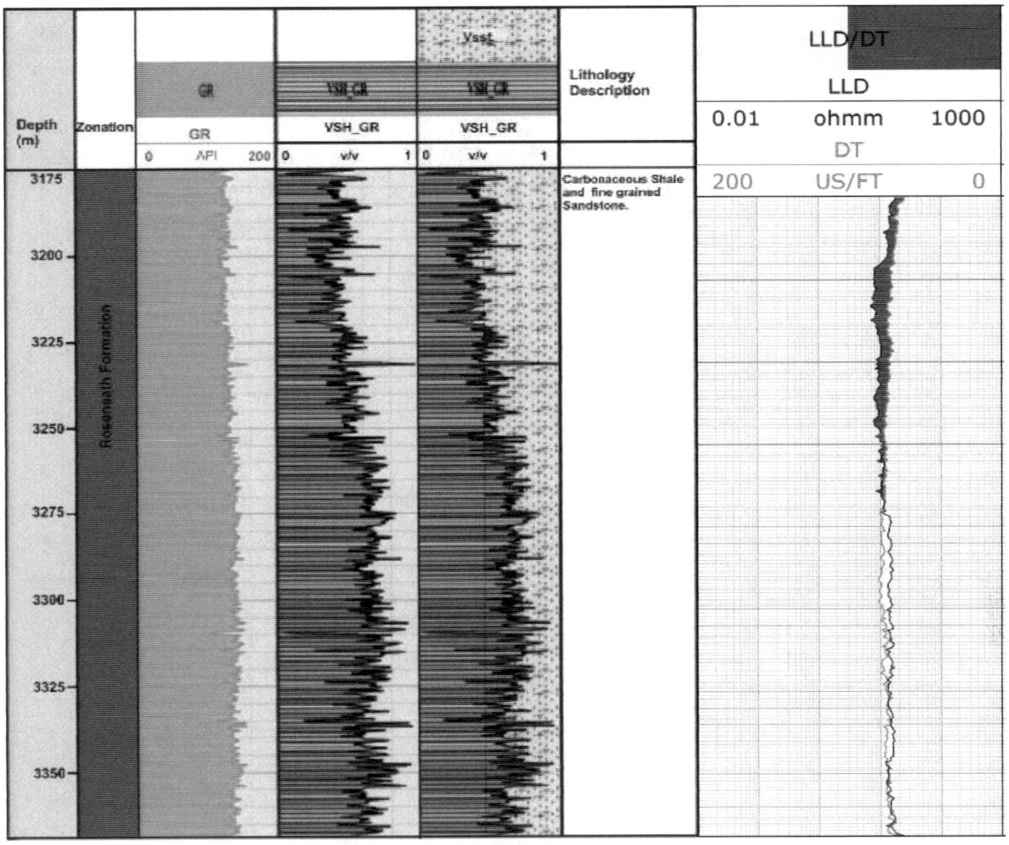

Figure 4: TOC calculated from resistivity/sonic relations.

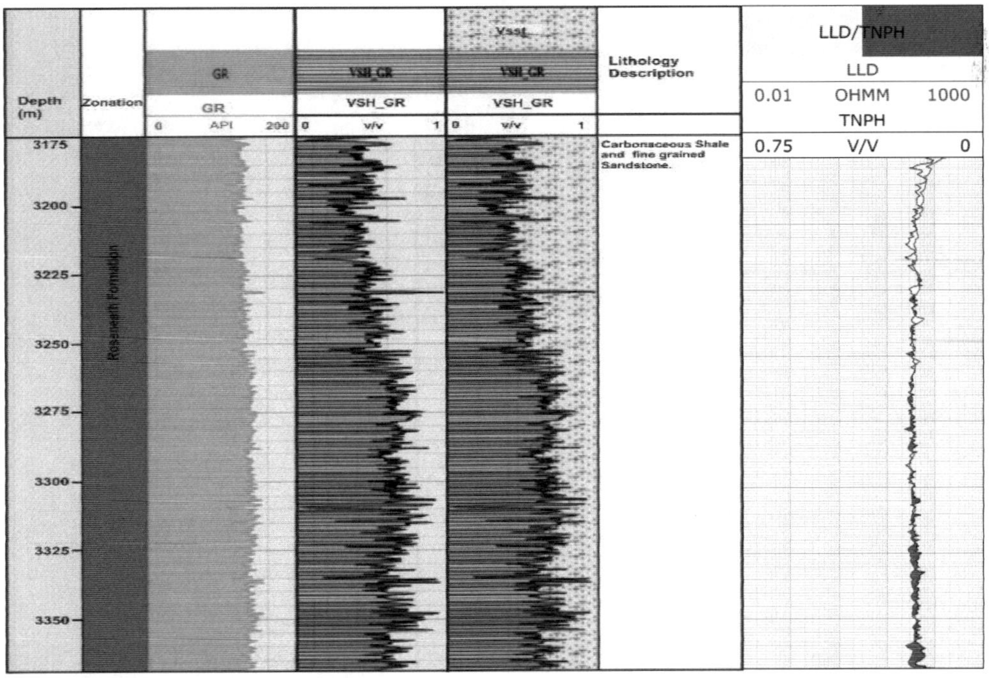

Figure 5: TOC calculated from Δ log R from neutron/resistivity overlay.

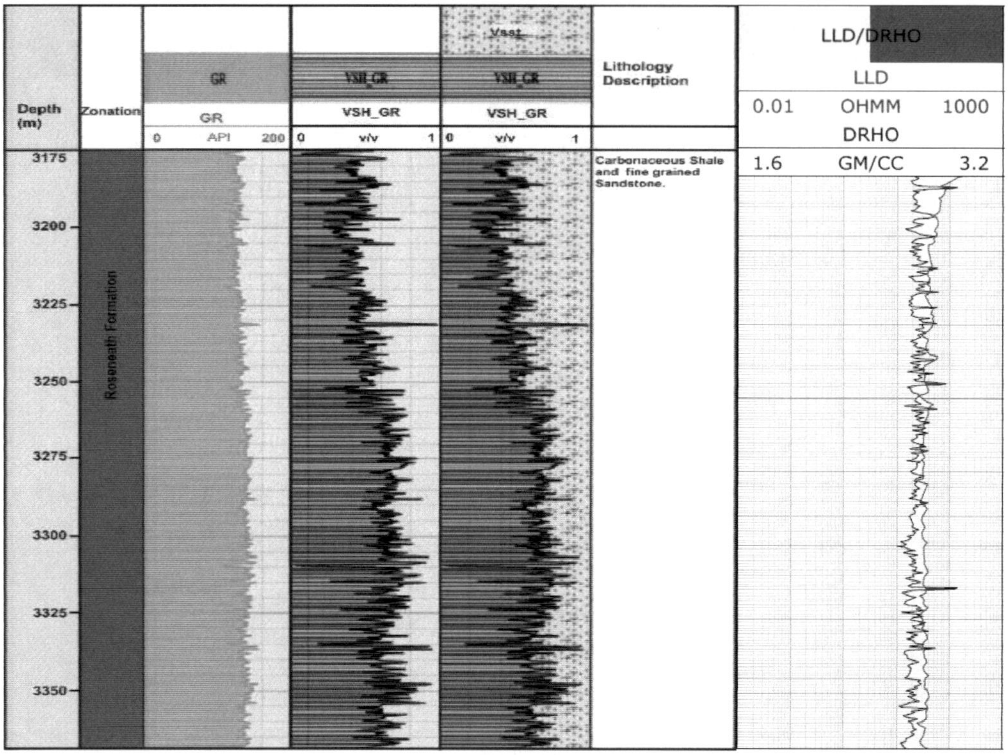

Figure 6: TOC calculated from Δ log R from density/resistivity overlay

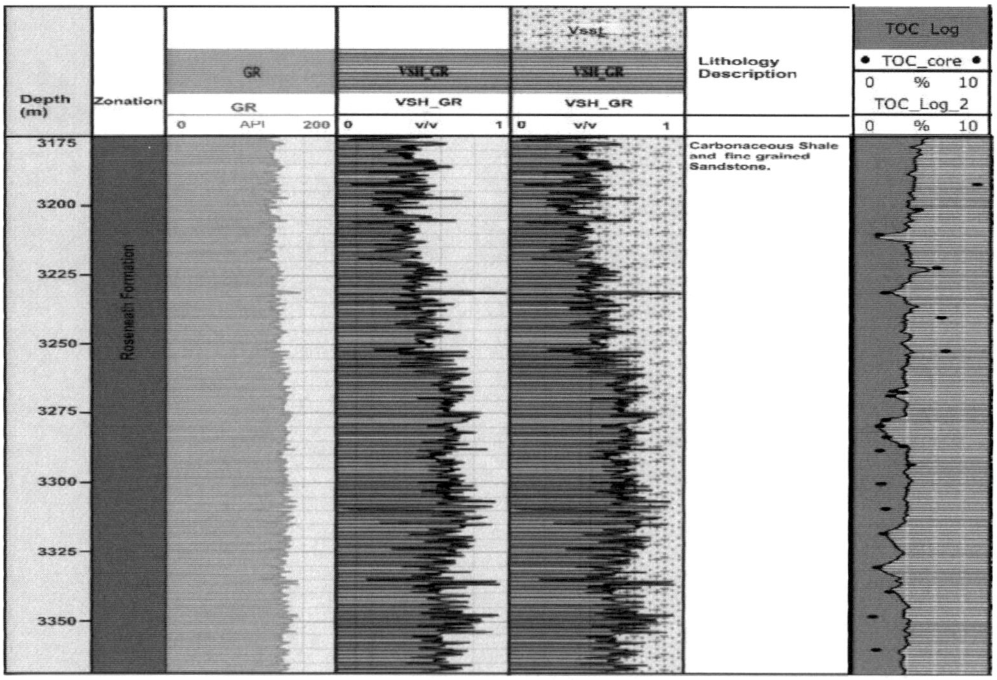

Figure 7: Showing promising correlation between TOC calculated from averaged Δ log R and TOC measured on samples.

5 Conclusion

The source rock evaluation from wireline logs is easy and quick method. The evaluation is based on response of each wireline logs (GR, CNL, DEN, DT,LLD) to increasing organic content. The increase of gamma ray, compressional slowness, neutron porosity, resistivity and decrease of density may be the consequence of increase in organic matter.

The comparative analyses indicated that the empirical relations proposed on density relations may be side specific and may not give promising results for other formation. According to PIRSA 2011, the Roseneath formation is kerogen type II and amount of TOC is 1 -5 wt% .The TOC from density relations vary from 0- 1.9 wt% with averaged 0.56 wt% for Roseneath formation which is inconsistent with sample data and published data. The TOC from sonic/resistivity, neutron/resistivity and density/resistivity overlay are 3.84 wt% , 3.68 wt% and 4.40 wt% averaged. While TOC from averaged of three Δ log R method is vary from 2-6wt% and 3.98 wt% averaged which is consistent with sample data. Therefore, the averaged TOC from Δ log R method is preferred which indicated that Roseneath formation has appreciable amount of organic matter. The log technique can be successful for quantification of organic richness if it is correlated with laboratory results. The result from Δ log R method showed good agreement with laboratory results. Δ log R is quick and reliable method to quantify organic richness. It is recommended to use average of Δ log R methods using different log suits for identification and quantification of organic richness before sampling of intervals.

The maturation history should be known before evaluating organic richness of formation.

6 Acknowledgment

We would highly thankful to DSD Australia for providing us well logs and geochemical data. The special thanks to Faculty of Department of petroleum engineering, in University Teknologi PETRONAS in Perak, Malaysia.

References

[1] Autric, A., and P. Dumesnil, 1985, Resistivity, radioactivity and sonic transit time logs to evaluate the organic content of low permeability rocks: The Log Analyst, v. 26, p. 36-45.

[2] Beers, R. F., 1945, Radioactivity and organic content of some Paleozoic shales: AAPG Bulletin, v. 29, p. 1-22.

[3] Dellenbach, J., J. Espitalie, and F. Lebreton, 1983, Source rock logging: Transactions of the 8th European SPWLA Symposium, paper D.

[4] Hunt, J.M., 1996. Petroleum Geochemistry and Geology, 2nd edition W.H. Freeman and Company, New York.

[5] Herron, S. L., L. Letendre, and M. Dufour, 1988, Source rock evaluation using geochemical information from wireline logs and cores (abs.), AAPG Bulletin, v. 72, p. 1007.

[6] Hood, A., Gutjahr, C.C.M., Heacock, R.L., 1975. Organic metamorphism and the generation of petroleum. AAPG Bulletin 59,
986– 996.

[7] Heath, R. 1989. Exploration in the Cooper Basin. The APPEA Journal, 29(1), 366–378

Meyer, B. L., and M. H. Nederlof, 1984, Identification of source rocks on wireline logs by density/resistivity and sonic transit time/resistivity crossplots: AAPG Bulletin v. 68, p. 121-129.

[8] Meissner.F.F.,1978, Petroleum geology of the Bakken Formation Williston basin, North Dakota and Montana, in The economic geology of the Williston basin: Montana geological Society. 19

[9] Mendelson.J.D.,1985, and M.N. Toksoz. 1985 Source rock characterization using multivariate analysis of log data. Transactions of the Twenty-sixth SPWLA Annual Logging Symposium,paper UU.

[10] Nixon, R.P., 1973, Oil source beds in cretaceous mowry Shale of north-western interior United States: AAPG Bulletin. V. 57.p.136-161.

[11] Passey, Q. R., S. Creaney, J. B. Kulla, F. J. Moretti, and J. D. Stroud, 1989, Well log evaluation of organic-rich rocks, 14th International Meeting on Organic Geochemistry, Paris, abstract 75.

[12] Rieke,H.H.,III,D.W.Oliver,W.H.Fertl, and NcCord, 1980, Successful application of carbon/oxygen logging to coalbed exploration, 55th Annual Fall Technical conference of the Society of petroleum Engineers , Dallas, Texas, September 21-24,1980.
[13] Tissot.B.P., and D.H Welte. 1984. Petroleum formation and occurance: New York. Springer-Verlag. 699p.

IOP Conf. Series: Earth and Environmental Science **88** (2017) 012021 doi:10.1088/1755-1315/88/1/012021

Accuracy Analysis of DEM Generated from Cokriging Interpolators

A Setiyoko[1] and A M Arymurthy[2]

[12]Faculty of Computer Science, University of Indonesia, Depok, Indonesia
[1]Remote Sensing Technology and Data Center-LAPAN, Jakarta Timur, Indonesia

andie.setiyoko@ui.ac.id

Abstract. DEM as a representation of the earth's surface has many functions for spatial analysis. DEM can be produced from several kinds of techniques such as satellite technology stereo optical or radar technology. Problems when using the optical stereo data is at the high point density level that is not distributed evenly. In regions with homogeneous character, the height point is becoming sparse. This will affect to DEM accuracy. In order to solve the problem, performing fusion techniques using interpolation method cokriging involving data points ALOS PRISM and SRTM height point was conducted. The sparse height point derived from ALOS PRISM on some object is expected to be enhanced by using SRTM data. There were several aspects to enhance the accuracy of DEM-derived from this process: the character of topography, land cover types, density in height point of the data and the precise type of interpolation method used.

1. Introduction

Digital Elevation Model (DEM) is very important for various purposes including geology, hydrology, environmental modeling and urban planning [1]. DEMs also support the study of rainfall and earthquake-induced landslides [2]. DEM is needed in a large number of applications, starting from virtual globes and visualization to engineering and environmental planning [3]. DEM serves as an input for much spatial analysis [4]. DEM and its derivation such as slope and aspect were very important input for spatial analysis, particularly for forest fire vulnerability. Rogeau and Armstrong conducted a research to quantify the effects of elevation, aspect, slope as variables on probabilities of burning [5]. On the other side, satellite image orthorectification process requires DEM data to correct error affected by image perspective (tilt) and relief (terrain) effects [1].

DEM as a representation of height information of landscape can be generated from multisource, such as stereo optical imagery, Synthetic Aperture Radar (SAR), land surveying, airborne light detection and ranging (LIDAR) [3]. Older methods of generating DEMs often involve interpolating digital contour maps that may have been produced by direct survey of the land surface. This method is still used in mountain areas, where interferometry is not always satisfactory. DEM generated from Cartosat-1 stereo data has a value of RMSE ranging from 1.29 m to 2.96 m [6].

The quality of a DEM is a measure of how accurate elevation is at each pixel (absolute accuracy) and how accurately is the morphology presented (relative accuracy). Several factors play an important role for quality of DEM-derived products: (1) terrain roughness; (2) sampling density (elevation data collection method); (3) grid resolution or pixel size; (4) interpolation algorithm; (5) vertical resolution; and (6) terrain analysis algorithm [7]. Based on the previous research it was observed that ordinary kriging is the best interpolator for Cartosat-1 DEM [6]. It has been observed also that interpolation method with the least error is universal kriging and interpolation method with the highest error is global polynomial in generating IRS 1C DEM [8]. In some specific objects, height points are not well generated. Lack of height point usually occurs in the low frequent object where the slope of the terrain is very high [9].

In order to solve the problem, [9] have conducted research robust stereo image matching for spaceborne imagery concluded that experimental results with Cartosat-1 images indicate that the aspect-based correlation and blunder detection works very efficiently and effectively in stereo image matching. There is various research have been conducted in DEM fusion. Papasaika *et al* have conducted research in Fusion of Digital Elevation Models using Sparse Representations and concluded that the DEM fusion achieved up to 43% improvement in RMSE [3]. Schindler *et al* have conducted research in Improving Wide-Area DEMs through Data Fusion – Chances and Limits [1]. The research has an output that the experiments confirm significant improvements are possible by fusion of existing DEMs – in the ALOS + SPOT case the RMSE was reduced by 29%. There are various alternate methods to solve the problem by using geostatistical theory. Common techniques are probabilistic interpolators: simple kriging, ordinary kriging, universal kriging, indicator kriging, probabilistic kriging, disjunctive kriging, cokriging and also deterministic interpolators: inverse distance weighted (IDW), global polynomial, local polynomial, radial basis functions (RBF) written by [10]. Setiyoko and Kumar have conducted research to interpolate height point derived from stereo imagery using both probabilistic interpolators and deterministic interpolators [6]. It's concluded that cokriging method as the probabilistic method has better accuracy.

Another method used to interpolate point is a multiple-point geostatistical simulation (MPS) that was conducted by [11] to interpolate bathymetry data. Tang *et al* proposed a FILTERISM method that combined traditional geostatistics and the MPS for the purpose of image fusion and super-resolution enhancement which was used to downscaling remote sensing data applied only to image fusion of multispectral and panchromatic bands [12]. Tang *et al* then improved the method as improved FILTERISM by combining with a geostatistic approach to derived digital elevation data fusion using multiple-point geostatistical simulation [13]. Even there was view research in the development of geostatistic, to apply cokriging on DEM interpolation based on multi-point is still challenging because it has many method setting to be implemented. The issue in fitting variogram, the best setting of neighborhood searching are the problem to be solved. Because to fit the best variogram requires an understanding of the assumptions in the underlying theory of random processes on which geostatistics is based [14]. In order to get the best result, understanding trend and characteristic of the data used are very critical to choose the best variogram and parameters of the searching neighborhood.

In this research, we still conducted cokriging interpolation technique to fuse ALOS PRISM height points and SRTM height points by applying best fitting variogram and searching neighborhood by considering trend and character of study area topographic. This research project is aimed to calculate the accuracy of DEM generated from various cokriging interpolations by using two kinds of height points sources: ALOS PRISM and SRTM. Combination both data in cokriging interpolation should increase the accuracy of DEM-derived.

2. Cokriging

Cokriging uses more than one variable types [15], compared to kriging that only uses one variable type. Models based on more than one variable of interest form the basis of cokriging. Cokriging could be used as a tool for image fusion [16]. Cokriging could predict unknown point by calculating the main variables of $Z(s)$, both autocorrelation of $Z(s)$ and cross-correlation between $Z(s)$ and other type of variables would help to make a better predictions [15]. Cokriging is more complex compared to kriging. Basically cokriging is based on kriging theoretical term. Kriging requires a semivariogram modeling including values for the parameters: nugget, sill, and range. The semivariogram is defined as [15]:

$$Y(s_i, s_j) = \frac{1}{2} var(Zs_i - Zs_j) \quad (1)$$

where var is the variance.

Ordinary cokriging expressed in the following simple mathematical formula [15]:

$$Z_1(s) = \mu_1(s) + \varepsilon_1(s) \qquad (2)$$
$$Z_2(s) = \mu_2(s) + \varepsilon_2(s)$$

where $Z_1(s)$ and $Z_2(s)$ are the variables of interest, decomposed into a deterministic trend $\mu_1(s)$ and $\mu_2(s)$ which have random autocorrelated errors form $\varepsilon_1(s)$ and $\varepsilon_2(s)$. There would be autocorrelation for each of them and cross-correlation between them. Variations on this formula form the basis for all of the different types of cokriging. The trend can be a simple constant; that is, $\mu(s) = m$ for all locations s. Understanding different type of cokriging seen in Table 1 [15].

Table 1. Cokriging Methods.

No	Method	Definition
1	Ordinary Cokriging	If $\mu_1(s)$ and $\mu_2(s)$ are unknown, then this is the model on which Ordinary Cokriging is based.
2	Simple Cokriging	Whenever the trend is completely known (i.e., all parameters and covariates known), whether constant or not, it forms the model for Simple Kriging.
3	Universal Cokriging	Trends that vary, and where the regression coefficients are unknown, form models for Universal Kriging.
4	Disjunctive Cokriging	The general unspecified transformations of the $Z(s)$, namely $f_i(Z(s_i))$ or the ith variable can be made. The Disjunctive Cokriging predictor is processed by performing functions of variables to predict at location s_0.

3. Methodology

Location of study area, lies in city of Bandung, West Java, Indonesia, having boundary coordinates 761716.36 E – 791911.41 E and 9255543.35 S – 9216181.95 S in the datum of WGS (World Geodetic System) 84 and the projection system is UTM (Universal Transverse Mercator) as seen in Figure 1, which has both plain and hilly area with covered by multi type of land cover objects.

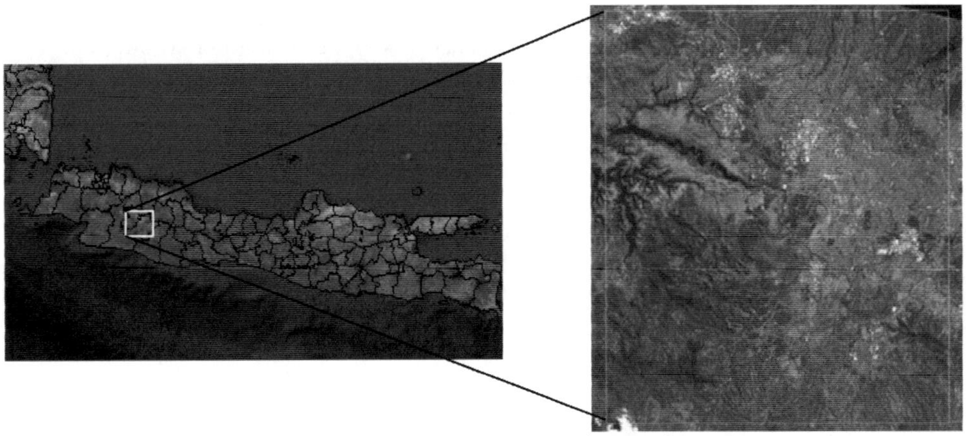

Figure 1. Study area of research experiment.

Data used in this research project are height points generated from ALOS PRISM stereo data and height points generated from Shuttle Radar Topography Mission (SRTM) seen in Figure 2. To validate the generated DEMs, high precision height points are used. Details of ALOS PRISM height points: generated from stereo panchromatic data; spatial resolution 2.5 m at nadir, observation date August 7th 2006, ellipsoid/datum EGM96, projection system UTM, zone no=48. Details of SRTM height points: generated from SRTM raster data version 4.1; horizontal datum WGS84; vertical datum EGM96 (Earth Gravitational Model 1996) ellipsoid; Spatial Resolution 3 arc-seconds for global coverage (90 meters); C-band Wavelength 5.6 cm. While high precision Points as reference points are the national base coordinate system used for national mapping, with the detail information as follows: Number of points 11 points, Horizontal datum WGS 84, Vertical datum EGM96.

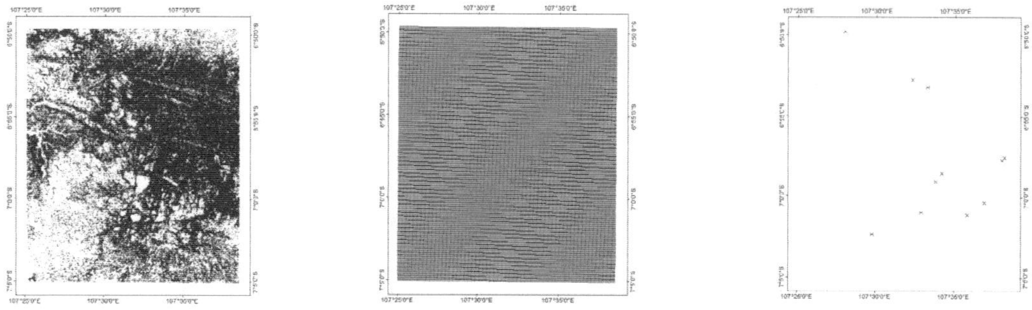

(a) (b) (c)

Figure 2. Height Points: (a) ALOS PRISM, (b) SRTM (c) Reference.

ALOS PRISM height points which contained 305882 points were generated from the interior and exterior orientation process based on satellite photogrammetric method. ALOS PRISM height points weren't well distributed in the study area. Dense points were located in high frequent objects such as buildings, road or settlement area, while sparse points found in a homogeneous object such as paddy field, forest, water body where matching points hardly to be generated in satellite photogrammetric processing. SRTM height points were derived from raster format of DEM by conducting raster to point processing. SRTM height points contained 83739 height points which were the same as the pixel number of the raster of the study area. One height point was a representation of each pixel value of DEM SRTM.

Both ALOS PRISM height points and SRTM height points were fused by implementing cokriging interpolation to generate DEM. ALOS PRISM height points defined as dataset one and SRTM height points defined as dataset two. By using module geostatistical analyst in ArcMap software, four techniques of cokriging (universal cokriging, ordinary cokriging, simple cokriging, and disjunctive cokriging) were applied. Four generated DEMs were expected to be generated from each technique. Accuracy analysis would be conducted by calculating a residual error based on the root mean square (RMSE) method of height information derived from interpolation method that has been applied to the datasets. In this analysis, reference height points were used as validation.

4. Result and Analysis

After performing interpolation techniques, four DEMs were generated. There are DEM derived from ordinary cokriging method (DEM OC), DEM derived from simple cokriging method (DEM SC), DEM derived from ordinary universal method (DEM UC), and DEM derived from ordinary disjunctive method (DEM DC) seen in Figure 3.

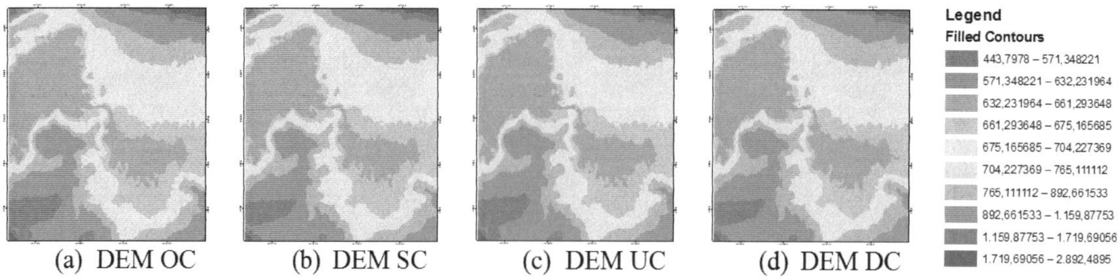

(a) DEM OC (b) DEM SC (c) DEM UC (d) DEM DC

Figure 3. DEMs generated from cokriging interpolation method.

In this experiment, analysis is conducted by performing RMSE quantitative analysis and visual comparation. General visualization of the four DEMs is resemblance one to another. In general, there is no difference when seen visually. Even so, if considered in more detail in some places there is a difference. DEM OC and UC there are similarities visually, while DEM SC and DC look difference. After validation analysis using height reference points, the calculation results obtained residual error for each DEM, as shown in Table 2.

Table 2. Residual Error (m) of Generated DEMs.

No	Reference Point	Character	Land Cover	ALOS Height Point	DEM UC	DEM OC	DEM SC	DEM DC	Average Residual Error
1	303	Plain	Settlement	Dense	1.50	1.50	1.70	0.92	1.41
2	172	Plain	Settlement	Dense	3.90	3.90	3.42	4.25	3.87
3	182	Plain	Settlement	Sparse	7.99	7.99	7.14	7.61	7.68
4	310	Plain	Settlement	Dense	5.70	5.70	5.38	6.03	5.70
5	045	Plain	Settlement	Dense	4.69	4.69	9.75	7.15	6.57
6	007	Plain	Settlement	Dense	2.49	2.49	2.14	2.93	2.51
7	237	Plain	Settlement	Dense	0.75	0.75	6.64	3.08	2.81
8	029	Plain	Settlement	Dense	4.18	4.18	3.44	4.39	4.05
9	028	Plain	Settlement	Dense	1.49	1.49	1.15	1.43	1.39
10	0282	Hilly	Plantation	Sparse	12.02	12.02	0.99	8.81	8.46
11	0283	Hilly	Lake Area	Dense	3.64	3.64	5.75	3.01	4.01
	Average Residual Error				4.40	4.40	4.32	4.51	

The accuracy of generated DEMs depends on character of the topographic surface, landcover type, density of height point, and interpolation method. Hilly surface affected the accuracy of the DEM, seen at the reference point number 0282, where the average of the highest residual error, which is 8.46 m. While in the plain area have a relatively lower error and the lowest at 303 points, with an average of 1.39 m. When considering the influence of the type land cover at the reference point in the plantation area has a residual error is higher than other regions, although there are variations in the residual error settlement but remained relatively smaller which is also influenced by the density of the ALOS PRISM height points. As seen reference points 182 and 0282 where the height point is sparse, residual error is relatively higher compared to other points. Simple cokriging method was resulting a better interpolated DEM, as seen in Table 2, DEM SC has a smallest average residual error, 4.32 m.

In order to see the connection between the residual errors every cokriging method used, in this study, the T-Test Analysis across methods was applied. The asymp. Sig. (2-tailed) value of the residual error between each method is higher than 0.05, so that each pair did not have a significant difference.

5. Conclusion

In general, the accuracy of the DEM from ALOS PRISM fusion product height points with SRTM height points by using interpolation method cokriging depends on several factors such as the character of topography, land cover types, densities in height point of the data and the precise type of interpolation method used. In this study, high precision of interpolated value is obtained at reference point 028 which is plain located on settlement area with a dense ALOS PRISM height point. While the highest accuracy is obtained at the reference point 237 with the same character with 028 points, which is obtained by using Universal Cokriging or Ordinary Cokriging. While the highest residual errors obtained at reference point 0282 which is located on hilly and covered by plantation land cover and contained sparse ALOS PRISM height point. Therefore land cover type could be added as a factor that affects the quality of DEM-derived products, as additional the several factors that play an important role in DEM quality [7]. It's occurred particularly for DEM generated from stereo imagery based on this research experiment. This additional factor of DEM error is the contribution for the research.

Further research will be done by conducting a study on the problem to increase the number of height points of a region mainly on homogeneous land cover.

References

[1] K. Schindler, H. Papasaika-hanusch, S. Schütz, and E. Baltsavias, "Improving Wide - Area DEMs Through Data Fusion - Chances and Limits," pp. 3–14.

[2] J. Iwahashi, I. Kamiya, and H. Yamagishi, "Geomorphology High-resolution DEMs in the study of rainfall- and earthquake-induced landslides : Use of a variable window size method in digital terrain analysis," *Geomorphology*, vol. 153–154, pp. 29–38, 2012.

[3] H. Papasaika, E. Kokiopoulou, E. Baltsavias, K. Schindler, and D. Kressner, "Fusion of Digital Elevation Models Using Sparse Representations," *U. Stilla al.*, pp. 171–184, 2011.

[4] P. C. Kyriakidis, A. M. Shortridge, and M. F. Goodchild, "Geostatistic for conflation and accuracy assesment for digital elevation models," *Int. J. Geogr. Inf. Sci.*, vol. 13, no. 7, pp. 677–707, 1999.

[5] M. Rogeau and G. W. Armstrong, "Forest Ecology and Management Quantifying the effect of elevation and aspect on fire return intervals in the Canadian Rocky Mountains," *For. Ecol. Manage.*, vol. 384, pp. 248–261, 2017.

[6] A. Setiyoko and A. Kumar, "COMPARISON ANALYSIS OF INTERPOLATION TECHNIQUES FOR DEM GENERATION USING CARTOSAT-1 STEREO DATA," *Int. J. Remote Sens. Earth Sci.*, vol. 9, no. 2, pp. 78–87, 2012.

[7] A. M. Elmoustafa, H. N. Farres, and M. M. Elfawy, "Effect of Elevation Data Accuracy on Storm Drainage Schemes , Lagos , Nigeria," *Nat. Resour.*, vol. 6, pp. 433–441, 2015.

[8] A. Setiyoko, "Analysis of spatial interpolation techniques for DEM generation using IRS-1C data," in *34th Asian Conference on Remote Sensing 2013, ACRS 2013*, 2013, vol. 2, pp. 789–794.

[9] V. N. Radhika, B. Kartikeyan, B. Gopala Krishna, S. Chowdhury, and P. K. Srivastava, "Robust stereo image matching for spaceborne imagery," *IEEE Trans. Geosci. Remote Sens.*, vol. 45, no. 9, pp. 2993–3000, 2007.

[10] P. A. Burrough and M. Rachael A., *Principles of Geographical Information Systems*, no. January 1998. OXFORD UNIVERSITY PRESS, 1998.

[11] S. K. Jha, G. Mariethoz, and B. F. J. Kelly, "Bathymetry fusion using multiple-point geostatistics: Novelty and challenges in representing non-stationary bedforms," *Environ. Model. Softw.*, vol. 50, pp. 66–76, 2013.

[12] Y. Tang, P. M. Atkinson, and J. Zhang, "Downscaling remotely sensed imagery using area-to-point cokriging and multiple-point geostatistical simulation," *ISPRS J. Photogramm. Remote Sens.*, vol. 101, pp. 174–185, 2015.

[13] Y. Tang, J. Zhang, L. Jing, and H. Li, "Digital Elevation Data Fusion Using Multiple-Point Geostatistical Simulation," *IEEE J. Sel. Top. Appl. EARTH Obs. Remote Sens.*, vol. 8, no. 10, pp. 4922–4934, 2015.

[14] M. A. Oliver and R. Webster, "A tutorial guide to geostatistics: Computing and modelling variograms and kriging," *Catena*, vol. 113, pp. 56–69, 2014.

[15] K. Johnston, J. M. Ver Hoef, K. Krivoruchko, and N. Lucas, "ArcGIS 9 ® Using ArcGIS Geostatistical Analyst," in *ESRI*, vol. 5, no. 2, 2001, pp. 131–166.

[16] N. Memarsadeghi, J. Le Moigne, N. Gsfc, and D. M. Mount, "Image Fusion Using Cokriging," *IEEE*, pp. 2–5, 1999.

Acknowledgments

This project was supported by Faculty of Computer, University of Indonesia and Remote Sensing Technology and Data Center, LAPAN Indonesia.

IOP Conf. Series: Earth and Environmental Science **88** (2017) 012022 doi:10.1088/1755-1315/88/1/012022

Application of Simultaneous Inversion Characterizing Reservoir Properties in X Field, Sabah Basin

Lee Ming Xiang[1], Luluan Almanna Lubis[1]
Geoscience Department, Universiti Teknologi PETRONAS, 32610 Tronoh, Perak Darul Ridzuan, Malaysia.

Email: mingxiang1006@gmail.com

Abstract. X field located between the East Baram Delta and Southern Inboard Belt, which is elongated northwest to southeast direction in Middle Miocene Sabah Basin. Previous studies were unable to differentiate the lithology and fluid distribution in the field by using conventional reservoir characterization approaches for example amplitude versus offset (AVO) analysis and post-stack deterministic inversion due to the low contrast in P-impedance between gas sand and shale. Thus, the goal of this approach is to obtain a reliable estimation of P wave velocity (V_P). S-wave velocity (V_S), and density (ρ). From these parameters, we predict the fluid and lithology properties for better reservoir characterization. A feasibility analysis was conducted and the results show that prediction of the hydrocarbon reservoir in X field is feasible using P-impedance and Vp/Vs. Consequently, the simultaneous inversion is carried out to derive the rock properties such as density, P-impedance, S-impedance, Lambda-rho (λρ) (Incompressibility), and Mu-rho (μρ) (Rigidity). The results from this study demonstrate the effectiveness of simultaneous inversion and the best parameters for enhancing the resolution and characterizing the hydrocarbon sand layer of the prospect.

1. Introduction

1.1 Background

Back in years, amplitude has only been used for prospect evaluation and reservoir characterization or development since the advent of bright spot Direct Hydrocarbon Indicator (DHI) technology. However not all bright spots are DHI, it can be happened when hard shale is overlying on the soft shale or brine sand, and when there is anisotropy caused by the presence shales or thinly bedded intervals [1]. Thus, it was best studied in AVO-inversion domain by classifying the AVO response from forward modelling and extract more information from the shear impedance seismic data at different offset that can highlight the fluid content and discriminate lithology in the reservoir [2].

Both AVO and inversion should work consistent with well log and vertical seismic profile (VSP) data. According to [3], elastic parameters can be obtained through prestack AVO inversion for reservoir characterization. Thus, the integration of seismic, geology, borehole, rock physics, and petrophysics are significant for the AVO inversion process. Inversion is a technique that convert the interface property or reflection coefficient from seismic data to a layer property which is known as P-impedance [4].

[1] Geoscience Department, Universiti Teknologi PETRONAS, Malaysia, Bandar Seri Iskandar, 32610, Tronoh, Perak Darul Ridzuan, Malaysia.

Information of internal rock properties such as lithology, porosity or fluid type could be extracted from the amplitudes of inverted data set. From the inverted data, it is more ideal for us to do stratigraphic interpretation and reservoir characterization.

1.2 Problem Statement

In earlier study, reservoir characterization using post stack deterministic method to estimate acoustic impedance was attempted, but this method was unable to discriminate the reservoir and non-reservoir lithology in the field as both gas sand and shale are sharing almost similar P-impedance response . Thus, in most wells, the fluid contacts were not seen although the pressure data suggested that the contacts were close to the amplitude shut-offs and flat spot that can found in the given seismic. Besides, the field has also polarity issue and amplitude which increase the challenges to characterize the reservoir *[1]*. The associated seismic data quality and amplitude of certain area of prospect were poor due to wipe out of amplitudes caused by gas leakage.

From previous case study, X field was being abandoned before due to lack of hydrocarbon indication and the negative results from previous drilling activities. Thus, a combination of geological concepts and modelling with good quality of seismic and careful geological and geophysical evaluation were required for successful exploration and development for the prospect area.

1.3 Objectives

The main objective of this research is to discriminate the lithology and maps the fluid distribution of the hydrocarbon bearing reservoir by using simultaneous inversion method. The subsequent objectives to be achieved while carrying the study are:

- Estimate the reservoir properties such as P-impedance, S-impedance, density (ρ), Vp/Vs, Lambda-rho (λρ) (Incompressibility), and Mu-rho (μρ) (Rigidity) from the inversion results.
- Enhance the prediction of actual reservoir's condition to reduce to risk of drilling a dry well.
- Determine the best parameter for inversion process.

1.4 Scope of Study

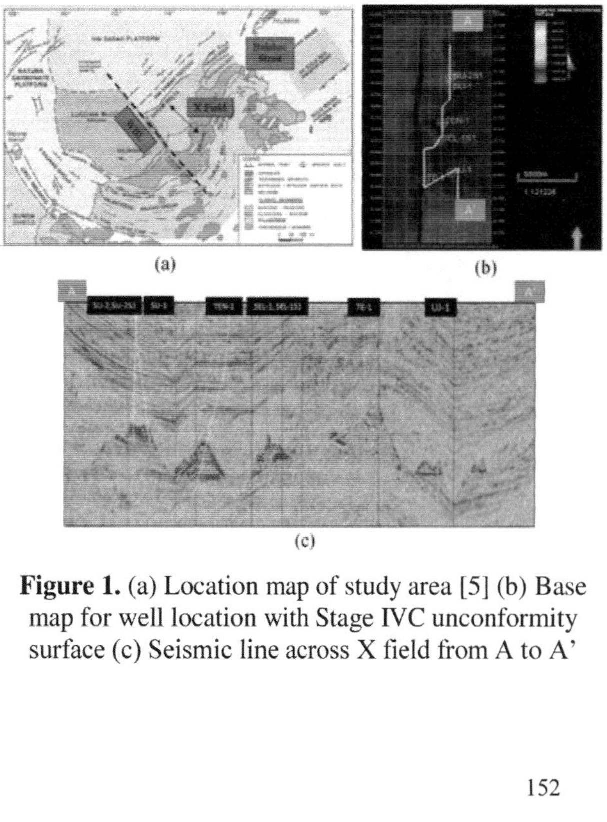

Figure 1. (a) Location map of study area [5] (b) Base map for well location with Stage IVC unconformity surface (c) Seismic line across X field from A to A'

The study includes the 3D angle stacks seismic and well logs data from X field. The 3D seismic data are about 300 km2 (925 inline and 1981 crossline) with angle stacks at different ranges, near (5°-15°), mid (15°-25°), and far (25°-40°). Check shot and well logs from 8 deviated wells which required for both geological and petrophysical analysis were complete and ready for the study. These datasets will be loaded into Schlumberger Petrel and Hampson Russell software for further analysis. In this case study, only 3 wells will be emphasized which are SU-1(deviated), SEL-1 (vertical) and UJ-1(vertical). Figure 1 shows the location map, base map and seismic cross section of the study area.

IOP Conf. Series: Earth and Environmental Science **88** (2017) 012022 doi:10.1088/1755-1315/88/1/012022

2. Methodology

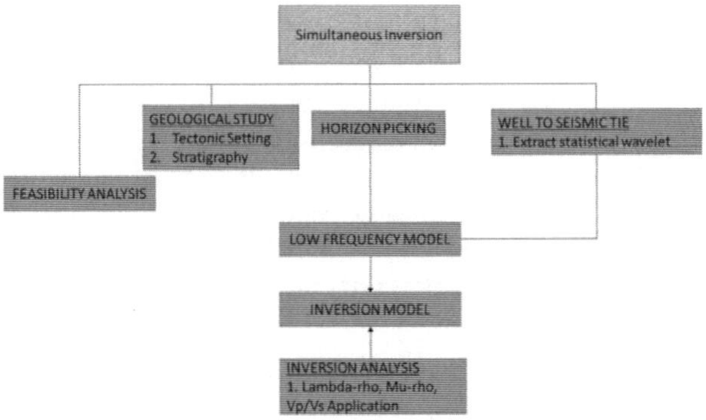

The workflow in figure 2 describes the steps and processes carried out to achieve the objectives of this research that include lithology and fluid distribution of the hydrocarbon bearing reservoir, reservoir properties estimation, and reservoir's conditions prediction. Each of the methods used is discussed in relation to its applicability to the study of reservoir characterization of X Field in Sabah Basin.

Figure 2. Simultaneous inversion workflow

2.1 Feasibility Analysis

The cross-plot approach is applied on the respective wells, to determine feasibility of the simultaneous seismic inversion process on the dataset, before proceeding to the full inversion workflow. This step could help us to understand the quality of both seismic and well logs data. Figure 3 represent the cross plot of Vp/Vs ratio versus P-impedance at the hydrocarbon zone of interest at well SU-1, SEL-1, and UJ-1. From the cross plots, separation between the points scattered can be observed, and make it possible to differentiate hydrocarbon from claystone and brine sand from these elastic rock properties.

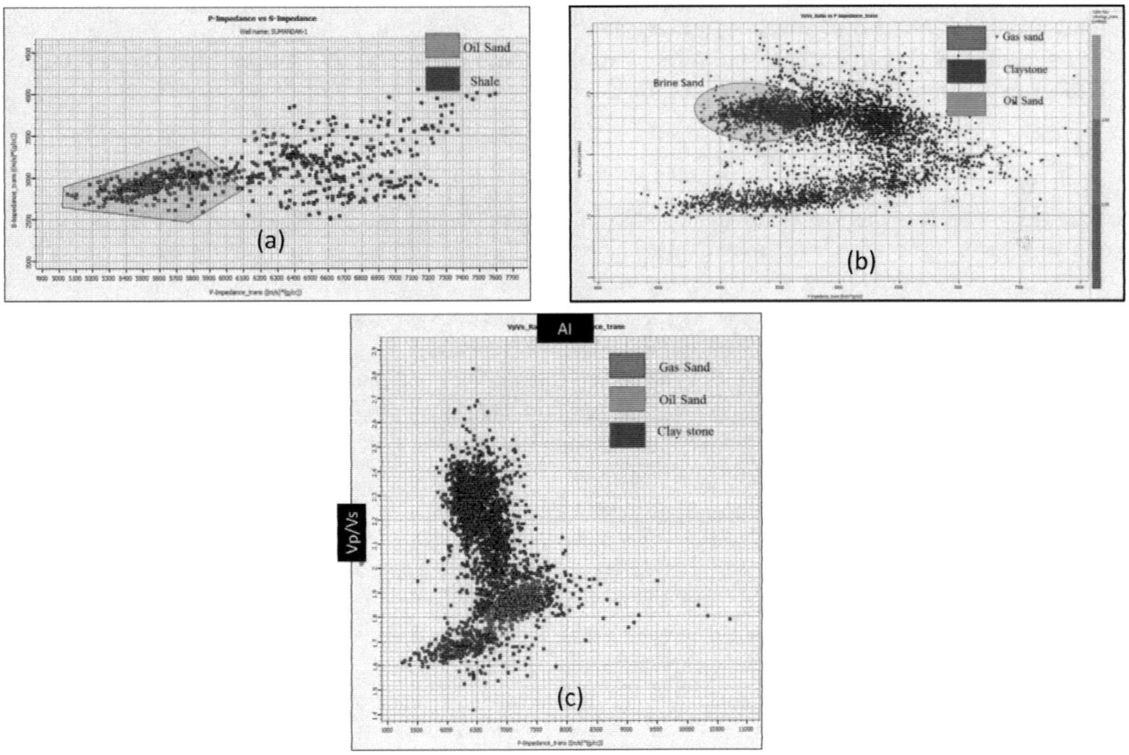

Figure 3. Cross plot of Vp/Vs ratio against P-impedance. (a) Well SU-1 (b) Well SEL-1 (c) Well UJ-1. (Red dots represent the gas sand, green dots represent oil sand, purple dots represent clay stone)

2.2 Wavelet Extraction and Well Tie

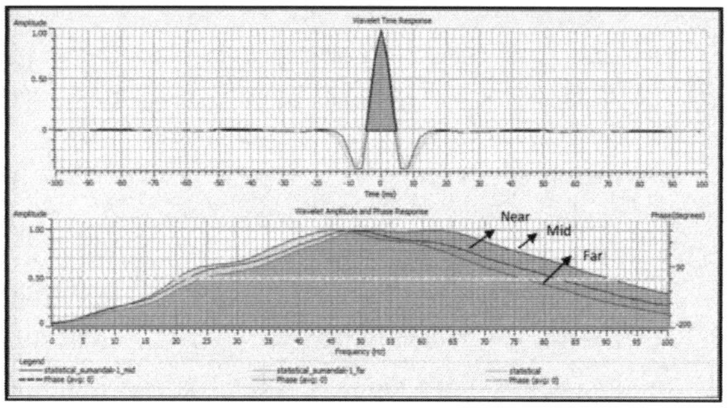

Figure 4. Extraction of statistical group wavelet

For a pre-stack data, it is highly expected that the seismic data will experience a continuous loss of high frequency energy from near to far offset [5]. Thus, each trace of CDP gather would have different wavelet representation. In this study, both statistical and respective well's wavelets from are extracted for comparison to determine which type of wavelet can generate better inversion results (figure 4).

For this study, well to seismic tie is important to examine the current depth-time conversion and optimizes it so that the generated zero-offset synthetic can optimally match the seismic. This step is crucial since the depth time curve derive only from sonic log is insufficient for the purpose.

2.3 Generate Low Frequency Model

A low frequency model (LFM) or initial model is basically an initial impedance volume to represent geological model of a field. The models cropped from 700ms to 2000ms are generated with the input of well logs data (P-wave velocity, S-wave velocity and density), angle gathers and the interpreted horizons. The simple model was built, with the values from the well logs are being extrapolated by using the log trend from all logs and the interpolation between the horizon of top reservoir and base reservoir is top lap to each other. This model able to provide data that will compensate for missing information in the seismic data at low end of spectrum. It can integrate the information and data from several sources, introduce trends, increase resolution, remove tuning and provide absolute results from the inversion. These models were then used as the input for inversion process.

2.4 Inversion Analysis and Modelling

After both wavelets and low frequency models have been obtained, the convolution process is carried out to generate PP synthetic trace by using Aki-Richard's equation (1), where, $c^1 = 1 + \tan^2\theta$, $c^2 = -8y^2 + \tan^2\theta$, $c^3 = -0.5\tan^2\theta + 2y^2\sin^2\theta$, $\gamma = V_S/V_P$, R_P= P-reflectivity, R_S= S-reflectivity, R_D= Density-reflectivity [6]. In this study, Hampson Russel's Simultaneous Inversion algorithm will be used and applied to the volume. Three inverted volumes will be generated namely, inverted P-impedance (Zp), inverted S-impedance (Zs) and density. The inverted results will be quality checked by using the cross-plot approach and compare the results with the well data at the bandpass filter with cut off frequencies of 10/15-70/90H. Relationships between P-impedance, S-impedance, and Lamė parameters were analysed for reservoir characterization.

$$R_{PP}(\theta) = c_1 R_P + c_2 R_S + c_3 R_D \quad (1)$$

3. Results and Discussion

3.1 Well Log Analysis

This is the first step to identify the reservoir of interest by considering the reservoir properties such as volume of shale, porosity and water saturation. The basis log used are gamma ray, density, neutron porosity, and resistivity logs. Whole formation was evaluated to identify the primary reservoir of this study. No core data available for calibration, input parameters were referred to the log analysis report from well SU-1.

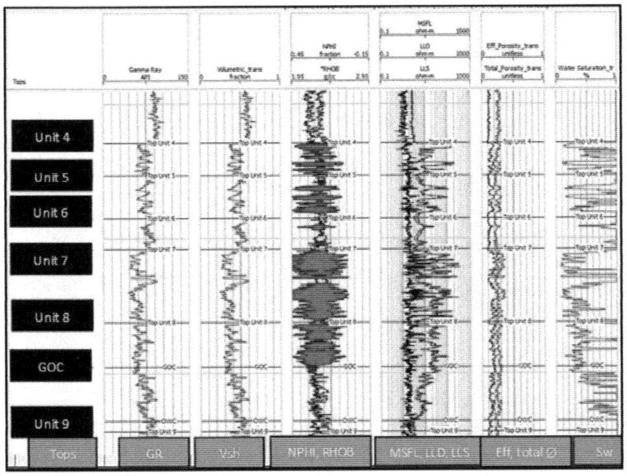

From this well log analysis, hydrocarbon zone can be indicated by density-neutron crossover. According to [7], presence of negative cross over for density and neutron log, with low gamma ray and high resistivity indicates the existence of hydrocarbon in a sandy reservoir. Figure 5 shows the petrophysical results for the control vertical well of this study SEL-1.

Figure 5. Results of petrophysical quick look from measured depth (MD) at well SEL-1

3.2 Inversion Analysis

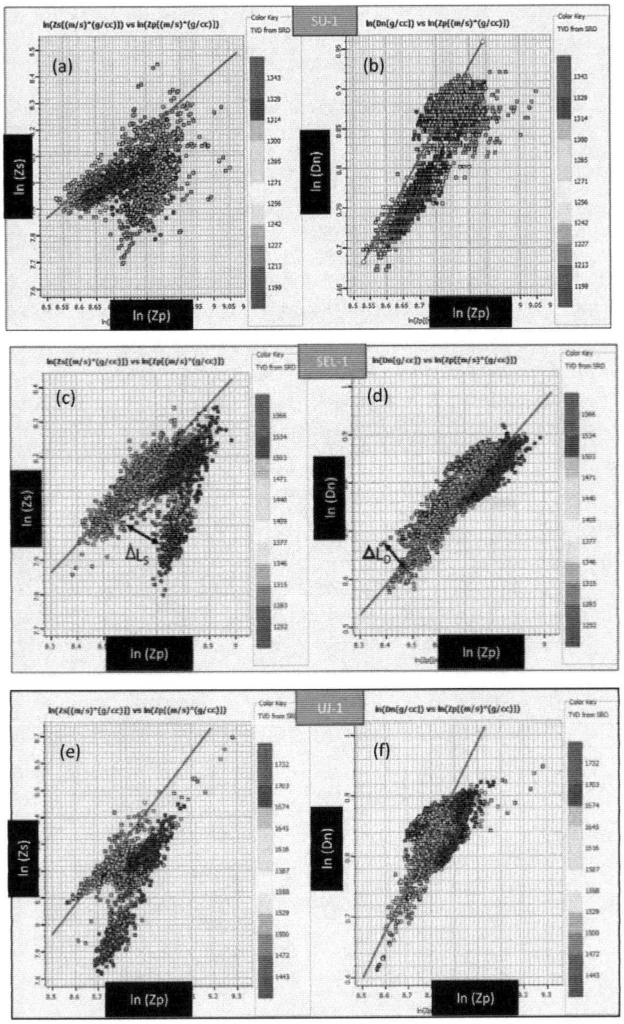

Figure 6. Cross plots of reservoir properties for ln Z_S versus ln Z_P (a, c and e), and ln Dn versus ln ZP (b, d and f) in well SU-1, SEL-1 and UJ-1 respectively

Pre-inversion analysis is a platform for the interpreter to test variety of inversion parameters on selected well before applying to the seismic volume. It allows us to measure success of the inversion and analyse different possible outcomes resulting from different type of wavelets.

Since inversion is a non-unique process as there are many geological answers consistent with input seismic data. Thus, cross plots with background trend information were derived from well logs in the model relating variables of Zp (P-impedance), Zs (S-impedance), and density on a log-log scale to reduce the non-uniqueness. Linear relationship between these variables are assumed to be in the absence of hydrocarbon [6]. Thus, in this case, we are looking for deviations away from a linear fit in logarithmic space to identify the desired fluid anomalies as shown in Figure 6.

From the figures, the red lines indicate the current linear trend, which has been manually adjust for better relationships of these variables within the interest reservoir interval. The colour scale of these cross plots was based on the true vertical depth (TVD) from the seismic reference datum (SRD).

Statistical and well group wavelets have been tested and compared to determine which is the best wavelet to reduce the uncertainty during the inversion process. Its impact on the inversion results also been examined in this research. The study was repeating with 50 iterations on the logs of inverting the synthetic using a constant value of γ =0.5 due to lower root mean square (RMS) error as supported by Hampson's research [6].

Table 1 is the summary of inversion results which compare both well and statistical group wavelets for respective wells and the combination of three wells. We can conclude that inversion by using statistical group wavelet is more realistic and applicable for reservoir characterization. Besides, log correlation factor play more significant role than the RMS error between original and inverted well logs for inversion process.

Table 1. Resulting error in the inversion curves for different group wavelets

Well	Group Wavelets	RMS Error				Correlation	Error
		Zp	Zs	Density	Vp/Vs		
SU-1	Well	481.655	239.237	0.171	0.160	0.984	0.179
	Statistical	502.828	251.595	0.174	0.163	0.978	0.205
SEL-1	Well	234.16	164.06	0.06	0.170	0.987	0.165
	Statistical	436.28	220.93	0.103	0.188	0.984	0.178
UJ-1	Well	329.712	245.873	0.094	0.120	0.977	0.218
	Statistical	227.392	170.484	0.066	0.093	0.964	0.266
SU-1, SEL-1, UJ-1	Well	294.304	333.342	0.122	0.152	0.636	0.772
	Statistical	664.945	351.088	0.170	0.180	0.982	0.188

3.3 Elastic Properties Analysis

Inversion process was applied to the whole seismic volume for modelling with the input of angle gathers, well logs, horizons, wavelets, as well as the initial models. Models with inverted elastic properties such as P-impedance, S-impedance, density, Vp/Vs and Lambda-mu-rho (LMR) between the target window of top reservoir to base reservoir were built. These models have a better image with clearer and details layering that allow us to easily differentiate the lithology and fluid present in our target reservoir.

These inverted results from each well were combined and few data slices created at the window targeted on top reservoir at a fixed window size of 40 ms. This allows interpreter to characterize the reservoir by using a base map with elastic properties as the reference parameters.

Figure 7 is the inversion result for the control well SEL-1 where arrows indicating the hydrocarbon zone. The arrow in red indicate the gas sand zone, while the green arrows refer to oil sand layers. From the figure, we can identify the gas oil contact from well SEL-1 due to the sudden change of elastic properties such as acoustic impedance, Vp/Vs, density and Lambda-rho from gas sand to oil sand. This proves the study can resolve the problem stated in [1] that the fluid contacts were not seen in most wells of X field. Besides that, we can also identify thin shale that lying in between the hydrocarbon sand by using the inverted elastic properties.

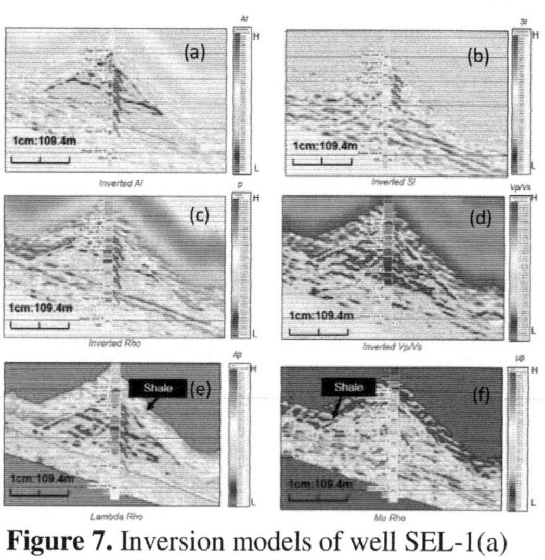

Figure 7. Inversion models of well SEL-1(a) Inverted AI (b) Inverted SI (c) Inverted Rho (d) Inverted Vp/Vs (e) Lambda-rho (f) Mu-rho

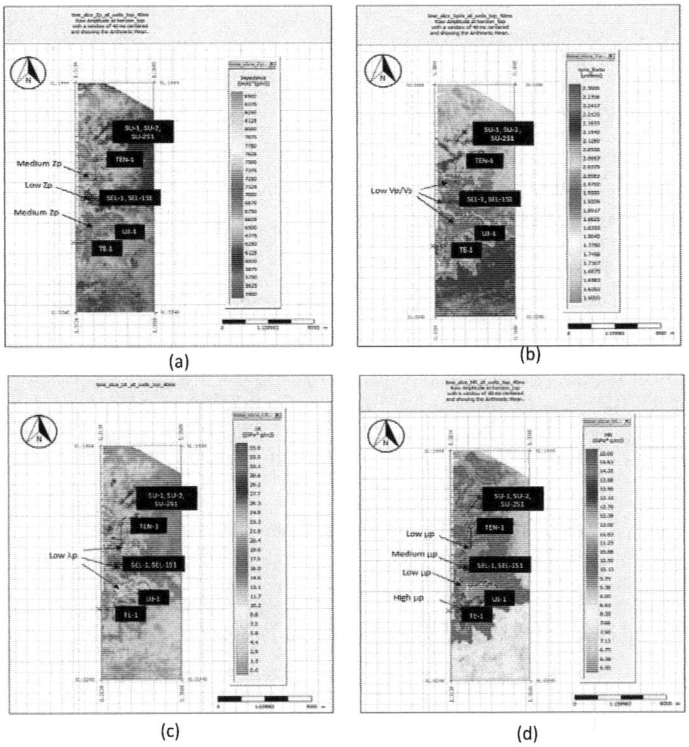

Figure 8. Data slice at top reservoir with window size of 40ms (a) Inverted P-impedance (b) Inverted Vp/Vs (c) Lambda-rho (d) Mu-rho

The inversion results for respective wells were then combined to generate few maps based on the distribution of elastic rock properties such as inverted P-impedance, inverted Vp/Vs, and Lambda-Mu rho across the survey area (figure 8). From these base maps, we can conclude that the area around well UJ-1, TE-1, SEL-1 and TEN-1 which gives low to medium acoustic impedance, low Vp/Vs, low λp, and low to high value of μpdue to presence of silty sand are indicating the region of interest in X Field.

4. Conclusions

In conclusion, this research successfully achieved the objectives and solved the problems with the inability to detect fluid contacts in most wells and to differentiate reservoir and non-reservoir lithologies from previous study. In this research, the gas oil contact was detected in well SEL-1 due to the viable contrast in elastic properties between gas sand layer and oil sand layer as shown in Figure 7. We also presented that statistical group wavelet and vertical well are yielding a better inversion result for lithology and fluid discrimination as summarized in Table 1. Besides, provided with current availability data and software, the application of simultaneous inversion and its innovative approach Lambda-Mu rho (LMR) have been demonstrated in this research that these methods are powerful tools in predicting the lithology and fluid distribution in "X" Field of Sabah basin for better reservoir characterization as supporting from the results shown in Figure 8. This finding was further supported by the research done by [8] and [9]. Their research proved that the resolution of simultaneous inversion allows them to characterize the reservoir better compared to the post-stack inversion.

5. Future Study

It is recommended to further extend this study to Elastic Impedance Inversion (EEI) in the future as it is more advance and most closely related to the reservoir properties. It will describe better for our seismic cube as it is generally sensitive to both fluid types and porosity. It is also better if the reservoirs can be characterized by using the improvement method to Goodway's and comparing these two approaches to conclude which method will be the best for modelling the reservoir in my study area. Besides, correction to deviated wells should be done before conducting the inversion as the anisotropy effect will affect the quality of inversion results.

IOP Conf. Series: Earth and Environmental Science **88** (2017) 012022 doi:10.1088/1755-1315/88/1/012022

6. References

[1] D. P. Ghosh and M. Brewer, Interpretation of Amplitudes, Kuala Lumpur: Petronas Carigali, 2007.

[2] D. J. Foster, R. G. Keys and F. D. Lane, "Interpretation of AVO anomalies," *Geophysics,* vol. 75, no. 5, pp. 75A3-75A13, 2010.

[3] D. Hou and J. Guo, "AVO inversion based on Bayesian theory in ray-parameter domain and its application," in *Society of Exploration Geophysicist Technical Programme*, 2016.

[4] L. A. Lubis, "Amplitude Variation with Offset, AVO," in *Rock Physics and AVO Analysis*, 2014.

[5] M. Sams, "Short Cource 5 Seismic Inversion," in *Petroleum Geoscience Conference & Exhibition*, Kuala Lumpur, 2013.

[6] D. Hampson and B. Russell, "Joint SImultaneous Inversion of PP and PS Angle Gathers," *CSEG Recorder,* vol. 38, no. 6, 2013.

[7] G. Asquith and D. Krygowski, Basic Well Log Analysis, Oklahoma: American Association of Petroleum Geologists, 2004.

[8] A. A. Yoong, L. A. Lubis and D. P. Ghosh, "Application of Simultaneous Inversion Method to Predict Lithology and Fluid Distribution in "X" Field, Malay Basin," in *IOP Conference Series: Earth and Environmental Sicence 38*, 2016.

[9] Petro-Explorers, "Integrated Reservoir Characterization Services," in *Power Point Slides*, 2015.

IOP Conf. Series: Earth and Environmental Science **88** (2017) 012023 doi:10.1088/1755-1315/88/1/012023

Facies Distribution and Petrophysical Properties of Shoreface-Offshore Transition Environment in Sandakan Formation, NE Sabah Basin

M. Firdaus A. Majid, M. Suhaili Ismail, A. Hadi A. Rahman, M. Azfar Mohamed

Department of Petroleum Geosciences, Universiti Teknologi PETRONAS, Seri Iskandar, Malaysia

fyrdaos@yahoo.com, suhaili.ismail@petronas.com.my, hadi_rahman@petronas.com.my, azfar.mohamed@petronas.com.my

Abstract. Newly exposed outcrop of Miocene shallow marine sandstone in Sandakan Formation, allows characterization of the facies distribution and petrophysical properties of shoreface to offshore transition environment. Six facies are defined: (1) Poorly bioturbated Hummocky Cross Stratified (HCS) sandstone (F1), (2) Moderately bioturbated HCS sandstone (F2), (3) Well bioturbated HCS sandstone (F3), (4) Poorly bioturbated Swaley Cross Stratified (SCS) sandstone (F4), (5) Interbedded HCS sandstone with sand-silt mudstone, (6) Heterolithic mudstone. The sedimentary successions were deposited in upper to lower shoreface, and offshore transition environment. Facies F3, F4 and F5 shows good reservoir quality with good porosity and fair permeability values from 20% to 21% and 14 mD to 33 mD respectively. While Facies F1 exhibits poor reservoir quality with low permeability values 3.13 mD.

1. Introduction

The shallow marine sandstone of Sandakan Formation is one of the potential hydrocarbon prospect in the Northeast (NE) Sabah Basin (Figure 1a). Shallow marine sandstone has been known as one of the prolific areas for the hydrocarbon reservoir.

Previous studies on the Sandakan Formation have primarily focused on the regional geology and depositional environment (Fitch, 1958; Haile & Wong, 1965; Noad, 1998; Staufer & Lee, 1972; Tulot, 2002), sequence stratigraphy correlation (Wong, 1992), geochemical element evaluation (Tarek, 2002), and stratigraphy succession and depositional framework (Chung, Sum, & Rahman, 2015). Generally, little attempt has been made to consider this formation as a potential reservoir (Teoh, 2007). But, in this paper, comprehensive studies were further refined on the facies distribution in conjunction with petrophysical properties. The aim of this study is to characterize the shoreface- offshore environment as a shallow marine sandstone reservoir on newly exposed outcrop in Sandakan Formation. The objective of this study are: (1) To describe the facies distribution and depositional processes, (2) To study the petrophysical properties of sandstone facies and associates with different facies, in order to establish the reservoir quality of the facies distribution.

2. Geological Setting

The NE Sabah Basin can be classified into two deep sedimentation areas and a platform area (Meng & Anuar, 1999). The deep sedimentation areas comprise the Central Sabah Sub-basin and the Sandakan Sub-basin (Figure 1b). While the platform area comprises the Kinabatangan area and the Dent Peninsula and consists of olistostromes and volcaniclastic units of the Segama Group. The western and southern margins of the basin consist of an arcuate crystalline basement–ophipolite belt. The northeastern and eastern boundaries are probably offshore in the Phillipines (Meng & Anuar, 1999).

The NE Sabah Basin consist of several pre-basin and basin units. The pre-basin deposits primarily constitute deep marine and older clastics (Crocker, Kulapis and Labang Formations), ophiolitic complex and deep marine sediments (Chert-Spillite, Sapulut, Trusmadi and East Crocker Formations), recrystallized massive, and oolitic mollusk-rich and algal limestone, the Madai-Baturong Limestone (Fontaine & Ho, 1989). Blocks of these rock-units, especially the ophiolite complex, Labang and Kulapis Formations are abundant in the olistostrome (Meng & Anuar, 1999).

The basin is controlled by two structural trends which are older WNW-ESE Sulu Trend, and the younger NE-SW trend. These structures are parallel with the orientation of the grabens formed by the opening of the southeastern Sulu Sea (Clennell, 1991, 1996; Hutchison, 1992). However, the WNW-ESE Sulu Trend might be reactivated in Late Miocene or Pliocene times as evidenced by similarly oriented faults, probably with strike slip components, displacing the Middle Miocene and older sediments (Meng & Anuar, 1999).

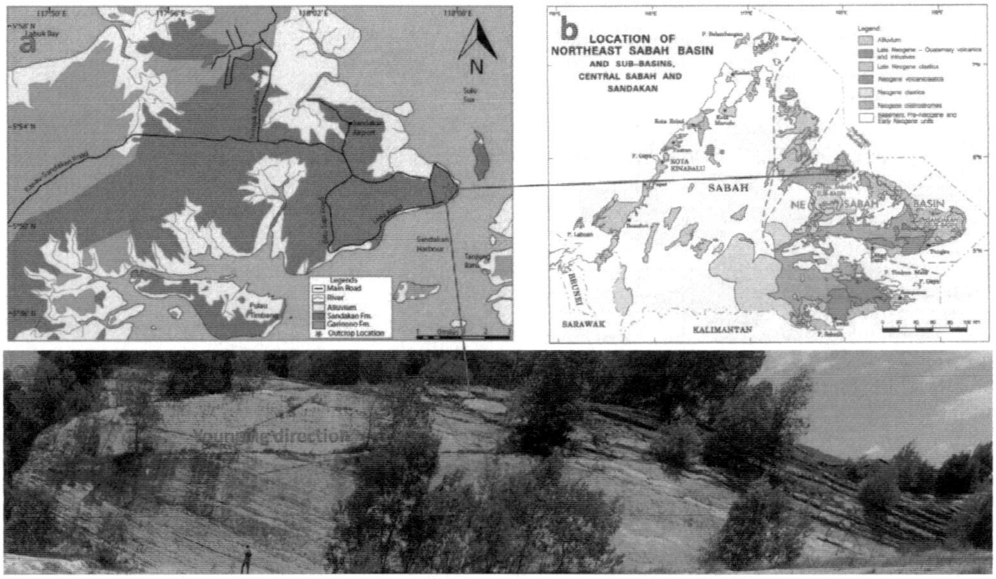

Figure 1: (a) The geological map of Sandakan Formation (modified from (Lee, 1970)). (b) Geological map of NE Sabah Basin (Meng & Anuar, 1999)). (c) Panorama view of outcrop exposed in Sandakan (Note that red arrow shows younging direction)

3. Methodology

In this study, fieldwork was conducted on outcrops exposed adjacent to the Jalan Buli Sim Sim, Sandakan (Figure 1c). Several data were collected such as detail sedimentary logging and sampling of rock units using hand drilling tools.

The sedimentary logging is described briefly with emphasis on the lithology, sedimentary structures, geometry, and bioturbation activity. These characteristics were used to describe facies types and develop the facies distribution.

Laboratory measurements were performed on one-inch-diameter plugs of sandstones based on the standard procedure. The plug samples were treated using solvent extraction units containing azeotrope solvents. The plugs then were placed in a humidity oven at 80 °C and 60% relative humidity. The plugs were dried until their weight were constant. The porosity was measured using helium gas porosimeter at ambient pressure, while permeability was measured using nitrogen gas as flowing fluid.

4. Results

4.1. Facies description

Six (6) facies type are identified in the patched outcrop with a total thickness about 41 m. These facies are differentiated based on its sedimentological characteristics. These facies are summarizing as follows (Figure 3):

4.1.1. Poorly Bioturbated Hummocky Cross Stratified (HCS) Sandstone (F1). This facies consists of fine grained sandstone, pale white in colour and well to moderately sorted. Some parts of the beds are slightly weathered. It is characterized by hummocky cross stratified structures. Conglomeratic lag and mud clast sparse the trough part of HCS which is usually lined the internal stratification including thin carbonaceous material. Thin mud lamination can be seen at lower part of sandstone and mud lens are rare. Both of top and base of the sandstone usually shows irregular contact. The facies is poorly bioturbated with (Bioturbation Index (BI): 1-2) that commonly characterized by *Ophiomorpha*.

4.1.2. Moderately Bioturbated Hummocky Cross Stratified (HCS) Sandstone (F2). The facies consists of fine grained sandstone, slightly grey in colour and well to moderately sorted. The thickness of this facies is ranging between 0.8 to 1.0 m. It is characterized by hummocky cross stratified structures. Conglomeratic lag and mud clast sparse the trough part of HCS which is usually lined the internal stratification including thin carbonaceous material (Figure 2a). Thin mud lamination stratified the upper part of sandstone and mud lens are common. The base of the sandstone shows irregular contact and certain sections shows erosive contact, while the top of sandstone shows sharp contact. The facies is moderately bioturbated (BI: 2-3) and is commonly filled by *Ophiomorpha* and *Chondrites*.

4.1.3. Poorly Bioturbated Swaley Cross Stratified (SCS) Sandstone (F3). This facies consists of fine grained sandstone, slightly grey in colour, well to moderately sorted. It is characterized by massive sandstone (10 m thick) with amalgamated HCS sandstone. Conglomeratic lag and mud clast sparse the trough part of HCS which is usually lined the internal stratification including thin carbonaceous material, coal clast, and cross cut structures (Figure 2b). Mud laminations and mud lens are common on this facies. Base and top of the sandstone shows irregular contact. The sandstone is sparsely bioturbated with BI: 0-1 characterized by burrow of *Ophiomorpha.*

Figure 2: Sedimentary features of outcrop.(a) Conglomeratic lag (pointed by red arrow) on HCS sandstone.(b) HCS structure with cross cut structure.(c) Interbedded HCS sandstone and sand-silt mudstone. d) Carbonaceous matter and mud lamination are thinning upward.

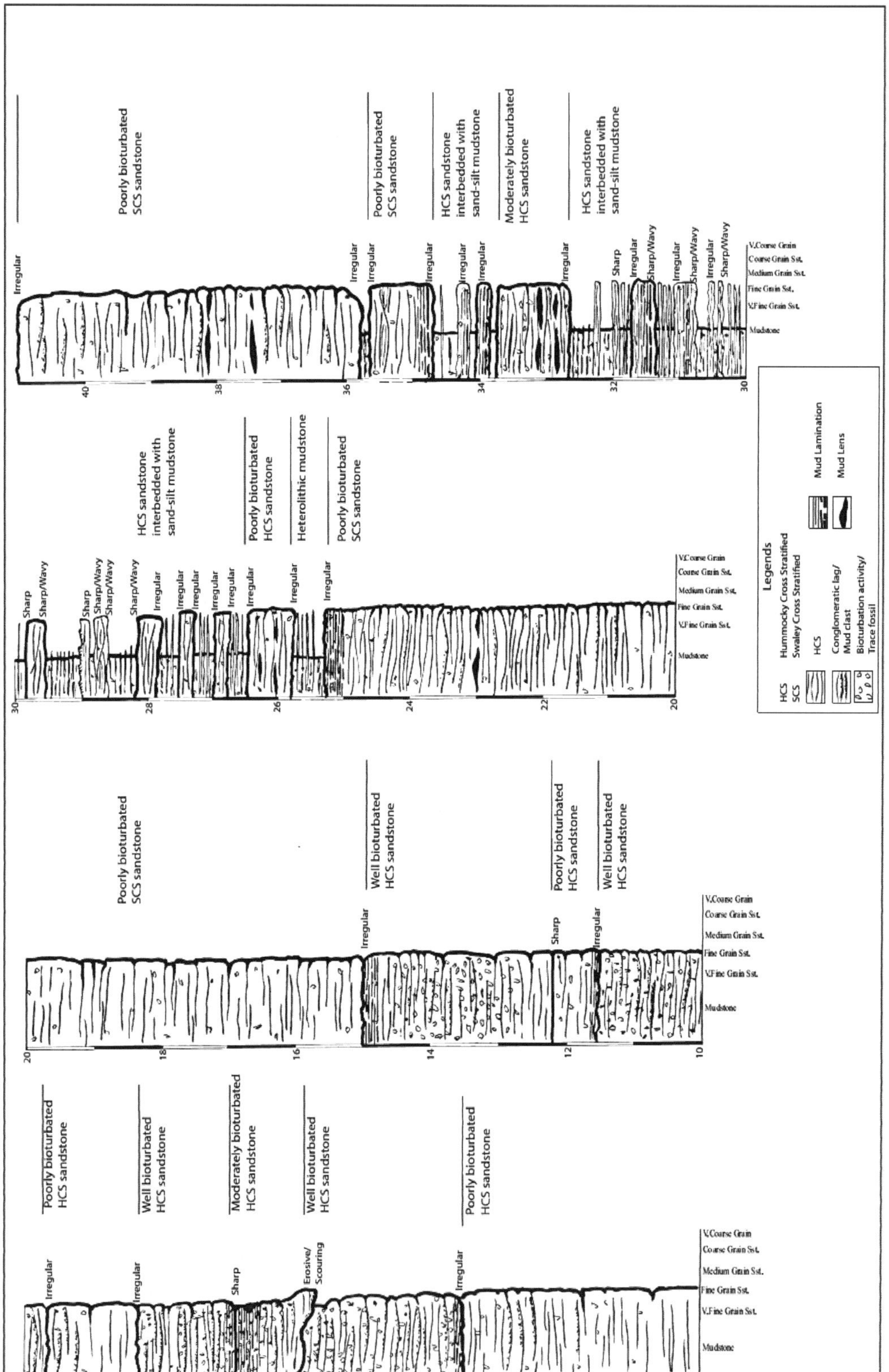

Figure 3: Sedimentary succession of the study area with 41 m thicks.

4.1.4. Well Bioturbated Hummocky Cross Stratified (HCS) Sandstone (F4). The facies consists of fine grained sandstone, slightly grey in colour and well to moderately sorted. The thickness of this facies ranges between 1.5 to 2.8 m.. It is characterized by low angle, hummocky cross stratified structures. Conglomeratic lag and mud clast sparse the trough part of HCS which usually lined the internal stratification including thin carbonaceous material. Certain sections of HCS structure were unclear due to intense bioturbation activity. Thin mud lamination stratified at upper and lower part of sandstone. The base of sandstone shows sharp contact, while the top of sandstone shows irregular contact. The facies is well bioturbated (BI: 3-4) that commonly filled by *Ophiomorpha*, *Chondrites*, *Skolithos*, and *Teichichinus*.

4.1.5. Interbedded Hummocky Cross Stratified (HCS) Sandstone with Sand-Silt Mudstone (F5). The facies consists of interbedded HCS sandstone and sand-silt mudstone (Figure 2c). The sandstone is characterized by fine grained, grey in colour, well to moderately sorted with thickness about 10-30 cm. HCS is low angle, cross cut, sparse with conglomeratic lag and mud clast at the trough part that align the stratification of HCS. Thin mud lamination thickness decreasing upward at certain section (Figure 2d) and presence of mud lens is common. Both of top and base of the sandstone usually shows irregular contact, but some part is also shows wavy-sharp and ripple top contact. The sand-silt mudstone is characterized by dark grey. mudstone alternating with thin lamination of sand-silt layer. The layer shows wavy-like structure, non-continuous lamination with thickness ranging between 2-10 cm. Flaser structure of thin sand-silt layer with thickness decreasing upward at some section of mudstone beds. The facies is poorly bioturbated (BI: 1-2) that commonly filled by *Ophiomorpha*.

4.1.6. Heterolithic Mudstone (F6). The muddy dominated facies consists of dark grey mudstone and finely laminated. Within the mudstone is locally alternating of fine grained layer sand-siltsone (1-2mm) and shows wavy-like lamination. Bioturbation within this facies is poor with BI: 1-2.

4.2. Petrophysical Properties

The results obtained from the conventional analyses (Table 1) shows the porosity, permeability and grain density. The grain density values ranged from 2.65 g/cm^3 to 2.7 g/cm^3. The mean value of 2.66 g/cm^3 is close to the grain density of quartz (2.64 g/cm^3). The porosity distribution for all samples ranges from 20.56% to 22.91% with a mean value of 21.85% and standard deviation value of 1.62. The permeability distribution for all samples ranges from 3.13 mD to 34.27 mD, with a mean value of 16.29 mD and standard deviation value of 0.501.

Table 1: Routine core analysis results.

Plug Number	Core Depth (m)	Pressure (psi)	He Porosity (% of Vb)	Klinkenberg Permeability (mD)	Grain Density (g/ml)
SOC 1/01	3.20	400	22.91	3.134	2.70
SOC 1/02	13.90	400	20.56	14.357	2.65
SOC 1/03	23.50	400	21.40	32.853	2.66
SOC 1/04	29.00	400	21.15	14.800	2.67

5. Discussion and conclusion

The hummocky cross stratified sandstone is primarily dominates the shoreface environment in Sandakan Formation, where these repetitive facies can only be characterized by distinguishing the bioturbation index. The facies is common for high energy setting environment which related to the wave and storm processes. The presence of HCS clearly points to a storm origin, related to the presence of high energy oscillatory and combined flows (unidirectional flow and combination of oscillatory) which is produced by storm event (Cheel, Leckie, & Wright, 1993; Dumas & Arnott, 2006). This facies is interpreted to have been deposited within lower to middle shoreface setting below the weather wave base. Conglomeratic lag and mud clast that deposited and lined the trough part of HCS are the result of oscillatory high energy that was able to mobilise the pebble clast. (Lambiase & Tulot, 2013) suggest that mud and carbonaceous that sparse in the lower shoreface sandstones of the Sandakan Formation originate from back-barrier mangrove swamps which deposited from suspended-sediment plumes. The swaley cross stratified sandstone facies is also the product of high energy oscillatory flow, deposited above fair weather wave base and as part of upper shoreface deposits. Different intensity of bioturbation (poor to moderate) in HCS sandstone is probably due to different in operation condition during deposition, substrate, salinity, oxygen level, sheltered environment and etc.

Interbedded HCS sandstone with sand-silt mudstone reflect a fluctuation and alternation of quiet water sediment fall out (low energy), with combined and pure oscillatory flows (high energy) (Buatois et al., 2012). The presence of HCS structure suggests that the sandstone body reflects deposition under high energy storm conditions (Arnott & Southard, 1990; Cheel et al., 1993; Dumas, Arnott, & Southard, 2005). The sand-silt mudstone interbedding is deposited from fall out suspension during a period of quiet, non-storm and fair weather conditions (Maceachern, Stelck, & Pemberton, 1999). The thin lamination of sand-siltstone is a result from rapid deposition by quasi-steady transitional flow of plug flow where traction-deposited bedform are unstable (Baas & Best, 2008). Therefore, the facies can be interpreted as part of lower shoreface to offshore transition zone.

In petrophysical properties, four distinct characteristics with facies affinities were observed. Facies F1 has good porosity and low permeability, with the value shows 22.91 % and 3.13 mD respectively. Facies F3 exhibits good porosity and fair permeability, with the value shows 21.40 % and 32.85 mD respectively. Facies F4 has good porosity and fair permeability, with the value shows 20.56 % and 14.36 mD respectively. Facies F5 exhibits good porosity and fair permeability, with the value shows 21.15 % and 14.80 mD respectively. The low values of permeability recorded in some samples may be due to the high content of mud.

In conclusion, six facies (Facies F1, F2, F3, F4, F5 and F6) were grouped according to lithology, sedimentary structures, geometry, and bioturbation activity. Based on facies description, the studied sedimentary successions were deposited in upper shoreface to lower shoreface and offshore transition zone environment. Facies F3, F4 and F5 shows good reservoir quality with good porosity and fair permeability values from 20% to 21% and 14 mD to 33 mD respectively. While Facies F1 exhibits poor reservoir quality with low permeability values 3.13 mD.

6. Acknowledgement

The author would to express the gratitude to University Teknologi PETRONAS for research funding.

7. References

[1] Arnott, R. W., & Southard, J. B. (1990). Exploratory flow-duct experiments on combined-flow bed configurations, and some implications for interpreting storm-event stratification. *Journal of Sedimentary Research, 60*(2).

[2] Baas, J. H., & Best, J. L. (2008). The dynamics of turbulent, transitional and laminar clay-laden flow over a fixed current ripple. *Sedimentology, 55*(3), 635-666.

[3] Buatois, L. A., Santiago, N., Herrera, M., PLINK-BJÖRKLUND, P., Steel, R., Espin, M., & Parra, K. (2012). Sedimentological and ichnological signatures of changes in wave, river and tidal influence along a Neogene tropical deltaic shoreline. *Sedimentology, 59*(5), 1568-1612.

[4] Cheel, R. J., Leckie, D. A., & Wright, V. (1993). Hummocky cross-stratification. *Sedimentology review, 1*, 103-122.

[5] Chung, K. W., Sum, C. W., & Rahman, A. H. A. (2015). Stratigraphic Succession and Depositional Framework of the Sandakan Formation, Sabah. *SAINS MALAYSIANA, 44*(7), 931-940.

[6] Clennell, B. (1991). The origin and tectonic significance of melanges in Eastern Sabah, Malaysia. *Journal of Southeast Asian Earth Sciences, 6*(3), 407-429.

[7] Clennell, B. (1996). Far-field and gravity tectonics in Miocene basins of Sabah, Malaysia. *Geological Society, London, Special Publications, 106*(1), 307-320.

[8] Dumas, S., & Arnott, R. (2006). Origin of hummocky and swaley cross-stratification—the controlling influence of unidirectional current strength and aggradation rate. *Geology, 34*(12), 1073-1076.

[9] Dumas, S., Arnott, R., & Southard, J. B. (2005). Experiments on oscillatory-flow and combined-flow bed forms: implications for interpreting parts of the shallow-marine sedimentary record. *Journal of Sedimentary Research, 75*(3), 501-513.

[10] Fitch, F. H. (1958). The geology and mineral resources of the Sandakan area and parts of the Kinabatangan and Labuk Valleys, North Borneo: US Government Printing Office.

[11] Fontaine, H., & Ho, W. (1989). Note on the Madai–Baturong Limestone, Sabah, East Malaysia; discovery of Caprinidae (rudists). *CCOP Newsletter, 14*(3), 27-32.

[12] Haile, N., & Wong, N. (1965). The Geology and Mineral Resources of the Dent Peninsula, Sabah, Malaysia: Borneo Reg. *Malaysia Geol. Survey Mem, 16*.

[13] Hutchison, C. (1992). The Southeast Sulu Sea, a Neogene marginal basin with outcropping extensions in Sabah. *Bulletin of the Geological Society of Malaysia, 32*, 89-108.

[14] Lambiase, J. J., & Tulot, S. (2013). Low energy, low latitude wave-dominated shallow marine depositional systems: examples from northern Borneo. *Marine Geophysical Research, 34*(3-4), 367-377.

[15] Lee, D. T. C. (1970). *Sandakan Peninsula, Eastern Sabah, East Malaysia*: US Government Printing Office.

[16] Maceachern, J. A., Stelck, C. R., & Pemberton, S. G. (1999). Marine and Marginal Marine Mudstone Deposition: Paleoenvironmental Interpretations Based on the Integration of Ichnology, Palynology and Forarniniferal Paleoecology. *SEPM Special Publication 64*, 205-225.

[17] Meng, L. K., & Anuar, A. (1999). Northeast Sabah Basin. In A. J. B. Zainul, A. Anuar & A. Ali (Eds.), *The petroleum geology and resources of Malaysia* (pp. 545): Kuala Lumpur: Petroliam Nasional Berhad (Petronas).

[18] Noad, J. J. (1998). *The sedimentary evolution of the tertiary of eastern Sabah, Northern Borneo.* (Unpublished Thesis), Birkbeck (University of London). (Noad, J. 1998. The sedimentary Evolution of the Tertiary of Eastern Sabah, Northern Borneo.)

[19] Staufer, P. H., & Lee, D. T. C. (1972). Sedimentology of the Sandakan Formation, East Sabah. *Geological Survey of Malaysia, 1*, 10-17.

[20] Tarek, A. (2002). *Geochemistry of Sandakan Formation, East Sabah.* (MSc. Unpublished), University of Malaya, Kuala Lumpur.

[21] Teoh, Y. J. (2007). Characteristics Of Sedimentary Facies And Reservoir Properties Of Some Tertiary Sandstones In Sabah And Sarawak, East Malaysia (MSc. Unpublished Thesis), Universiti Sains Malaysia, Penang.

[22] Tulot, S. (2002). Sedimentology and Reservoir Properties of Sandakan Formation Shorface Sandstones, Sabah, Malaysia. (MSc. Unpublished Thesis), University Brunei Darussalam, Brunei.

[23] Wong, R. H. (1992). *Sequence stratigraphy of the Middle Miocene–Pliocene southern offshore Sandakan part of the South China Sea.* Paper presented at the Proceedings symposium on the Tectonic framework and energy resources of the western margin of the Pacific Basin, 29th November-2nd December.

A Review on the research and practice of city sustainable development indicators and indices

Ning Xiang[1,2,3]

[1] China National Institute of Standardization, No.4 Zhichun Road, Haidian District, Beijing 100191, P R China
[2] Key Laboratory for Solid Waste Management and Environment Safety, Tsinghua University, Zhongguancun Street, Haidian District, Beijing 100084, China
[3] Supported by the Foundation of Basic Research Fund for Central Public Scientific Institution（512016Y-4496）

Xning_pku@163.com

Abstract. City sustainable development indicators and indices have become a hot issue in academic research and practical application, alongside the high-speed worldwide urbanization and driven by the actual managing demand. This article is aimed at a clear understanding of the progress in relevant research and practice. This is done by collecting common indicators and indices for city sustainable development and making comparison of the assessment process and contents, so as to find out main obstacles for the development of this research field and explore the direction for efforts to be made next step. The article divides these indicators and indices into two categories: ① indicators serving as single index which can provide an explicit description on the relationship between economic activities and environmental carrying capacity, but have a narrow scope of assessment and use complicated methods to collect and calculate data; ② indices based on indicator systems which can represent multiple processes, could reflect the view of strong sustainability and are easy to use, but can hardly depict the responding relationship between social, environmental and economic changes for city sustainable development or assure the scientific rigor of weight setting. Practices on indicators and indices for city sustainable development was summarized, and its problems were reviewed with China being representative of transitioning countries. According to the review, great progress has been achieved in the research and practice of indicators and indices for city sustainable development, but consistency of theories, rationality of indicators and scientific rigor of methodology are to be improved significantly.

1. City sustainable indicators and indices as an emerging important issue

1.1. Cities are a key area promoting the sustainable development of humanity

Areas where urban population is concentrated, industry and commerce are developed and residents focus on non-agricultural population are featured by element concentration, systematic development, man-made environment domination and overflow effect. At present, urbanization is advanced rapidly across the world, especially in developing countries. Cities gather more than 50% of world population, and the number is expected to reach 6.3 billion, out of the approximately 9.3 billion world population, by 2050[1], as shown in Figure 1. Along with the trend comes the rapidly increasing impact of cities in resource and environmental protection and sustained economic growth. Cities take away 75% of the world's natural resources, cover 3% of land, discharge 50% of wastes and 70% of greenhouse gases and produce over 80% of GDP[1-5].

As urbanization proceeds at a high speed, lots of problems in the governance of city sustainable development have started to emerge—economic growth is weakly driven, economic development model is to be transformed urgently; social inclusiveness isn't strong, and equalization of public services is to be improved urgently; resources are used less efficiently, and eco-environment is to be improved; cities are expanded blindly, urban planning and construction and urban governance capacity are to be improved earnestly. Evidently, cities have become a key pushing ahead the sustainable development of human society.

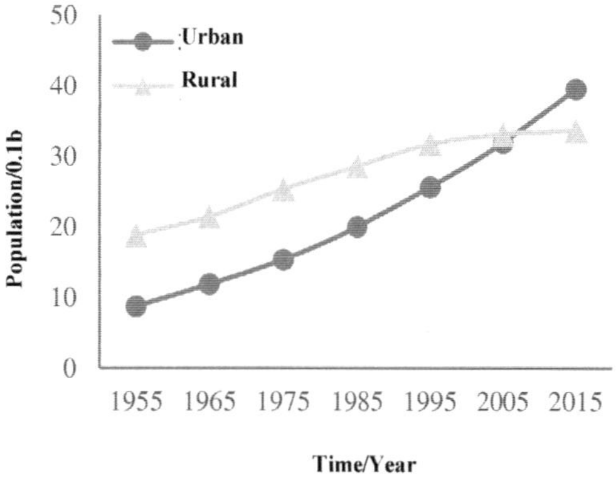

Figure 1. World Population Change, 1955-2015

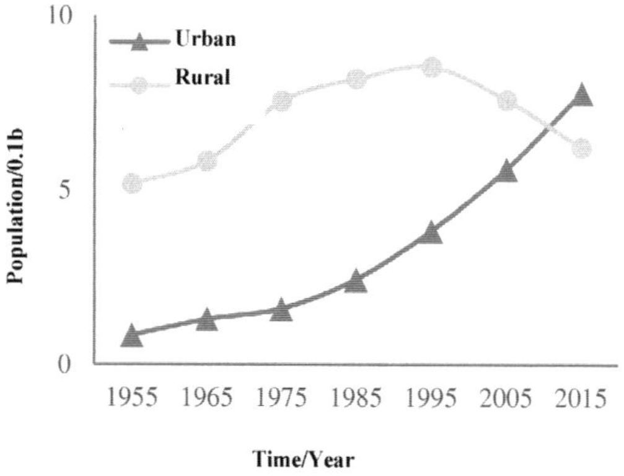

Figure 2. Chinese Population Change, 1955-2015

1.2. City sustainable development indicators and indices are an important instrument supporting urban governance

Intended for city sustainable development, urban activities shouldn't exceed the necessary limit while satisfying the production and living needs of residents. As an important issue worldwide, city sustainable development challenges the management capability and wisdom of all governments. In September, 2015, the United Nations approved 17 sustainable development goals, of which goal No. 11 was to "make cities and human settlements inclusive, safe, resilient and sustainable", targeting urban development.

Indicators are "pointers capable of reflecting or measuring some conditions, which are helpful in translating information in a more intelligible manner and can describe complicated conditions in a concise way". This is taken from reference[6]. A comprehensive evaluation index may be acquired from a mixture of indicators through single index integration. Indicators have the functions of organizing and planning regulation and control, verifying validity and conducting social propaganda [7].

Researching and practicing city sustainable development indicators can guide the formulation of urban development strategies, monitor the implementation effect of public policies and promote through social propaganda information transparency and public supervision. A set of global indicators will be adopted to implement and assess each sustainable development goal and specific targets, according to the *2030 Agenda for Sustainable Development*. In 2012, International Organization for Standardization (ISO) approved to build ISO/ TC 268 Technical Committee for the Sustainable Development of Cities, which has released ISO 37120 Sustainable Development of Communities: Indicators for City Services and Quality of Life and is developing indicators for city resilience and smart cities. It's thus clear that the research and development of indicators for city sustainable development has drawn universal attention and received vigorous support from international organizations.

To sum up, indicators for city sustainable development are an important instrument supporting city governance and have become a hot issue in academic research and management practice. This paper is designed to explore the research progress of indicators for city sustainable development, learn about their application and review existing problems with the focus on China and thereby, present pertinent directions and suggestions of improvement.

2. Research of indicators and indices for city sustainable development

The establishment of sustainable development indicators was a process of constant evolution. With regard to academic research, the sustainable development evaluation methodology based on environmental impact assessment emerged in the middle of the 20th century; subsequently, multiple integrated environmental and economic assessment techniques sprang up over time, such new methods as lifecycle analysis (LCA) and synthetic environmental and economic accounting (SEEA) system were worked out, the scope of evaluation expanded from environmental damage and resource use to multiple services of ecosystem; theories and indicators relating to city sustainable development, such as ecological city, compact city and low-carbon city were put forward; when it came to the 21st century, apparently more attention was paid to social sustainability. In relation to management practice, developed countries and some developing countries have adopted, since the 1980s, a great mixture of city development policies, such as ecological city, livable city and low-carbon city, which focused on the coordination between society, environment and economy; as a result, they have accumulated application cases and experience of indicators for city sustainable development. In a word, obvious progress has been achieved in the research and application of indicators for city sustainable development in recent years.

The following part contains a summarization of relevant studies starting with the design thought of indicators for city sustainable development. These indicators are divided into two classes, with the first class based on single index and the second class based on indicator system[8]. They have different features in methodology and process: the former are designed to target relatively single objects, whose calculating process is supported by a complete set of conversion coefficient database obtained via long-term experiments and surveys, and the latter target diversified objects, whose reference values are selected in a more subjective manner.

2.1. Indicators serving as single index

These indicators are designed mainly to represent the relationship between economic activities and eco-environmental support levels, based on single, universal computational items, and they can translate, through conversion factor, objects of different categories into computational item values and thereby, sum them to get a single index value. For instance, ecological footprint (EF) takes the area of productive land as the computational item and the genuine progress indicator (GPI) takes the currency as the computational item. Table 1 presents a summarization of ordinary indicators based on single index.

Single-index indicators are advantageous in that the index value has a relatively more intuitional physical significance and thus, can provide a clear description on the relationship between economic activities and natural environment's support levels.

Table 1. City sustainable development evaluation technique based on single index.

Technique Name	Assessment Scope & Computing Method
Ecological Footprint	$EF=P/Y_N \times YF \times EQF$, P is the amount of a product harvested or waste emitted, Y_N is the average yield for P, and YF and EQF are the yield factor and

(EF)[9]	equivalent factor respectively, for the land use type in question.
Genuine Progress Indicator (GPI)[10]	$GPI=C_{adj}+G_{nd}+W-D-E-N$ C_{adj} is personal consumption expenditure adjusted by income unfairness, G_{nd} is government expenditure except defense, W is the non-market contribution of welfare, D is personal expenditure on defense, E is the cost of environmental degradation and N is the depreciation of natural capital stock.
Genuine Saving (GS)	GS=Gross domestic savings- fixed capital consumption (depreciation)+ educational spending- air pollution cost- water pollution cost- consumption of non-renewable resources- CO_2 loss cost
Life Cycle Assessment (LCA)	Overall load of environmental impact= ∑(output quantity of the nth type of environmental loads× characterization factor of the mth environmental impact type of the nth environmental loads/ reference value of the mth environmental impact type, weighted average Environmental impact types include consumption of non-renewable resources, greenhouse effect, acidification effect, forming of photochemical smoke and human health damage.
Material Flow Analysis (MFA)	Total quantity of regional material input=Direct material input+ regional hidden flow Total quantity of regional material output= Regional processing output+ regional hidden flow Total quantity of regional material consumption= Total material demand- export and its hidden flow Net increase of regional material stock= Total material demand- total material input Measuring unit is fresh weight of materials, and conversion methods include material weight conversion coefficient and the standard coal conversion coefficient of energy materials.

Single-index indicators are disadvantageous in that ① the scope of evaluation is small, and processes that can be taken into consideration are fewer than the indicators based on index system. In particular, it's more difficult to reflect social sustainability and the equity principle. For instance, EF can measure environmental sustainability alone; ② The view of strong sustainability isn't adopted. For instance, SEEA converts natural and social costs into currency amount and sums it with total income, total expenditure or total savings and thus, doesn't reflect fully natural environment ' s significance in limiting city development; ③ The scientificity of conversion factor is in dispute. For instance, converting natural and social costs into currency amount using SEEA must be on the premise of several assumptions; and ④ High requirements are raised for data collection and computing methods are complicated, which together increase the cost of evaluation and make it hard to generalize.

2.2. Indices based on indicator systems

Indicators based on index system can obtain single-dimensional index or comprehensive index through combining and integrating multiple indicators that represent different objects or processes, and relatively more influential indexes include Green City Index (GCI), City Development Index (CDI) and Human Development Index (HDI); see Table 2 for more details. The generalized flow for building an indicator system is to establish a conceptual model of city sustainable development, set an assessment framework and select indicators. Ordinary assessment frameworks are the three-pillar framework, city sector constitution, city metabolism model, life cycle theory and city sustainable development policies and goals.

These indicators are advantageous because they can give consideration to all dimensions or processes of city sustainable development at the same time, and make it convenient for conducting the horizontal comparison of single indicators and experience reference. Theoretically, these indicators can express from the perspective of strong sustainability. In addition, they are easy to generalize because they can be increased and modified according to local characteristics and data availability, and their data collection and computing methodology are relatively easier.

Their disadvantages are: ① variable selection and indicator weight are more random than the indicators serving as single index discussed in subsection 2.1; ② composite index value has a non-intuitional physical significance; ③ limited to people's knowledge level, the indicator system evaluation technique can hardly capture the complex nature of the city system and is difficult to give an accurate description on the responding relationship in changes of social situation, natural environment and economic activities; and ④ when indicators are converted into indices, the weighted mean method is adopted generally, in which environmental quality isn't taken as the fundamental limiting condition and the view of strong sustainability isn't reflected.

Table 2. City sustainable development evaluation technique based on the indicator systems.

Technique Name	Assessment Scope & Computing Method
Green City Index (GCI)[11]	GCI= CO_2 emissions+ energy+ building+ land utilization+ transport+ water+ health+ waste management+ air quality+ environmental governance
City Development Index (CDI) [11]	GCI= (Infrastructure index+ waste index+ well-being index+ education index+ product index)/5; Here, infrastructure= 25× water supply+ 25× sewage discharge+ 25× electricity supply+ 25× communication; Waste= Wastewater treatment× 50+ regular solid waste disposal× 50; Well-being= (Life expectancy- 25) × 50/60+ (32-child mortality) × 50/31.92; Education= Literacy rate× 25+ comprehensive enrollment rate× 25; Product= (log city product-4.61)×100/5.99
Environmental Performance Index (EPI) [11]	EPI=f (Environmental health, ecological vitality), where 21 indicators represent 9 themes, and the weighted mean method is used for weighting
Human Development Index (HDI) [11]	HDI (1990-2009)= 1/3(life expectancy)+ 1/3(education index)+ 1/3(parity purchase power-adjusted per capita GDP); here, educational index is the integration of adult literacy rate and enrollment rate; HDI (2010-)= (Life expectancy) $^{1/3}$+ (educational index) $^{1/3}$+ (parity purchase power-adjusted per capita GDP) $^{1/3}$; here, educational index is the integration of the upper limit of average schooling years and the schooling year expectancy.
Happy Planet Index (HPI) [11]	HPI= Happy life expectancy/ ecological footprint, Here, happy life expectancy is a product of life expectancy and happy experience obtained from questionnaire form
Well-being Index (WI) [11]	WI= (Ecosystem well-being index+ human well-being index)/2; Here, Ecosystem well-being index= (Land+ water+ air+ species and gene+ natural resources)/5; Human well-being index= (Health and population+ wealth+ knowledge

	and culture+ community+ fairness)/5
Sustainable Society Index (SSI) [11]	SSI= Human health+ environmental health+ economic health, or SSI= (Basic needs+ health+ personal and social development+ natural resources+ climate and energy+ transport+ economy)/7
Sustainable Chinese Cities [12]	R=(Average life expectancy+ ratio of educational spending to GDP under municipal administration+ per capita GDP under municipal administration+ urbanization rate+ ratio of added value of non-agricultural industries+ centralized treatment rate of domestic sewage+ harmless treatment rate of domestic garbage+ integrated utilization rate of industrial solid wastes+ standard attainment rate of industrial wastewater emission+ SO_2 emissions per unit of industrial output value+ smoke emissions per unit of industrial output value+ green coverage ratio of built-up areas)/ 12
Development Satisfaction[13]	Development satisfaction= f(development sustainability, development coordination, development level); Here, development sustainability is represented by 3 themes and 13 indicators, development coordination by 4 themes and 16 indicators and development level by 3 themes and 17 indicators. The analytic hierarchy process is used to determine the weight, and weighted mean is adopted.
City Development Level Index [14]	Development level index= f(environmental sustainability index, economic sustainability index, social sustainability index); Here, environmental sustainability index= f(air quality, water environment quality, acoustic environment quality, environmental control, eco-environment); Economic sustainability index= f(gross economic strength, economic structure, economic effectiveness, economic extroversion, economic prosperity, economic intensiveness); Social sustainability index= f(population index, dwelling level, infrastructure level, educational level, cultural life conditions, medical conditions, social security status, living level, technological level); Weighted mean

3. Practice of indicators for city sustainable development: taking China as example

Many international organizations and city management agencies have designed and practiced indicators for city sustainable development. These indicators are applied largely in three circumstances: action plans of international organizations designed to promote city sustainable development, local sustainable development strategies made by Chinese and foreign cities based on *The 21st Century Agenda* and city development plans relating to sustainable development. In the course of the formulation and implementation of these plans and strategies, assessment of city sustainable development serves as important basis and reference for identifying problems, setting goals, making strategies, conducting supervision and checking results. At present, no indicator has been universally accepted in practice and these indicators have poor political influence. Parris et al. concluded there are

IOP Conf. Series: Earth and Environmental Science **88** (2017) 012024 doi:10.1088/1755-1315/88/1/012024

three reasons[15]: fuzziness of concept, multiple goals of sustainable development measurement and confusions in the terminology, data and methodology of measurement.

Over more than 30 years of rapid economic development, China has highly representative problems in city development. Figure 2 shows the changes in urban and rural distribution of Chinese population between 1955 and 2015; over the 60 years alone, China's urbanization rate soared from 10% to 50%, and its urban population accounted for as much as 54.77% in 2014. By 2030, China will have had another 310 million urban residents, adding its urban population to over 1 billion and urbanization rate to 70%. Apparently, China needs urgently to intensify city sustainability so as to cope with the social, environmental and economic challenges arising in the high-speed urbanization process.

The Chinese government attaches great importance to city sustainable development. In 2015, the Central Urban Work Conference was held again, after 36 years, presenting overall requirements of "building harmonious, livable and viable modernized cities with their respective characteristics, and opening up a characteristic path for city development". Starting from the problems they concern respectively, departments of the central government have released multiple development plans and policies relating to sustainable development and applicable to the urban scale, and have formulated indicators serving as a technical instrument for policy implementation. Seen from Table 3, the great quantity of urban development plans have different themes and goals, involving a huge number of indicators with different themes and levels. Based on the feasibility research conducted by competent departments, such city assessment indicators have a foundation in the effectiveness, usability, feasibility and replicability for reflecting specific problems. Nevertheless, their relationship with city sustainable development assessment isn't clear enough, with great differences in the structure of indicator system and involving redundant, overlapping indicators. Due to the fragmented urban assessment activities, indicators are more expensive to use and more difficult to generalize, basic-level departments receive a bigger pressure and many government resources are consumed internally.

Table 3. Examples of city sustainable development plans & policies of China's central government.

No.	Theme	Competent Authorities	Relevant Document	Type & Quantity of Indicators
1	Sponge City	MOHURD[a]; MWR[b]; MOF[c]	Guiding Opinions of the General Office of the State Council on Advancing the Construction of Sponge Cities (GBF [2015] No. 75)	18 norms
2	Eco-city	MEP[d]	Notice on Issuing the Work Appraisal Plan for the Establishment of National Eco-counties and Eco-cities (HB [2005] No. 137)	34 indicators, inc. 6 qualitative
3	Circular Economy Demonstration City	NDRC[e]	Interpretation of the Evaluation Content for the Construction of National Circular Economy Demonstration Cities (2013)	67 core indicators, inc. 14 qualitative and 8 characteristic
4	Water Saving City	MOHURD[a]; NDRC[e]	Measures for the Declaration and Assessment of National Water Saving Cities (JC [2012] No. 57)	24 indicators, inc. 10 qualitative
5	Health City	National Patriotic Health Campaign Committee	Health City Standard (2014 Version)	40 norms

6	National Forest City	SFA [f]	Indicators for the Assessment of National Forest Cities (LY/T 2004-2012)	40 indicators, inc. 21 qualitative
7	Garden City/ Ecological Garden City	MOHURD[a]	Measures for the Declaration and Assessment of National Garden Cities, Standard for National Garden Cities (JC [2010] No. 125)	62 core indicators, inc. 11 qualitative and 4 negative; 68 improvement indicators, incl. 15 qualitative and 7 negative
8	"China Human Settlement Prize"	MOHURD[a];	Assessment Indicator System of China Human Settlement Prize (for Trial Implementation) (JC [2010] No. 120)	64 indicators, inc. 22 qualitative and 1 comprehensively negative
9	Environmental Protection Model City	MEP [d]	Assessment Indicators for National Environmental Model Cities and Detailed Rules of Implementation (Stage VI) (HB [2011] No.3)	26 norms

[a] MOHURD is short for the Ministry of Housing and Urban-Rural Construction of P.R. China.
[b] MWR is short for the Ministry of Water Resources of P.R. China.
[c] MOF is short for the Ministry of Finance of P.R. China.
[d] MEF is short for the Ministry of Environmental Protection of P.R. China.
[e] NDRC is short for the National Development and Reform Commission of P.R. China.
[f] SFA is short for the State Forestry Administration of P.R. China.

4. Brief summary

Cities have become an important focal point advancing the sustainable development of human society. For the purpose of city sustainable development, the three pillars of society, environment and economy must be sustainable at the same time and the view of strong sustainability should be adopted[8]. Indicators are an important instrument supporting the governance of city sustainable development.

These indicators can be divided into two classes. On the one hand, indicators serving as single index can provide a clear description on the relationship between economic activities and the environment, but are applicable to a small scope and request complicated methods for collecting and calculating data. On the other hand, indicators based on the indicator system can represent multiple processes, reflect the view of strong sustainability and adopt simple methods to collect and calculate data, but they present a responding relationship, not clear enough, among changes in social situation, natural environment and economic activities and the scientificity of weight setting can be hardly assured. These indicators are practiced in the city development strategies and policies relating to sustainable development by international organizations, central governments and municipal governments. However, there is no such indicator commonly accepted by academia or in practice. For instance, many government organs of China have had management practice based on city sustainable development assessment, but with great differences in the selection of assessment themes and indicators and the methodology of calculation. When selecting indicators, they always highlight excessively local advantages and stress departmental benefits, which is more random and less comprehensive.

To sum up, great progress has been achieved in the research and application of indicators for city sustainable development in terms of theory presentation, indicator establishment, assessment methodology and empirical verification. Nevertheless, theoretical consistence, indicator rationality, methodology scientificity are still to be improved significantly. On the whole, the assessment contents about assessment indicators and indices for city sustainable development are decentralized, and the assessment methods are diversified. The lack of a unified theoretical paradigm and a method system is a key technology obstacle for relevant practice.

In future work regarding city sustainable development indicators and indices, the focal point is to build a consensus on the concepts as well as the construction method of indicators and indices. On the one hand, standardized and process-based institutional arrangements and management tools could be used to strengthen communication between interested parties at all levels to reach a consensus on the concept of city sustainable development and gradually form city sustainable development indicators with stronger universality. On the other hand, research on city classification needs to be strengthened to find out similarities and differences of challenges in sustainability facing various cities and enhance pertinence of indicators and indices for city sustainable development, so as to work out appropriate city development goals and strategies according to local conditions.

References

[1] UNDESA (United Nations Department of Economic and Social Affairs) 2015 *World Urbanization Prospects: The 2014 Revision* ST/ESA/SER.A/366
[2] Camagni R, Capello R and Nijkamp P 1998 *Ecol. Econ.* **24** 103-18
[3] Dodman D 2009 *Environ. Urban.* **21** 185-201
[4] Finco A and Nijkamp P 2001 *J. Environ. Pol. Plan.* **3** 289-309
[5] UNEP (United Nations Environment Programme) 2011 *Global Initiative for Resource Efficient Cities?*
[6] Sustainable Development Strategy Research Team of the 21st Century Agenda Management Center and CAS Institute of Geographic Sciences and Natural Resources Research 2007 *Sustainable Development Status and Trend in China* (Beijing: Social Sciences Adademic Press (China)) (in Chinese)
[7] Stockmann R and Meyer W 2012 *Evaluation* (Beijing: People's Publishing House) (in Chinese)
[8] Mori K and Christodoulou A 2012 *Environ. Impact Asses.* **32** 94-106
[9] Ewing B, Reed A, Rizk S M, Galli A, Wackernagel M and Kitzes J 2008 *Calculation Methodology for the National Footprint Accounts, 2008 Edition* (Oakland: Global Footprint Network)
[10] Yang Y C 2009 *Environ. Sci. Manage* **34** 23-5 (in Chinese)
[11] Huang L, Wu J and Yan L 2015 *Landscape Ecol.* **30** 1-19
[12] Urban Sustainability Research Team of CAS Institute of Urban Environment 2010 *China Sustainable City Report 2010* (Beijing: Science Press) (in Chinese)
[13] Tursun H, Wang H D, Wang L H and Peng Y D 1997 *China Popul. Resour. Env* **7** 46-50 (in Chinese)
[14] Xu X Q and Zhang J J 2001 *Acta Geogr. Sin.* **56** 54-63 (in Chinese)
[15] Parris T M and Kates R W 2003 *Annu. Rev. Env. Resour.* **28** 559-86

Influence of Saturant on Seismoelectric Coupling Response of Porous Media

Shiferaw Regassa Jufar[1], Ulugbek Djuraev[1], Sonny Irawan[1], Luluan Almanna Lubis[2]

[1]Petroleum Engineering Department, Univerisiti Teknologi PETRONAS, Malaysia
[2]Geosciences Department, Universiti Teknologi PETRONAS, Malaysia

E-mail: shiferaw.jufar@utp.edu.my

Abstract. The seismoelectric coupling phenomenon in porous medium saturated with gas and water was studied numerically. The numerical model was formulated based on Pride's theoretically developed seismoelectric coupling coefficient that describes conversion of seismic energy into electromagnetic radiation. The conversion is complex function of physical, electrical and poroelastic properties of the porous medium and the saturant. In this study, a porous medium that is partially saturated with water is considered. Pore spaces that are not saturated with water are considered to be filled with gas. The gas phase is methane. The water saturation was varied from 0.2 to 0.9. The physical and electrical properties of both phases are determined at a temperature of 325 °K. The porosity and permeability of the porous medium is 0.2 and 500 mD, respectively. The porous medium in this study is isotropic. Zeta potential was estimated using an empirically derived correlation. At 0.001 mol/lit salinity, zeta potential is 66.7 mv. The numerical study result shows that at a fixed seismic wave frequency, lower water saturation results in higher seismoelectric coupling coefficient. When water saturation increases, the seismoelectric coupling coefficient decreases. However, the rate of decline is lower at higher seismic frequency. The rate of decline at 10 kHz is 1.38 nV/Pa compared to 4.35 nV/Pa at 100 kHz. Higher frequencies result in stronger nonlinearities, which causes higher mechanical energy dissipation. As such, the result obtained in this study is consistent with what is normally expected when mechanical wave propagates through a porous media.

1. Introduction

When a porous rock is saturated with an electrolyte, an electric double layer is formed on the interface between the solid and the fluid (Zhu et al., 2013). When a seismic wave passes through a saturated porous rock, it creates charge separation in the electrical boundary layer. Subsequently, the charge separation leads to formation of an electric dipole. These dipoles in turn emit an electromagnetic wave, which can be measured to infer the properties of the fluid, flow and the porous rock. Zhu et al. (2000) experimentally investigated seismoelectric conversion in fluid-saturated porous media. They concluded that the mechanism of the seismoelectric conversion is completely different from the piezoelectric effect of quartz grains. They also determined the seismoelectric sensitivity with respect to salinity of the saturant. It was shown that the amplitude of seismoelectric signals increases as the saturant conductivity decreases. They demonstrated that seismoelectric measurements could be an effective means of obtaining transport coefficients such as hydraulic permeability and other porous rock properties. Theoretically, the seismoelectric is related to the fluid conductivity, rock porosity, pore size, etc. (Pride et al., 1996).

The objective of this paper is to investigate the possibility of applying seismoelectric conversion phenomenon to monitor saturation changes and fluid propagation pattern in a reservoir. Seismic waves and electromagnetic waves are applied to investigate the subsurface. Seismic waves have fair resolution as well as good penetration depth, however, impedance contrast between reservoir fluids are small. As such, they are insensitive to change in fluid content of a reservoir. On the other hand, electromagnetic waves have good electrical conductivity determination, but due to large wave length, they have poor resolution. In the present paper, electromagnetic wave generated due to seismic wave excitation, a phenomenon called seismoelectric conversion, will be examined to monitor changes in saturation in gas-water system. Successful implementation of the method should exploit the merit of using seismic and electromagnetic methods individually. Accordingly, the main goal of this paper is to provide some insight into how change in water saturation affects the seismoelectric response in gas-water system so that the seismoelectric phenomenon can be used to monitor and map the desaturation pattern of the porous media.

2. The Seismoelectric Phenomenon

Ivanov (1939) and Frenkel (1944) are the pioneers of the seismoelectric method, as they were the first ones to record seismoelectrically induced signals and to develop the electrokinetic theory of the seismoelectric phenomenon. Ivanov (1939) measured an electric field generated due to seismic wave passing through the earth without the application of external voltage. Frenkel (1944) developed set of equations to explain the phenomenon using the concept of electric double layer. Their works were extended and widely implemented in Soviet Union, whereby a great number of theoretical and field works were conducted using seismoelectric method for mineral exploration and to determine the weathered zone thickness to assist in interpretation of seismic data (Revil et al., 2015). Years later, Thompson et al. (1993) and Thompson et al. (2007) conducted a study, they adopted seismoelectric method to use for exploration of hydrocarbon reservoirs. As Thompson and Gist concluded, the signals induced due to the seismoelectric effects were detected at the depth of 300 meters, however, they proposed that this method could be successfully applied to much greater depths (several kilometers). A great deal of theoretical and field works done aspire us to attempt to study the seismoelectric method for monitoring saturation changes in gas-water system. The seismoelectric method has been studied for several decades. However, a comprehensive theory of the phenomenon was developed by Pride in the early '90s (Pride, 1994). His work has led to numerous breakthroughs and the phenomenon is being studied for various applications. In this paper, we adopted Pride's model to investigate the effect of water saturation in partially saturated porous media.

3. Methodology

3.1. Pride's Model for Partially-Saturated Medium

Pride (1994) developed a complete set of equations that characterizes the coupling between seismic and electromagnetic wave propagation by combining Biot's poroelasticity theory and Maxwell's equations via frequency-dependent seismoelectric coupling shown in Eq. 1 (Jouniaux et al., 2012). The coupling equation is considered to be quite complex, since it takes into account multiple variables that affect both seismic and electromagnetic fields.

$$L(\omega) = L_0\left[1 - i\frac{\omega}{\omega_c}\frac{m}{4}\left(1 - 2\frac{\tilde{d}}{\Lambda}\right)^2\left(1 - i^{3/2}\tilde{d}\sqrt{\frac{\omega\rho_f}{\eta_f}}\right)^2\right]^{-1/2} \tag{1}$$

where L_0 is the low-frequency electrokinetic coupling, m is a dimensionless number, $\tilde{d}$ is the Debye length, Λ is a geometry term of the porous material, ω_c is the transition angular frequency, and ω is the angular frequency. Additionally, ρ_f and η_f denote the density and dynamic viscosity of the fluid mixture, respectively.

Due to complexity of Pride's coupling coefficient, Walker and Glover (2010) proposed a simplified version of the equation assuming that the Debye length can be neglected when compared to the characteristic pore size. Moreover, they modified the following parameters:

$$m = 8\left(\frac{\Lambda}{r_{eff}}\right)^2 \tag{2}$$

where r_{eff} is effective pore radius. The transition angular frequency is given by equation 3.

$$\omega_c = \frac{8\eta_f}{\rho_f r_{eff}^2} \tag{3}$$

And thus, the simplified Pride's electrokinetic equation for fully-saturated porous medium becomes

$$L(\omega) = L_0\left[1 - 2i\frac{\omega}{\omega_c}\left(\frac{\Lambda}{r_{eff}}\right)^2\right]^{-1/2} \tag{4}$$

where

$$L_0 = -\frac{1}{F}\frac{\varepsilon_f \varepsilon_o \zeta}{\eta_f} \tag{5}$$

Early theoretical developments accounted for seismoelectric effects that occur only in fully-saturated conditions (Ivanov, 1939; Frenkel, 1944). However, the practical potential of using the seismoelectric method is realized when it takes into account both fully- and partially-saturated conditions in porous media. Therefore, in the numerical study presented in this paper, we adopted the simplified version of Pride's seismoelectric coupling coefficient modified by Walker and Glover (2010) and extended it for partially saturated conditions with water and gas saturating the porous medium.

Table 1. Parameters used in the numerical study.

	Notation	Value	Unit
Water density	ρ_w	988	kg/m^3
Gas density	ρ_g	1.1829	kg/m^3
Water dynamic viscosity	η_w	0.000894	$Pa * s$
Gas dynamic viscosity	η_o	0.0000181	$Pa * s$
Porosity	ϕ	20	%
Tortuosity	α_∞	4	m^2
Zeta-potential	ζ	66.7	mV
Formation factor	F	18.28	–
Effective pore radius	r_{eff}	80	μm
Pore geometry length	Λ	8E-6	m

Seismoelectric coupling is directly dependent on reservoir rock and fluid properties and the electric double layer (Jouniaux et al., 2012). The zeta-potential, which is an electric potential on the slipping plane within the double layer, is a key element in the seismoelectric phenomenon (Vinogradov et al., 2010). It is directly dependent on brine concentration, and this in turn, has a substantial effect upon the magnitude of the seismoelectric amplitudes (Zhu et al., 2016). Additionally, recent studies revealed that seismoelectric effects are also greatly affected by changing fluid saturations in a porous medium (Revil et al., 2015). In this study, the zeta-potential was estimated using an empirically derived correlation (Vinogradov et al., 2010), which links zeta-potential to the saturant salinity. At 0.001 mol/lit salinity, zeta potential is 66.7 mV. Water is assumed to be the wetting phase in the porous medium, and thus varying water saturation does not affect both the electric double layer and the zeta-potential. The frequency in the present study is varied between 10 kHz to150 kHz. Brine concentration was varied from 0.001 mol/l to 0.1 mol/l. Water saturation was varied from 0.2 to 0.9. The remaining parameters used in our calculations are listed in Table 1 that is shown below.

4. Results and Discussion

As discussed in the introduction section, the zeta-potential strongly affects the seismoelectric coupling coefficient L. The charge density and fluid properties in the electric double layer in turn affects the zeta-potential. As such, the zeta-potential varies with salinity. Fig. 1 shows variation of the seismoelectric coupling coefficient with salinity at various water saturations and constant frequency of 10 kHz. In semi-log plot, the coupling coefficient varies with salinity linearly. The model we used in this study is valid for salinity values as high as 0.4 mol/l. In Fig. 1 salinity is varied from 0.001 mol/l to 0.1 mol/l. When water saturation is fixed at 0.2 and salinity increases from 0.001 mol/l to 0.1 mol/l, the seismoelectric coupling coefficient decreased from 5.47 nV/Pa to 2.35 nV/Pa, which is about 57 % drop. Under the same change in salinity, but at water saturation 0.9, the seismoelectric coupling coefficient decreased by about the same percentage from 2.56 nV/Pa to 1.10 nV/Pa. Even though the general trend shows that the coupling coefficient declines with salinity, fitting the semi-log plot to a straight line correlation shows that the rate of decline is larger at lower water saturation. At saturation 0.2, the decline rate is 0.678 compared to 0.317 at saturation 0.9. Hence, at fixed water saturation, the change in seismoelectric coupling coefficient as salinity changes is significant when water saturation is lower. In hydrocarbon reservoirs, salinity changes with depth and we may not expect sudden jumps in concentration gradient. The observation shown here can be useful during water injection where higher water saturation zones near the injector well prevails. When the porous medium is at higher water saturation, the seismoelectric coupling coefficient seems to be less sensitive to variation in salinity. As such, the contrast between zones with different water saturation can be observed. In general, the seismoelectric coupling coefficient contrast is significant at lower salinities.

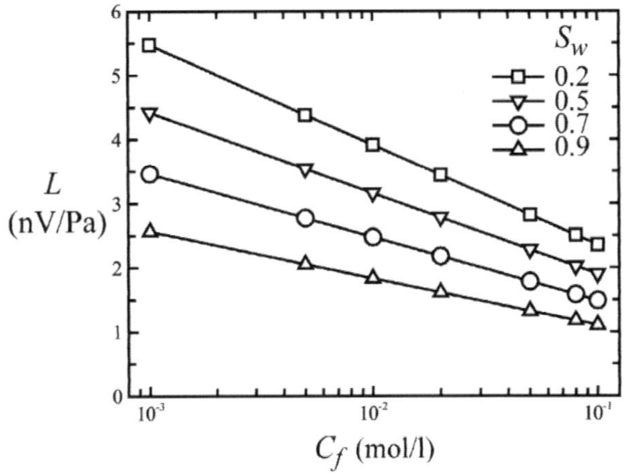

Figure 1. Effect of salinity on the seismoelectric coupling coefficient at f = 10 kHz.

Figure 2 shows variation of the seismoelectric coupling coefficient with frequency on a linear plot. The salinity of the water is fixed at 0.001 mol/l and frequency is varied from 10 kHz to 150 kHz at various water saturations. At a given frequency, the seismoelectric coupling coefficient is higher at lower water saturation. However, at constant water saturation the seismoelectric coupling coefficient is higher at lower frequency. When frequency increases, the seismoelectric coupling coefficient decreases and beyond some frequency values it starts to level off. Higher frequencies are associated with larger dissipation due to nonlinearities. Hence, the conversion of mechanical energy from the seismic source to the electromagnetic radiation becomes less efficient. Such characteristics is manifested by the trend shown in Fig. 2. The important observation from Fig. 2 is that at fixed frequency, the seismoelectric coupling coefficient is sensitive to the fluid content of the porous medium. The frequency range indicated here is much larger than those normally used in seismic surveys. However, since lower frequencies give off higher electromagnetic radiation, the magnitude of the signals should be much larger than what is shown in Fig.2. Therefore, the fluid content and the saturation of a porous medium may be recognized by measuring the amplitude of the seismoelectric response. However, it should be mentioned that the signal to noise ratio during the measurement of this electrokinetic effect is low. With the development of sensitive instrumentation, reliable measurement and post-processing, interpretation may become routine and its sensitivity to fluid saturation, porosity, permeability and other rock and fluid properties can be exploited for practical application.

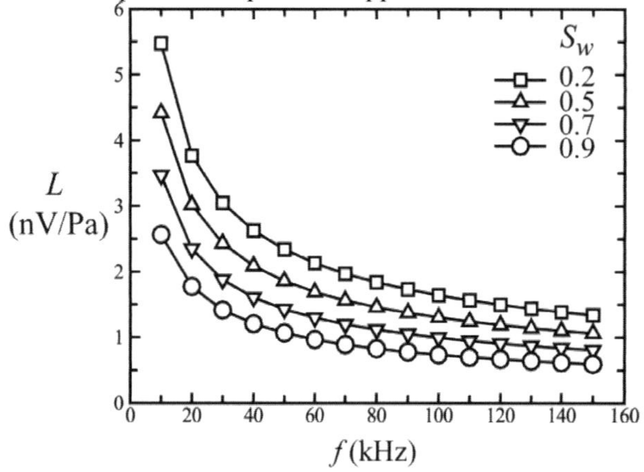

Figure 2. Effect of frequency on the seismoelectric coupling coefficient at C_f=0.001 mol/l.

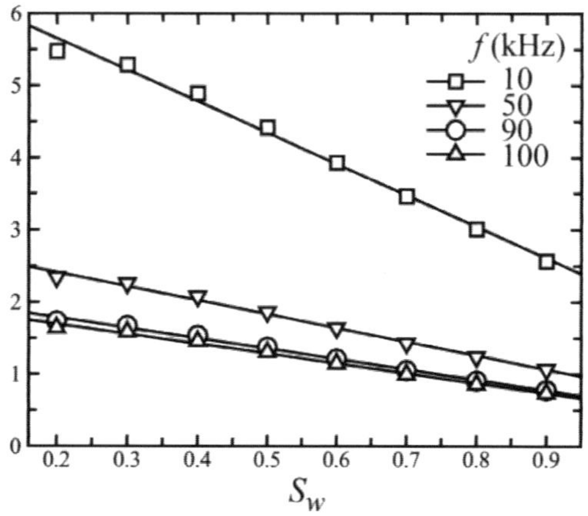

Figure 3. Effect of water saturation on the seismoelectric coupling coefficient at C_f =0.001 mol/l.

Effect of water saturation on the seismoelectric coupling coefficient at 0.001 mol/l salinity and various frequencies is presented in Fig. 3. When frequency is fixed at 10 kHz and water saturation is increased from 0.2 to 0.9, the seismoelectric coupling coefficient decreased by about 2.91 nV/Pa from 5.47 nV/Pa to 2.56 nV/Pa. Similarly, for the same change in saturation but at salinity frequency 100 kHz, the seismoelectric coupling coefficient decreased by about 0.91 nV/Pa from 1.64 nV/Pa to 0.73 nV/Pa. Although the seismoelectric coupling coefficient declines when water saturation increases at a given frequency, the rate of decline is lower at higher frequency. Table 2 summarizes the rate of decline at various frequencies.

As can be seen from Fig.3, the dependence of the coupling coefficient on saturation can be exploited to map saturation changes in a reservoir during hydrocarbon production. The contrast is significant at lower frequencies. But the challenge is the inherent difficulty in measuring the seismoelectric signals. With the advent of proper measurement techniques, the phenomenon should be capable of monitoring the saturation profile of a reservoir during hydrocarbon production. As such, various fluid injection and production processes can be monitored and optimized for better performance.

Table 2. Rate of decline of the seismoelectric coefficient at various frequencies

Frequency (kHz)	Rate of Decline (nV/Pa)
10	4.35
50	1.93
90	1.45
100	1.38

5. Conclusion

In this paper, we have numerically studied effect of change in saturation in gas-water saturated porous medium by extending Pride's seismoelectric coupling model. The result of the numerical study shows that the seismoelectric coupling coefficient is sensitive to changes in water saturation. However, the response is salinity and frequency dependent. The effect of salinity is stronger at lower water saturations. At higher water saturation, the response is less sensitive to salinity. At the given frequency, lower water saturation results in higher seismoelectric coupling coefficient. However, when water saturation remains constant and frequency is increased, the seismoelectric coupling coefficient declines faster until certain values of frequency beyond which the response seems to level off. The characteristics described here is based on the range of frequency studied in this paper, i.e. 10 kHz to 150 kHz. The study also shows that the seismoelectric coupling coefficient is sensitive to the change in saturation characteristics of the porous media in gas-water system. In general, when water saturation increases at constant salinity, the seismoelectric coupling coefficient decreases. However, the rate of decline is frequency dependent. The rate of decline is higher at lower frequencies. Regardless of frequency, the change in saturation results in change in the seismoelectric coupling coefficient. Apparently, the dependence of the coupling coefficient on saturation can be exploited to map saturation changes in a reservoir during hydrocarbon production. However, from practical point of view measuring seismoelectric response is inherently difficult due to small signal-to-noise ratio. Nevertheless, with the advent of improved measurement techniques and processing algorithms, the method should be capable of monitor change in saturation characteristics of a reservoir. Such possibility allows reservoir engineers to understand details of fluid propagation behavior in a reservoir. Such understanding is very helpful in maximizing sweep efficiency.

6. Acknowledgement

The authors would like to thank Shell – TU Delft – UTP Collaboration for providing funding.

7. References

[1] Zhu Z and Toksöz M 2013 Experimental measurements of the streaming potential and seismoelectric conversion in Berea sandstone *Geophys. Prospect.* **61(3)** 688-700.

[2] Zhu Z, Haartsen M, and Toksöz M 2000 Experimental studies of seismoelectric conversions in fluid-saturated porous media *J. Geophys. Res.:Solid Earth*, **105** 28055-28064.

[3] Pride S and Haartsen M 1996 Electroseismic wave properties *J. Acoust. Soc. Am.* **100** 1301-1315.

[4] Liu D, Castanier L, and Brigham W 1992 Displacement by foam in porous media *SPE ATCE*.

[5] Farajzadeh R, Andrianov A, Krastev R, Hirasaki G, and Rossen W 2012 Foam–oil interaction in porous media: Implications for foam assisted enhanced oil recovery *Adv. Colloid Interface Sci.* **183** 1-13.

[6] Ivanov A 1939 Effect of electrization of earth layers by elastic waves passing through them *Doklady Akademii Nauk SSSR* **24** 42-45.

[7] Frenkel J, 1944 On the theory of seismic and seismoelectric phenomena in a moist soil J. Phys. J. *Phys. (Soviet)* **8** 230-241.

[8] Revil A, Jardani A, Sava P, and Haas A 2015 The Seismoelectric Method: Theory and Application *Wiley*.

[9] Thompson A and Gist G 1993 Geophysical applications of electrokinetic conversion *The leading edge* **12** 1169-1173.

[10] Thompson A, Sumner J, and Hornbostel S 2007 Electromagnetic-to-seismic conversion: a new direct hydrocarbon indicator *The Leading Edge* **26** 428-435.

[11] Pride S 1994 Governing equations for the coupled electromagnetics and acoustics of porous media *Phys. Rev. B* **50** 15678.

[12] Jouniaux L and Bordes C 2012 Frequency-dependent streaming potentials: a review *Geophys J Int.* **2012**.

[13] Walker E and Glover P 2010 Permeability models of porous media: Characteristic length scales, scaling constants and time-dependent electrokinetic coupling *Geophysics* 75 235-246.

[14] Vinogradov J, Jaafar M, and Jackson M 2010 Measurement of streaming potential coupling coefficient in sandstones saturated with natural and artificial brines at high salinity *J. Geophys. Res.* **115**.

[15] Zhu Z, Toksöz M, and Zhan X 2016 Seismoelectric measurements in a porous quartz-sand sample with anisotropic permeability *Geophys. Prospect* **64** 700-713.

An analysis of the characteristics of Indonesian industrial sectors: 2005-2010

Ubaidillah Zuhdi

Faculty of Management and Economics, Gdansk University of Technology, Gdansk 80-233, Poland

zuhdi@zie.pg.gda.pl

Abstract. The purpose of the current study is to analyze the characteristics of Indonesian industrial sectors from 2005 through 2010. The study employs the analysis instruments from the Input-Output (IO) analysis, namely the indices of the power of dispersion, and the sensitivity of dispersion. For 2005 and 2010, the study focuses on nine and seventeen industries, respectively. The results show that industry 3, manufacturing, placed the quadrant I on the analysis period. The fact shows that, from 2005 through 2010, the industry had a strong influence on the Indonesian economic activities, and received high impacts from the external aspects. In other words, the industry has great potency for the Indonesian economy. From the results one can also see that sector 4, electricity and gas, included in the quadrant I in 2010. This fact explains that the sector has also great potency for the economic activities of Indonesia. Thus, the Indonesian government should prioritize the industries development.

1. Introduction

Industrial sectors have important roles on the economy of one country. The roles can be seen on the micro and macro aspects. On the former aspect, the roles can be observed from the economic interactions between industries and households. On the other hand, using the aggregate level, the latter one is represented by the interactions between industries and government.

One can argue that the roles itself cannot be separated from the unique characteristics of industrial sectors because it might describe how they behave in the economic system. Therefore, an analysis of the characteristics of industries can also be viewed as an effort to know the roles. Further, the proposals for improving the economic conditions of analyzed country might be generated from the analysis.

The examples of previous studies which conduct the analysis are [1], [2], [3], [4], [5], [6], [7], and [8]. The mentioned previous studies, however, focus on the specific industries. In other words, the research focuses on the analysis for the whole sector of specific country is still needed. The research is required in order to know the whole view regarding the characteristics of the industries of analyzed country so the suggestions for improving its economic conditions can be made properly. The current study is conducted in order to fulfill the gap.

The purpose of the study is to analyze the characteristics of Indonesian industrial sectors from 2005 through 2010. The period of analysis is aligned with the newest data. One can argue that, in both years, the Indonesian government politically had transitional situations because the presidential elections were conducted in 2004 and 2009. The study employs the Input-Output (IO) analysis as an analysis instrument because it is a suitable tool in describing the characteristics.

The rest of this paper is described as follows. Section 2 scientifically explains the methodology of the current study. Section 3 explores the results of calculations. The discussions for the results are also conducted on the section. The final section, section 4, gives the conclusions of the study, and suggestions for the further researches.

2. Methodology

This part scientifically explores the methodology of the current study. The first step of the methodology is to explain the data sources of the study. The data sources are the Indonesian IO tables for 2005 and 2010. The tables are obtained from [9] and [10], respectively. The former table consists of nine industrial sectors, and uses the producer's prices. On the other hand, the latter one has seventeen industries, and utilizes the basic prices. Because of the data characteristic, the years of data describe the analysis period of the study, namely from 2005 through 2010. The second step is to describe the industrial sectors of Indonesia used in this study. Tables 1 and 2 show the industries for 2005 and 2010, respectively.

Table 1. Indonesian industrial sectors used in this study, 2005.

Sector Number	Sector Name
1	Agriculture, livestock, forestry, and fishery
2	Mining and quarrying
3	Manufacturing industry
4	Electricity, gas, and water supply
5	Construction
6	Trade, hotel, and restaurant
7	Transport and communication
8	Financial, real estate, and business services
9	Services

Source: [11]

Table 2. Indonesian industrial sectors used in this study, 2010.

Sector Number	Sector Name
1	Agriculture, forestry, and fishing
2	Mining and quarrying
3	Manufacturing
4	Electricity and gas
5	Water supply, sewerage, waste management, and remediation activities
6	Construction
7	Wholesale and retail trade; repair of motor vehicles and motorcycles
8	Transportation and storage
9	Accommodation and food service activities
10	Information and communication
11	Financial and insurance activities
12	Real estate activities
13	Business activities
14	Public administration and defence; compulsory social security
15	Education
16	Human health and social work activities
17	Other services activities

The third step is to conduct the calculations in order to identify the characteristics of Indonesian industrial sectors on the analysis period. The methods applied in the calculations are the (1) index of the power of dispersion, and (2) index of the sensitivity of dispersion. Both methods are the analysis tools from the IO analysis. The former index is used to analyze the strength of one particular sector in influencing entire industries. A stronger influence is aligned with the higher index value. The detail of the index is explained by [12] as follows:

$$\text{Index of the power of dispersion by sector } = \frac{b_{*j}}{\overline{B}} \qquad (1)$$

where the numerator is each sum of column in the table of inverse matrix coefficients while the denominator explains the mean value of the entire vertical sum in the table of inverse matrix coefficients. More specifically, the equations of numerator and denominator are described as follows:

$$b_{*j} = \sum_{i}^{n} b_{ij} \qquad (2)$$

$$\overline{B} = \frac{1}{n}\sum_{j} b_{*j} = \frac{1}{n}\sum_{i}\sum_{j} b_{ij} \qquad (3)$$

where b_{ij} and n are the Leontief inverse value from sector i to sector j, and total number of analyzed industries, respectively.

The latter index is used to analyze the sensitivity of the specific sector to the external influences. A greater sensitivity is aligned with the higher index value. More specifically, one particular sector is called more sensitive to the influences from the external aspects if it has a greater index value. The detail of the index is explained by [12] as follows:

$$\text{Index of the sensitivity of dispersion by sector } = \frac{b_{i*}}{\overline{B}}. \qquad (4)$$

In this index, the numerator is each sum of row in the table of inverse matrix coefficients while the denominator describes the mean value of the entire horizontal sum in the table of inverse matrix coefficients. Further, the equations of the numerator and denominator of the index are explained as follows:

$$b_{i*} = \sum_{j}^{n} b_{ij} \qquad (5)$$

$$\overline{B} = \frac{1}{n}\sum_{i} b_{i*} = \frac{1}{n}\sum_{i}\sum_{j} b_{ij}. \qquad (6)$$

In order to get a compatibility sense with the previous index, equation (5) is slightly modified from the original source. More specifically, the part explains the total number of discussed industries, n, is added into the equation. As with the previous explanation, b_{ij} is the value of Leontief inverse from sector i to sector j. The next step is to analyze the characteristics of Indonesian industrial sectors on the analysis period. Conclusions of this study, and suggestions for further researches are described on the final step.

3. Results and analysis

Figures 1 and 2 plot the analyzed sectors, and combine both indices used in this study in one chart for 2005 and 2010, respectively. More specifically, the horizontal axis of the chart explains the values of the index of the power of dispersion while the vertical axis places the values of another index. The chart has four quadrants. Each discussed sector has a specific quadrant in the chart.

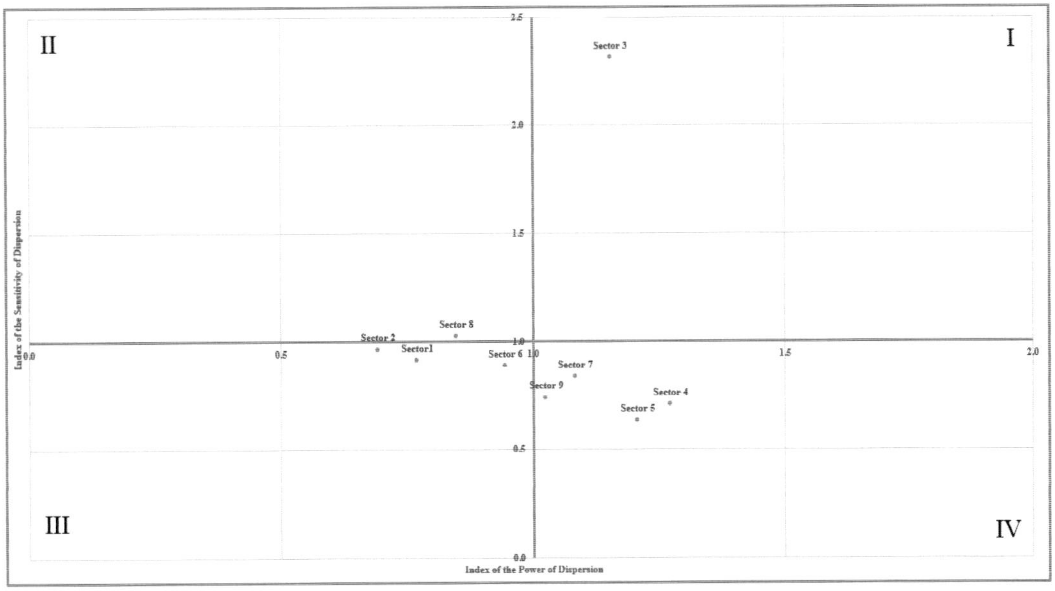

Figure 1. The quadrants for Indonesian industrial sectors based on the indices of the power of dispersion, and the sensitivity of dispersion, 2005 (Source: [11]).

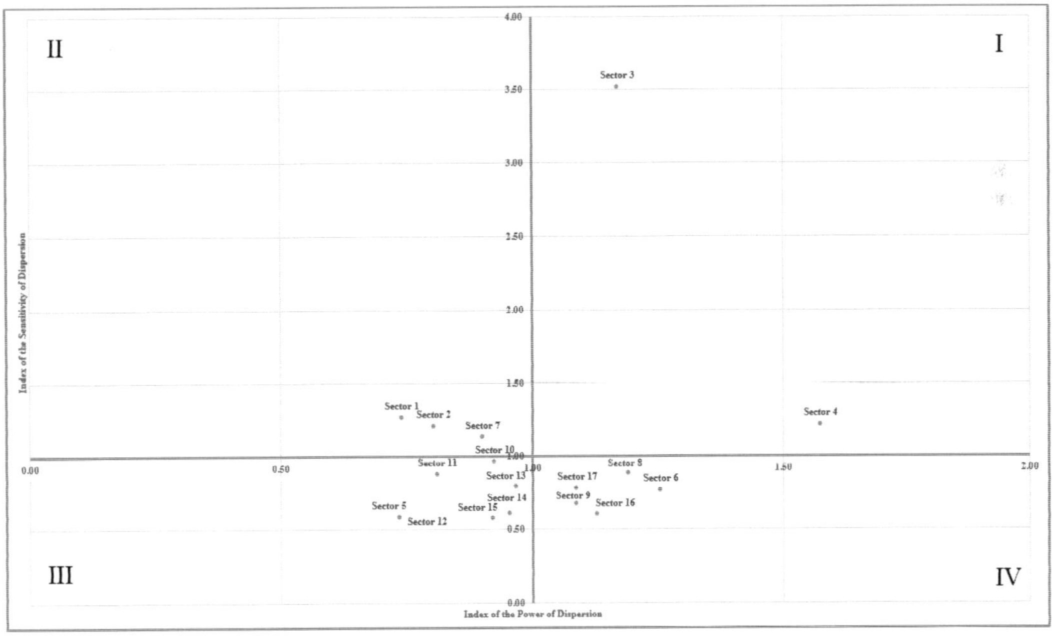

Figure 2. The quadrants for Indonesian industrial sectors based on the indices of the power of dispersion, and the sensitivity of dispersion, 2010.

Each quadrant has unique characteristics. More specifically, the quadrant I is a place where the values of both indices are more than one. In other words, the sectors include in this quadrant are those most affected by the external aspects as well as have strong influences on the entire industries. The opposite phenomena can be observed on the sectors which include in the quadrant III. On the other hand, quadrant II is an area where the value of the index of the power of dispersion is less than one while the value of another index is more than one. One can argue that the industries include in this quadrant are those which have weak influences on the entire industries but they get high impacts from the changes of external aspects. The opposite characteristics are owned by the sectors which include in the quadrant IV.

Based on the information in both figures, one can say that sector 3 placed the quadrant I on the analysis period. From tables 1 and 2, this is a manufacturing sector. The fact shows that, from 2005 through 2010, the sector had a strong influence on the Indonesian economic activities, and received high impacts from the external aspects. Sectors 5 and 12 in the latter figure have close positions because the indices values of both industries in 2010 were similar. Tables 3 and 4 summarize the quadrants of analyzed industries in 2005 and 2010, respectively.

Table 3. The quadrants of Indonesian industrial sectors, 2005.

Sector Number	Sector Name	Quadrant
1	Agriculture, livestock, forestry, and fishery	III
2	Mining and quarrying	III
3	Manufacturing industry	I
4	Electricity, gas, and water supply	IV
5	Construction	IV
6	Trade, hotel, and restaurant	III
7	Transport and communication	IV
8	Financial, real estate, and business services	II
9	Services	IV

Table 4. The quadrants of Indonesian industrial sectors, 2010.

Sector Number	Sector Name	Quadrant
1	Agriculture, forestry, and fishing	II
2	Mining and quarrying	II
3	Manufacturing	I
4	Electricity and gas	I
5	Water supply, sewerage, waste management, and remediation activities	III
6	Construction	IV
7	Wholesale and retail trade; repair of motor vehicles and motorcycles	II
8	Transportation and storage	IV
9	Accommodation and food service activities	IV
10	Information and communication	III
11	Financial and insurance activities	III
12	Real estate activities	III
13	Business activities	III
14	Public administration and defence; compulsory social security	III
15	Education	III
16	Human health and social work activities	IV
17	Other services activities	IV

4. Conclusions and further researches

This study analyzes the characteristics of Indonesian industrial sectors from 2005 through 2010. The study employs the analysis instruments from the IO analysis, namely the indices of the power of dispersion, and the sensitivity of dispersion. For 2005 and 2010, the study focuses on nine and seventeen industries, respectively.

The results show that industry 3, manufacturing, placed the quadrant I on the analysis period. The fact shows that, from 2005 through 2010, the industry had a strong influence on the Indonesian economic activities, and received high impacts from the external aspects. In other words, the industry has great potency for the Indonesian economy. Thus, the Indonesian government should prioritize the industry development.

From the results one can also see that sector 4, electricity and gas, included in the quadrant I in 2010. This fact explains that the sector has also great potency for the economic activities of Indonesia. Therefore, the sector enhancement should be considered by the Indonesian government.

Comparing with the previous study, the deeper understanding regarding the characteristics of Indonesian industrial sectors is obtained from the current study. However, the study uses the aggregated industries. In other words, the study is still far from describing the whole perspective about the characteristics. The perspective is needed in order to know better the conditions of Indonesian national economy so the comprehensive policies for enhancing those in the future can be developed. Therefore, as a further research, the study proposes the same analysis for the disaggregated Indonesian industries.

The other suggested further research from the study is to conduct the international comparison using the same analysis. The comparison can be done among developing as well as developed-developing countries. The comparison might describe the characteristics of the industries of compared countries so the similarities and differences among those can be investigated. The similar concept can be applied in order to compare the characteristics of the industrial sectors of regional organizations.

References

[1] Ilhan B and Yaman H 2011 A comparative input-output analysis of the construction sector in Turkey and EU countries *Engineering, Construction and Architectural Management* **18** 248–65

[2] Ofori G 2015 Nature of the construction industry, its needs and its development: a review of four decades of research *Journal of Construction in Developing Countries* **20** 115–35

[3] Zuhdi U 2014 An input-output approach to analyze the ways to increase total output of energy sectors: the case of Japan *IOP Conference Series: Earth and Environmental Science* **19** 012015

[4] Zuhdi U and Prasetyo A D 2014 Examining the total output changes of ICT sectors of Japan: an approach of input-output *Procedia-Social and Behavioral Sciences* **109** 659–63

[5] Zuhdi U 2014 Analyzing the role of creative industries in national economy of Japan: 1995-2005 *Open Journal of Applied Sciences* **4** 197–211

[6] Zuhdi U 2015 The dynamics of Indonesian creative industry sectors: an analysis using input-output approach *Journal of the Knowledge Economy* **6** 1177–90

[7] Zuhdi U 2015 An application of input-output analysis in analyzing the impacts of final demands changes on the total outputs of Japanese energy sectors: a further study *Journal of Physics: Conference Series* **622** 012041

[8] Zuhdi U 2014 Analyzing the impacts of final demand changes on total output using input-output approach: the case of Japanese ICT sectors *IOP Conference Series: Earth and Environmental Science* **19** 012016

[9] BPS-Statistics Indonesia 2015 *Statistical Yearbook of Indonesia 2015* [online] https://www.bps.go.id/website/pdf_publikasi/Statistik-Indonesia-2015_rev.pdf (accessed March 30, 2017)

[10] BPS-Statistics Indonesia 2016 *Statistical Yearbook of Indonesia 2016* [online] https://www.bps.go.id/website/pdf_publikasi/Statistik-Indonesia-2016--_rev.pdf (accessed December 30, 2016)

[11] Zuhdi U 2016 The Indonesian economy in 2005: an analysis using the input-output approach *Proceedings of 20th EBES Conference-Vienna* **3** 1825–42

[12] Japanese Ministry of Internal Affairs and Communications n.d. *Chapter IV: Coefficients for Input-Output Analysis and Computation Methods* [online] http://www.soumu.go.jp/main_content/000327480.pdf (accessed March 30, 2017)

IOP Conf. Series: Earth and Environmental Science 88 (2017) 012027 doi:10.1088/1755-1315/88/1/012027

An analysis of the characteristics of Japanese industrial sectors from 2005 through 2011

Ubaidillah Zuhdi

Faculty of Management and Economics, Gdansk University of Technology, Gdansk 80-233, Poland

zuhdi@zie.pg.gda.pl

Abstract. The purpose of the current study is to analyze the characteristics of Japanese industrial sectors from 2005 through 2011. The study employs the analysis tools from the Input-Output (IO) analysis, namely the indices of the power of dispersion, and the sensitivity of dispersion. The study focuses on the thirteen industries. The characteristics are described by the quadrants of analyzed charts. The results show that industries 3 and 9, manufacturing, and transport and postal services, respectively, placed the quadrant I on the analysis period. The fact shows that, from 2005 through 2011, these industries had strong influences on the Japanese economic activities, and received high impacts from the external aspects. In other words, the industries have great potency for the Japanese economy. Thus, the Japanese government should prioritize the industries development. From the results one can also view that, from 2005 through 2011, almost all Japanese industries had no the quadrant change. The fact describes that the characteristics of almost all Japanese industrial sectors on the period of analysis did not change.

1. Introduction

Industrial sectors have notable roles on the economy of one country. The roles can be observed on the micro and macro aspects. On the former aspect, the roles can be viewed from the economic interactions between industries and households. On the other hand, using the aggregate level, the latter one is described by the interactions between industries and government.

One can say that the roles itself cannot be separated from the unique characteristics of industrial sectors because it might explain how they behave in the economic system. Therefore, an analysis of the characteristics of industries can also be seen as an effort to know the roles. Further, the suggestions for improving the economic conditions of analyzed country might be generated from the analysis.

The examples of previous studies which discuss the analysis are [1], [2], [3], [4], [5], [6], [7], [8], [9], and [10]. The mentioned previous studies, however, focus on the particular industries. In other words, the research focuses on the analysis for the whole sector of specific country is still needed. The research is needed in order to obtain the whole view regarding the characteristics of the industries of analyzed country so the suggestions for enhancing its economic conditions can be made properly. The current study is conducted in order to fulfill the gap.

The purpose of the study is to analyze the characteristics of Japanese industrial sectors from 2005 through 2011. The period of analysis is aligned with the newest data. One can argue that the economic situations of Japan in both years were different because the earthquake and tsunami were happened in Japan in 2011. The study employs the Input-Output (IO) analysis as an analysis tool because it is a suitable instrument in explaining the characteristics. The rest of this paper is explained as follows. Section 2 scientifically explores the methodology of the current study. Section 3 describes the results of calculations. The discussions for the results are also done on the section. The next section, section 4, gives the conclusions of the study, and suggestions for the future researches.

2. Methodology

This part scientifically describes the methodology of the current study. The first step of the methodology is to explore the data sources of the study. The data sources are the Japanese IO tables for 2005 and 2011. The tables are obtained from [11] and [12], respectively. Both tables consist of thirteen industrial sectors, and use the producer's prices. The second step is to describe the industrial sectors of Japan used in this study. Table 1 shows the industries. The difference between the industries on both data resources can be seen on sector 9. More specifically, the name of the sector on the former data source is "Transport" while on the latter one is "Transport and Postal Services".

The third step is to conduct the calculations in order to know the characteristics of Japanese industrial sectors on the analysis period. The methods utilized in the calculations are the (1) index of the power of dispersion, and (2) index of the sensitivity of dispersion. Both methods are the analysis instruments from the IO analysis. The former index is used to analyze the strength of one specific sector in influencing entire industries. A stronger influence is aligned with the higher index value. The detail of the index is described by [13] as follows:

$$\text{Index of the power of dispersion by sector} = \frac{b_{*j}}{\overline{B}} \quad (1)$$

where the numerator is each sum of column in the table of inverse matrix coefficients while the denominator shows the mean value of the entire vertical sum in the table of inverse matrix coefficients. More specifically, the equations of numerator and denominator are explained as follows:

$$b_{*j} = \sum_{i}^{n} b_{ij} \quad (2)$$

$$\overline{B} = \frac{1}{n}\sum_{j} b_{*j} = \frac{1}{n}\sum_{i}\sum_{j} b_{ij} \quad (3)$$

where b_{ij} and n are the Leontief inverse value from sector i to sector j, and total number of analyzed industrial sectors, respectively.

Table 1. Japanese industrial sectors used in this study.

Sector Number	Sector Name
1	Agriculture, forestry, and fishery
2	Mining
3	Manufacturing
4	Construction
5	Electricity, gas, and water supply
6	Commerce
7	Finance and insurance
8	Real estate
9	Transport and postal services ("Transport" for 2005)
10	Information and communications
11	Public administration
12	Services
13	Activities not elsewhere classified

The latter index is applied to analyze the sensitivity of the specific sector to the external influences. A greater sensitivity is aligned with the higher index value. More specifically, one specific sector is called more sensitive to the influences from the external aspects if it has a greater index value. The detail of the index is described by [13] as follows:

$$\text{Index of the sensitivity of dispersion by sector} = \frac{b_{i*}}{\overline{B}}. \quad (4)$$

In this index, the numerator is each sum of row in the table of inverse matrix coefficients while the denominator explains the mean value of the entire horizontal sum in the table of inverse matrix coefficients. Further, the equations of the numerator and denominator of the index are described as follows:

$$b_{i*} = \sum_{j}^{n} b_{ij} \quad (5)$$

$$\overline{B} = \frac{1}{n}\sum_{i} b_{i*} = \frac{1}{n}\sum_{i}\sum_{j} b_{ij}. \quad (6)$$

In order to get a compatibility sense with the previous index, equation (5) is slightly modified from the original source. More specifically, the part describes the total number of discussed industries, *n*, is added into the equation. As with the previous explanation, b_{ij} is the value of Leontief inverse from sector *i* to sector *j*. The next step is to analyze the characteristics of Japanese industrial sectors on the analysis period. Conclusions of this study, and suggestions for further researches are explored on the final step.

3. Results and analysis

Figures 1 and 2 plot the discussed sectors, and combine both indices used in this study in one chart for 2005 and 2011, respectively. More specifically, the horizontal axis of the chart describes the values of the index of the power of dispersion while the vertical axis places the values of another index. The chart has four quadrants. Each discussed sector has a particular quadrant in the chart.

Each quadrant has unique characteristics. More specifically, the quadrant I is a place where the values of both indices are more than one. In other words, the industries include in this quadrant are those most affected by the external aspects as well as have strong influences on the entire industries. The opposite phenomena can be viewed on the sectors which include in the quadrant III. On the other hand, quadrant II is an area where the value of the index of the power of dispersion is less than one while the value of another index is more than one. One can argue that the industries include in this quadrant are those which have weak influences on the entire industries but they receive high impacts from the changes of external aspects. The opposite characteristics are owned by the industries which include in the quadrant IV.

Based on the information in both figures, one can say that sectors 3 and 9 placed the quadrant I on the analysis period. From table 1, the sectors are manufacturing, and transport and postal services industries, respectively. The name of sector 9 for 2005 is transport. The fact shows that, from 2005 through 2011, the industries had strong influences on the Japanese economic activities, and received high effects from the external aspects. Therefore, the Japanese government should prioritize the industries developments on the future based on the fact.

Tables 2 and 3 summarize the quadrants of analyzed sectors in 2005 and 2011, respectively. Based on the information in these tables, one can argue that, from 2005 through 2011, almost all Japanese industries had no the quadrant movement. The movement on the analysis period was owned by the sector 7, finance and insurance. More specifically, on the analyzed period, the movement of the sector was happened from quadrant II to quadrant III. The fact explains that the characteristics of almost all Japanese industrial sectors on the period of analysis did not change. Obviously, although this is an interesting fact, the further analysis is needed in order to explain whether the Japanese economic activities on the period had significant changes or not. One of the reasons is the earthquake and tsunami were happened in Japan in 2011.

IOP Conf. Series: Earth and Environmental Science **88** (2017) 012027 doi:10.1088/1755-1315/88/1/012027

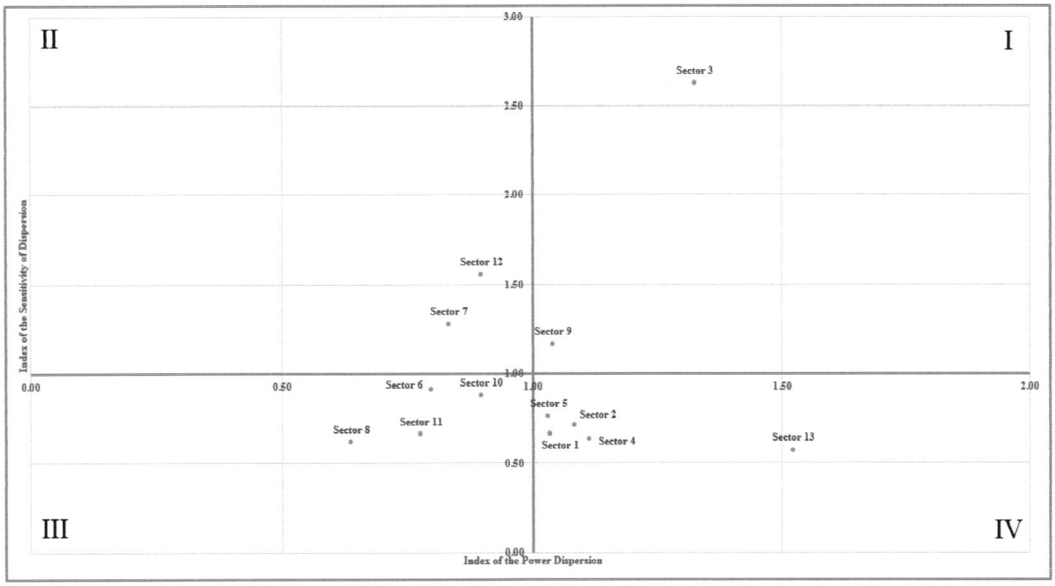

Figure 1. The quadrants for Japanese industrial sectors based on the indices of the power of dispersion, and the sensitivity of dispersion, 2005.

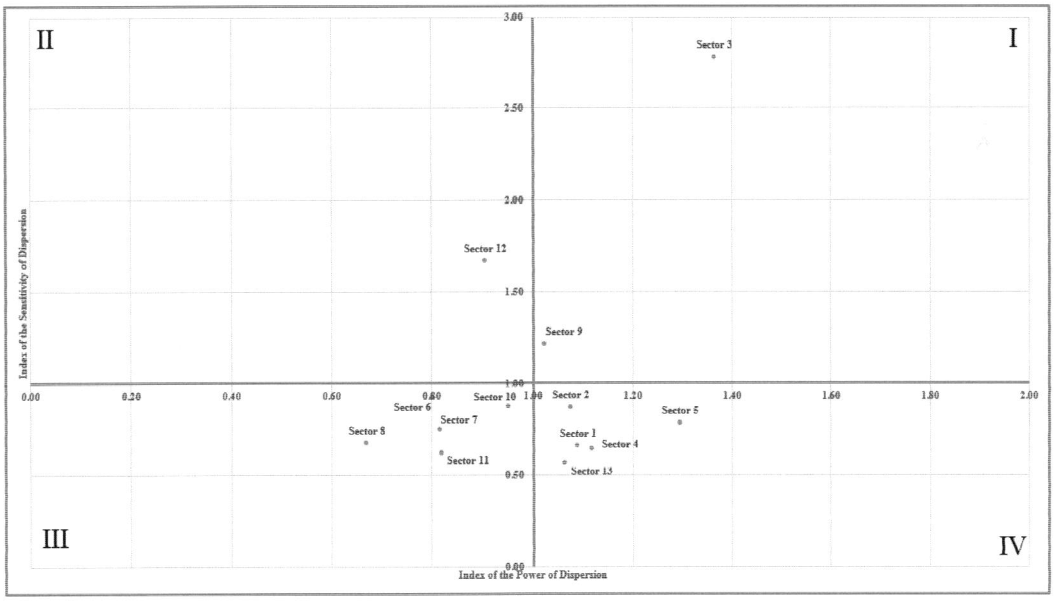

Figure 2. The quadrants for Japanese industrial sectors based on the indices of the power of dispersion, and the sensitivity of dispersion, 2011.

Table 2. The quadrants of Japanese industrial sectors, 2005.

Sector Number	Sector Name	Quadrant
1	Agriculture, forestry, and fishery	IV
2	Mining	IV
3	Manufacturing	I
4	Construction	IV
5	Electricity, gas, and water supply	IV
6	Commerce	III
7	Finance and insurance	II
8	Real estate	III
9	Transport	I
10	Information and communications	III
11	Public administration	III
12	Services	II
13	Activities not elsewhere classified	IV

Table 3. The quadrants of Japanese industrial sectors, 2011.

Sector Number	Sector Name	Quadrant
1	Agriculture, forestry, and fishery	IV
2	Mining	IV
3	Manufacturing	I
4	Construction	IV
5	Electricity, gas, and water supply	IV
6	Commerce	III
7	Finance and insurance	III
8	Real estate	III
9	Transport and postal services	I
10	Information and communications	III
11	Public administration	III
12	Services	II
13	Activities not elsewhere classified	IV

4. Conclusions and further researches

This study analyzes the characteristics of Japanese industrial sectors from 2005 through 2011. The study employs the analysis tools from the IO analysis, namely the indices of the power of dispersion, and the sensitivity of dispersion. The study focuses on the thirteen industries. The characteristics are described by the quadrants of analyzed charts.

The results show that industries 3 and 9, manufacturing, and transport and postal services, respectively, placed the quadrant I on the analysis period. The name of sector 9 for 2005 is transport. The fact shows that, from 2005 through 2011, these industries had strong influences on the Japanese economic activities, and received high impacts from the external aspects. In other words, the industries have great potency for the Japanese economy. Thus, the Japanese government should prioritize the industries development.

From the results one can also view that, from 2005 through 2011, almost all Japanese industries had no the quadrant change. The fact describes that the characteristics of almost all Japanese industrial sectors on the period of analysis did not change. The quadrant movement on the analysis period was owned by the sector 7, finance and insurance. More specifically, on the analyzed period, the movement of the sector was happened from quadrant II to quadrant III.

The understanding regarding the characteristics of Japanese industrial sectors is obtained from the current study. However, the study utilizes the aggregated industries. In other words, the study is still far from explaining the whole view about the characteristics. The view is needed in order to know better the conditions of Japanese national economy so the comprehensive policies for improving those in the future can be developed. Therefore, as a further research, the study proposes the same analysis for the disaggregated Japanese industries.

The other suggested further research from the study is to make an international comparison using the same analysis. The comparison can be done among developed as well as developed-developing countries. The comparison might explore the characteristics of the industries of compared countries so the similarities and differences among those can be investigated.

References

[1] Ilhan B and Yaman H 2011 A comparative input-output analysis of the construction sector in Turkey and EU countries *Engineering, Construction and Architectural Management* **18** 248–65
[2] Ofori G 2015 Nature of the construction industry, its needs and its development: a review of four decades of research *Journal of Construction in Developing Countries* **20** 115–35
[3] Zuhdi U 2014 An input-output approach to analyze the ways to increase total output of energy sectors: the case of Japan *IOP Conference Series: Earth and Environmental Science* **19** 012015
[4] Zuhdi U and Prasetyo A D 2014 Examining the total output changes of ICT sectors of Japan: an approach of input-output *Procedia-Social and Behavioral Sciences* **109** 659–63
[5] Zuhdi U 2014 Analyzing the role of creative industries in national economy of Japan: 1995-2005 *Open Journal of Applied Sciences* **4** 197–211
[6] Zuhdi U 2015 The dynamics of Indonesian creative industry sectors: an analysis using input-output approach *Journal of the Knowledge Economy* **6** 1177–90
[7] Zuhdi U 2015 An application of input-output analysis in analyzing the impacts of final demands changes on the total outputs of Japanese energy sectors: a further study *Journal of Physics: Conference Series* **622** 012041
[8] Zuhdi U 2014 Analyzing the impacts of final demand changes on total output using input-output approach: the case of Japanese ICT sectors *IOP Conference Series: Earth and Environmental Science* **19** 012016
[9] Zuhdi U, Mori S and Kamegai K 2014 Analysis of influences of GDP and ICT on Indonesian industrial structural changes using statistical analysis: 1990-2005 *Journal of Finance and Accountancy* **17** 1–19
[10] Zuhdi U 2015 An analysis of the role of information and communication technology sectors on Japanese national economy from 1995 through 2005: an application of multiplier analysis *IOP Conference Series: Earth and Environmental Science* **23** 012014
[11] Japanese Ministry of Internal Affairs and Communications 2009 *Introduction: The Japanese Economy and the 2005 Input-Output Tables* [online] http://www.soumu.go.jp/main_content/000327476.pdf (accessed 26 April 2016)
[12] Japanese Ministry of Internal Affairs and Communications 2016 *2011 Input-Output Tables for Japan* [online] http://www.soumu.go.jp/main_content/000443188.pdf (accessed December 30, 2016)
[13] Japanese Ministry of Internal Affairs and Communications n.d. *Chapter IV Coefficients for Input-Output Analysis and Computation Methods* [online] http://www.soumu.go.jp/main_content/000327480.pdf (accessed March 30, 2017)
[14] Zuhdi U 2015 An application of multiplier analysis in analyzing the role of information and communication technology sectors on Indonesian national economy: 1990-2005 *IOP Conference Series: Earth and Environmental Science* **23** 012015
[15] Zuhdi U, Mori S and Kamegai K 2014 Statistical analysis of influences of ICT on industrial structure changes from 1985 through 2005: the case of Japan *Journal of Computers* **9** 1291–9
[16] Zuhdi U, Mori S and Kamegai K 2013 Analysis of influences of ICT on structural changes in Japanese commerce, business services and office supplies, and personal services sectors using multivariate analysis: 1985-2005 *The Asian Journal of Technology Management* **6** 102–11
[17] Ohno K 2006 *The Economic Development of Japan: The Path Traveled by Japan as a Developing Country* [online] http://www.grips.ac.jp/forum/pdf06/EDJ.pdf (accessed 6 May 2017)

The Scenario of the Potential Analysis Alternative Energy in Order to Strengthening District's Energy Resilience (The Case Study in South Sumatera Province)

Muhammad Ferry Muhrom[1], Tb Ronny Rahman Nitibaskara[1], Herdis Herdiansyah[2], Ravita Sari[3]
[1] National Resilience Studies, School of Strategic and Global Studies, Universitas Indonesia, Salemba, Indonesia
[2] School of Environmental Science, Universitas Indonesia, Salemba, Indonesia
[3] Center of Strategic and Defense Studies, Universitas Indonesia, Salemba, Indonesia
E mail: herdis@ui.ac.id

Abstract: The current development of fossil energy, which is the driving force of the economy in Indonesia, is a non-renewable energy and is in need to know when it will be exhausted so it may be replaced with renewable energy. Many powerplant systems in Indonesia are still using conventional system that utilizes fossil energy as the primary energy in the process of electricity generation. The occurrence of electrical energy crisis is marked by several electricity blackout phenomena in some areas in South Sumatera province alternately, which is the proof that the installed power capacity has exceeded the capacity of generation power. Interconnection among several islands, namely Java Island, Sumatera Island, and Bali Island which has been interconnected with closed loop system through transmission network has not been able to overcome the electrical energy crisis. This paper aims to create alternative energy potential scenarios in the province of South Sumatera in sequence/ranking by using quantitative methods with sequential explanatory model formulated in the determination of alternative energy strategies then analyzed by using *Analytical Hierarchy Process*(AHP) method. The simulation results from this research indicate that geothermal energy potentials get the highest value so that it becomes the priority of alternative energy strategy in South Sumatera Province.

Keywords: *Alternative Energy, Renewable Energy, Energy Security, Resilience, Analytical Hierarchy Process* (AHP).

1. Introduction

Indonesia potentially has alternative energy in addition to fossil energy, so far to strengthen the renewal of Indonesia's energy through a policy tool that encourages new energy alternatives by suppressing petroleum subsidies and enabling new energy businesses, the issue of fuel price hike is often used as a political issue [1]. A shock will occur as predicted seeing how vital the role of fuel is in meeting the daily needs of life, as it has happened before, as a result of the increase in fuel triggering inflation against all basic commodities and ultimately will be costly for the people. The influence of energy on people's activities are: (1) As input for fertilizer factory, (2) Influence on farmers, (3) Influence on transportation and communication [2]. Therefore, development on renewable and non-renewable energy sources become important to be researched, and on non-renewable energy is needed to know when it is exhausted so that it can be immediately replaced with renewable energy.

Indonesia possesses "New Energy Source" which existence is located within the archipelago. This New Energy Source not only associated with amount and capacity, but also with its diversity, so this alternative energy sources could be Primary Energy Source as the replacement of Fossil Energy, to be used optimally in the future [3].

A huge challenge from the energy mix is the high dependence on fossil fuels (especially petroleum), which means the low share of renewable energy utilization. Excessive dependence on fossil fuels also raises questions on the sustainability aspect, and the rise in the price of Fuel Petroleum (BBM) in the country that has been closed with government subsidies, which consequently put the burden on the state budget (Budget Revenue and Expenditure) continues to grow. By 2014, the fuel subsidy burden amount to Rp 199.9 trillion from the budget devoted to energy subsidies, which is equivalent to about 2.5% of GDP and 25% of the total budget [4].

2. Theoretical Review

Alternative energy is spread throughout the region of Indonesia, including the province of South Sumatra, where there is a diversity of potential alternative energy resources. Energy potential in South Sumatra that can be renewed as an alternative energy source such as Geothermal, Water (micro hydro), Wind, and the Sun is large enough Scattered almost throughout the region of South Sumatra, as an area with the potential of large and varied energy resources, the province of South Sumatra needs an alternative priority energy strategy in managing energy potential to support the national energy security of Indonesia [5].

Furthermore, with the increasing demand for electrical energy in Indonesia currently unbalanced with the availability of electricity supply, or the crisis of electrical energy, it is a problem that cannot be avoided. Phenomenon of the electricity energy crisis such as electricity outages in Java-Bali and blackouts in rotation in some areas in South Sumatera province are signs that the supply of electricity in both the interconnection and conventional systems is no longer able to meet the increasing electricity demand of society and industry [6]. This is because the installed power capacity has exceeded the power generation capacity and this condition is aggravated by the price of fuel oil, which has a high selling price, resulting in power plants in the province of South Sumatra and other provinces in Indonesia that use diesel as primary energy for the process of voltage generation to experience an increase in operational costs. This condition is a challenge to the energy security of the region

There are two concepts of energy resistance that is 5S and 4A. Indonesia's energy security adheres to the 4A which are availability, accessibility, cost (Affordability), and usage (Acceptability) which then adapted to national conditions and capabilities. Energy security is a condition of ensuring the availability of energy and access to energy at affordable prices in the long term by keeping an eye on the environmental aspect as well [7].

3. Methodology

This paper uses a quantitative method with the Sequential explanatory model [8]. Quantitative method is used in determining the priority for alternative energy strategy. Population in this research is energy stakeholder in South Sumatera Area Energy and Mineral Resources Department of South Sumatera Province, PT. PERTAMINA (Persero) RU3 Palembang, PT. PLN Persero WS2JB Palembang, Ministry of Education and Training, PT. Pusri Palembang, and Academics. The technique for collecting data is by using primary data with resource person expert on energy and secondary data through data from documentation and publication of Ministry of Energy and Mineral Resources, PT. Pertamina (Persero), National Energy Council (DEN), Electronic Media, and relevant print media.

South Sumatera region contained a several potentials of renewable and non- renewable resources that can be the replacement for the fossil fuel energy such as, geothermal energy, coal, natural gas, water, wind, solar and peat, whereas the alternative energy like coal, natural gas, and peat is not a sustainable and environmental friendly. For that matters, in this paper we are only focused on the analysis of the geothermal, water, wind, and solar energy that can be managed and utilized as the source of energy which is sustainable along with the priority of environmental concept in its usage. Analytical Hierarchy Process (AHP) is used to analyze alternative energy potentials that influence the utilization and management of alternative energy potentials of strength, weakness, opportunity and threat factors. [9].

Modeling as shown below:

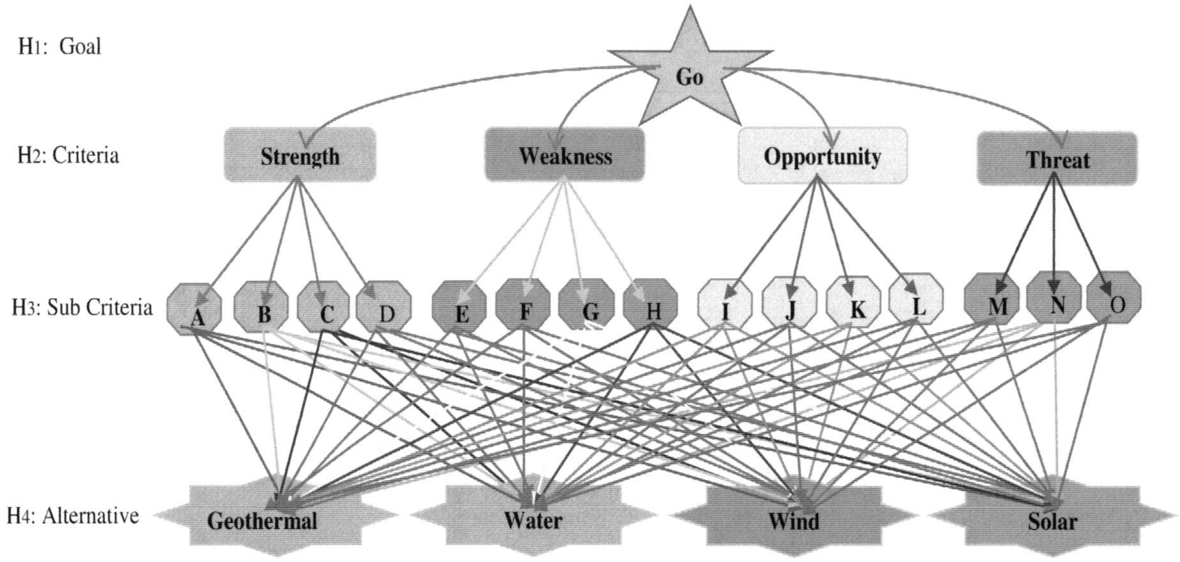

Picture 1. AHP Formulation of the Renewable Alternative Energy Strategy

Description: A = Low generating cost, B = No pollution and greenhouse gas emissions, C = Reduced oil imports, D = Ongoing utilization, E = High investment cost, F = inadequate infrastructure, G = localizes utilization, H = Cannot be traded before being converted, I = Increase national resilience, J = The demand of electricity and the growth of electrical energy, K = dependence on the fuel that causes National energy security problem, L = Reduce Co2 emissions, M = Lack of Human Resource particularly in the district, N = Deprivation of the infestation for renewable energy (because the rate of return for the capital investment is quite low and uncertain), O = the possibility of the regulation area which is different in accordance to the central government policy [10].

4. Result and Discussion

South Sumatera Province, which is part of Indonesia territory, has potential in alternative energy sources such as geothermal energy, water energy, wind energy, and solar energy which is immense enough and is spread all over South Sumatera region. The potential of this alternative energy can be utilized as primary energy source in South Sumatera province to strengthen the resilience of regional energy [11].

The AHP analysis uses pairwise comparison matrix based on several criteria [12], with several assumptions such as:

a. Analysis of the power aspects of the renewable energy selection strategy

The sub criteria studied in this aspect include: low cost generation factor (A), least pollution factor and greenhouse gas emissions generated at each plant (B), factors for reducing oil imports (C), and sustainable energy utilization factors (D).

Table 1. Local priority and global aspect of power

Priority of alternative strategy	A	B	C	D
Mass	0.085	0.140	0.233	0.542
Geothermal	0.074	0.066	0.096	0.074
Water	0.171	0.150	0.161	0.171
Wind	0.284	0.240	0.277	0.284

Solar	0.471	0.544	0.466	0.471
Total	**1.000**	**1.000**	**1.000**	**1.000**

Priority of alternative strategy	A	B	C	D	Weight
Geothermal	0.006	0.009	0.022	0.040	0.078
Water	0.015	0.021	0.038	0.093	0.166
Wind	0.024	0.034	0.065	0.154	0.276
Solar	0.040	0.076	0.109	0.255	0.480
Total	**0.085**	**0.140**	**0.233**	**0.542**	**1.000**

According to table 1, for low cost generating factors, the alternative strategy chosen is the alternative priority of the solar strategy. The energy priority that produces the lowest pollution is solar energy. Factors that can reduce oil imports are big priority solar energy, while for the use of renewable energy more priority to solar energy. Overall priority of renewable energy strategy alternative based on power aspect is more toward the alternative of solar energy that is equal to 0,480.

b. Analysis of the weakness aspects of renewable energy selection strategy

Sub criteria studied in this aspect include: factors that require expensive investment costs (E), factors lacking adequate infrastructure (F), local utilization factors (G), and non-selling factors before conversion (H).

Table 2. Local priority and global weakness aspect

Priority of alternative strategy	E	F	G	H
Mass	0.042	0.133	0.268	0.558
Geothermal	0.598	0.492	0.544	0.558
Water	0.209	0.309	0.240	0.259
Wind	0.133	0.136	0.150	0.112
Solar	0.060	0.063	0.066	0.071
Total	**1.000**	**1.000**	**1.000**	**1.000**

Priority of alternative strategy	E	F	G	H	Weight
Geothermal	0.025	0.065	0.146	0.311	0.547
Water	0.009	0.041	0.064	0.145	0.259
Wind	0.006	0.018	0.040	0.063	0.126
Solar	0.002	0.008	0.018	0.039	0.068
Total	**0.042**	**0.133**	**0.268**	**0.558**	**1.000**

Based on table 2, the alternative priority which has an expensive investment cost is geothermal energy. Based on the inadequate infrastructure factor that is owned by the province of South Sumatera is geothermal. Utilization of the local area is more geothermal and energy factors that cannot be traded before the conversion is also a geothermal alternative. Priority strategy to be done on the weakness aspect is an alternative aspect of geothermal strategy.

c. Analysis of opportunity aspect in renewable energy selection strategy

Sub criteria which is reviewed in this aspect is: An increase of national resilience factor (I), Electricity demand and the growth of electrical energy factor (J), Dependence on fuel causing national energy security problems factor (K), and reducing emission CO_2 factor (L).

Table 3. Local priority and Global opportunity aspect

Priority of alternative strategy	I	J	K	L	Weight
Geothermal	0.004	0.005	0.022	0.033	0.064
Water	0.011	0.014	0.036	0.051	0.112
Wind	0.017	0.037	0.061	0.114	0.228
Solar	0.038	0.057	0.141	0.360	0.596
Total	**0.071**	**0.112**	**0.259**	**0.558**	**1.000**

In table 3, the biggest opportunity based on an increased national resilience is the use of solar energy selection strategy. According to The Demand for electricity and the growth of electrical energy, solar energy alternatives are still a selected priority. Factors dependence on fuel prefer to alternative solar energy strategy alternatives, and to reduce CO2 emissions, the alternative of preferred energy strategy is solar energy. Overall, based on the opportunities that have been made on the utilization of renewable energy, the alternative strategy chosen is more to the selection of solar energy with a value of 0.596.

d. Analysis of threat aspects on renewable energy selection strategy

The sub criteria studied in this aspect are: the limited factors of human resources, especially in the region (M), the less desirable investment factor of the renewable energy sector (N), and the possible factors for the emergence of local regulations that are out of sync with central government policy (O).

Table 4. Local Priorities and Global Aspects of Threats

Priority of alternative strategy	M	N	O	Weight
Geothermal	0.047	0.139	0.347	0.533
Water	0.021	0.090	0.149	0.260
Wind	0.013	0.033	0.089	0.135
Solar	0.006	0.013	0.054	0.073
Total	**0.087**	**0.274**	**0.639**	**1.000**

Based on table 4. the threat aspect, the biggest threat to human resource constraints is on geothermal energy strategy. Less desirable investment in strategic priorities lies in the use of alternative geothermal strategies, as well as for the threat of emergence of local regulations that does not have synchronization with the central government. The greatest threat posed to the priority of alternative strategies lies in geothermal energy.

Overall, among the merging of strengths, weaknesses, opportunities, and threats that exist in the selection of renewable alternative energy can be seen in table 5.

Table 5. The selection of alternative priorities of renewable energy strategy of South Sumatra province

Priority of alternative	Geothermal	Water	Wind	Solar

strategy				
Strength	0.078	0.166	0.276	0.480
Weakness	0.547	0.259	0.126	0.068
Opportunity	0.064	0.112	0.228	0.596
Threats	0.533	0.260	0.135	0.073
Total	**1.221**	**0.796**	**0.766**	**1.216**

The priority matrix of energy alternative strategy above shows that geothermal energy is the main priority with value 1,221, the second alternative energy is solar energy with value 1,216, the third alternative is water energy with value 0,796, alternate to four that is wind energy with value 0,766.

5. Conclusion

According to the result of the analysis, the priority for alternative energy strategy based on the Strength aspect is solar energy with value of 0.480, while the priority for alternative energy strategy based on the weakness aspect is geothermal energy with value of 0.547. Priority on the alternative energy on the opportunity aspect is solar energy with value of 0.596 and on the threat aspect is geothermal energy with value of 0.533. Based on the final weighting, the result for the priority of alternative energy strategy that can be done in South Sumatera, as an added value of strength, weakness, opportunity and threat aspects, is geothermal energy with total value of 1.221.

6. References

[1]. Usman, Wan (2013). *Ketahanan Energi*, Jakarta
[2]. Kotarumalos, NurAisyah (2010). Menuju Ketahanan Energi Indonesia: Belajar dari Negara Lain, Jurnal Peneliti Pusat Penelitian Sumber Daya Regional Lembaga Ilmu Pengetahuan Indonesia, Vol. 3 - No. 1/2010-03.
[3]. Dewan Energi Nasional (2015).*Ketahanan Energi Indonesia*, Jakarta: Sekretariat Jenderal Dewan Energi Nasional.
[4]. Global Subsidies Intiative (2014). Tinjauan Subsidi Energi di Indonesia, *Edisi 1. Volume 1.Maret 2014.*
[5]. H. Padmadinata Utama (2006). *Master Plant Provinsi Sumatera Selatan Sebagai Lumbung Energi Nasional 2006-2025*, Jakarta: Pemerintah Provinsi Sumsel, Badan Pengkajian dan Penerapan Teknologi, Universitas Sriwijaya.
[6]. PT. PLN Persero (2014*). Rencana Usaha PenyediaanTenagaListrik (RUPTL) 2015-2024*, Jakarta.
[7]. Badan Intelijen Negara (2014). *Ketahanan Energi Indonesia 2015-2025 Tantangandan Harapan*, Jakarta: CV Rumah Buku.
[8]. Creswell, John. (2015). *Riset Pendidikan (Perencanaan, Pelaksanaan, dan Evaluasi Riset Kualitatif & Kuantitatif),* Yogyakarta.
[9]. Saaty, Thomas L, (1991), *Pengambilan Keputusan Bagi Para Pemimpin*. TerjemahanLiana Setiono, PT Pustaka Binaman Pressindo, Jakarta.
[10]. Lemhanas (2012). Pengembangan Energi BaruTerbarukan (EBT), *JurnalKajian Lemhanas,* Edisi 1*4.*
[11]. Bappeda Provinsi Sumatera Selatan (2014).*Rencana Pembangunan Jangka Menengah Daerah (RPJMD) Provinsi Sumatera Selatan Tahun 2013-2018,* Palembang: Pemerintah Provinsi Sumatera Selatan.
[12]. Usman, Wan dkk (2003). *Daya Tahan Bangsa*, Jakarta: Program Studi Pengkajian Ketahanan Nasional Universitas Indonesia.

The Model of Gas Supply Capacity Simulation In Regional Energy Security Framework: Policy Studies PT. X Cirebon Area

Nuryadin[1], Tb Ronny Rahman Nitibaskara[1], Herdis Herdiansyah[2], Ravita Sari[3]
[1] National Resilience Studies, School of Strategic and Global Studies, Universitas Indonesia, Salemba, Indonesia
[2] School of Environmental Science, Universitas Indonesia, Salemba, Indonesia
[3] Center for Strategic and Defense Studies, Universitas Indonesia, Salemba, Indonesia
E mail: herdis@ui.ac.id

Abstract: The needs of energy are increasing every year. The unavailability of energy will cause economic losses and weaken energy security. To overcome the availability of gas supply in the future, planning is crucially needed. Therefore, it is necessary to approach the system, so that the process of gas distribution is running properly. In this research, system dynamic method will be used to measure how much supply capacity planning is needed until 2050, with parameters of demand in industrial, household and commercial sectors. From the model obtained PT.X Cirebon area in 2031 was not able to meet the needs of gas customers in the Cirebon region, as well as with Business as usual scenario, the ratio of gas fulfillment only until 2027. The implementation of the national energy policy that is the use of NRE as government intervention in the model is produced up to 2035 PT.X Cirebon area is still able to supply the gas needs of its customers.
Keyword: Natural Gas, Energy Resilience, Planning, System Dynamic, Simulation.

1. Introduction

Oil and gas sector is an important strategic commodity for the country and affects the livelihood of many people. The need for the oil and gas sector continues to increase every year. For that the government needs to do strategic step through the utilization of natural gas either LNG (Liquified Natural Gas) or LPG (Liquified Petroleum Gas) as alternative energy of oil substitute. The existence of natural gas as an energy substitute for oil has an important role in reducing global warming by reducing carbon dioxide emissions. The government since 2007 began implementing the kerosene conversion program to LPG [1]. The goals of the program are: energy diversification by reducing dependence on petroleum especially kerosene, reducing misuse of subsidized kerosene, and providing practical and clean fuel for households and micro-enterprises.

Land conversion program to LPG is implemented based on Law no. 22/2001 on Oil and Gas, Presidential Regulation no. 5 of 2006 on National Energy Policy, Government Regulation No.104 of 2007 on the Provision, Distribution and Pricing of LPG Tube 3 Kg, and Regulation of Minister of Energy and Mineral Resources No. 26 of 2009 on the Supply and Distribution of LPG [1]. The kerosene to LPG conversion program needs to be anticipated considering that currently around 60% of Indonesia's LPG is supplied through imports [2]. With the import of LPG, the kerosene-to-gas conversion program will make a new problem: the shift of kerosene subsidy to gas subsidy. For 2017 the government plans to set a subsidy of Rp 20 trillion for LPG procurement subsidies [3].

The use of natural gas as fuel is more economical if we see it from the price when compared with other energy sources including LPG. Indonesia's gas reserves of around 151,3 TSCF with annual output reach 3,11 TSCF (Trillion Standard Cubic Feet), it is estimated that 49 years Indonesia still has reserve to production [4]. Nationally, the use of natural gas is mostly used by industry and PLN, for industries 19% and PLN 14%, while for households only 0.02% [5]. Cirebon Regency and City is one of the areas that have gas needs in three sectors including household, commercial and industrial sectors. Up to 2015 the number of household customers for Cirebon is 12,867 subscribers and 199 commercial sectors and industrial sector 69 customers. To meet these needs Pertamina allocated to PT. X Cirebon area of 10 MMSFCD (Million Standard Cubic Feet per Day) or 10,000 MMBTU (Million British Thermal Unit) if 1 MMBTU equivalent to 28.3205m^3 then the gas quota for PT.X Cirebon area is about 283.205 m^3 / day or 103,369,825 million m^3 of gas for each year [6].

The need for gas either industrial sector, household or commercial will increasingly influence PT.X Cirebon area as one of gas distributor in determining its supply capacity. To overcome the problem one of them is by making a model of gas supply capacity planning PT.X Cirebon area to meet the gas supply in the future. In this research, a gas supply model will be developed to determine the fulfillment ratio up to 2050, and will be scenario using Business as Usual (BaU) scenario of National Energy Council and government intervention in national energy policy that is NRE (New and Renewable Energy).

2. Literature review

The International Energy Agency (IEA) defines energy security as an uninterruptible energy source at an affordable price, and a state is said to have energy resistance if it has 90 days for energy supply [7]. Energy Resilience can be defined as the ability to supply energy to maintain economic growth and performance with reasonable price stability [8]. Energy is the lifeblood of the world economy, cut off the flow of energy and the economic will die. Energy is needed to produce food, manufacture goods, power machines and appliances, transport raw materials and finished product and provide heat and light [9].

On the other hand, the UN Economic Commission for Europe (UNECE) Energy Security Forum defines energy security as the availability of energy supply that can be used, at the point of final consumption, in sufficient quantities and timeliness, thus encouraging energy efficiency, economic development has no effect on the state social condition [10]. Energy resistance is disrupted if the energy supply is not balanced with the energy needs of the community. Threats to safety and energy security are influenced by several aspects, including political stability, manipulation of energy supply, competition and attack on production facilities, natural disasters and accidents. Soaring prices, limited access to energy sources, energy supply safeguards have a role in enlarging the threat of energy security.

Energy security is a concept whereby a country can defend itself and do development by prioritizing the security and availability of adequate energy reserves at affordable prices, whether oil or other types of energy variations. This is increasingly important with the fact that the economic and political dynamics also affect the energy supply that is crucial to the development activities of a country. Matters affecting the sustainability of energy reserves include the availability of energy reserves, price fluctuations, the threat of terrorism, the domestic instability of energy-exporting countries, the existence of war, geopolitical competition, to energy maps by the world's major energy-consuming countries [11].

The concept of energy resilience involves two dimensions, namely the dimensions of a country's independence to meet its energy needs derived from domestic energy resources, and the global interferential dimension in which the energy fulfillment of each country cannot be separated from the world's energy supply originating from particularly resource-rich exporting countries oil and gas. The availability of energy supply becomes a significant problem firstly, if the energy supply decreases, it will cause an increase in energy price which resulted in the decrease of energy purchasing power. This will impact on the collapse of economic activity and is destructive to the production and consumption activities of society. Secondly, with the discovery of new energy supplies, this can delay the scarcity of energy that may occur and secure energy reserves in a certain period. Supply plays a very important role, as the demand for energy as a primary commodity tends to remain constant and inelastic [12]. The Asia Pacific Energy Research Center states that a country can be said to have achieved energy security if it has fulfilled the 4 A indicator, namely: Accessibility (Access to energy sources), Availability (Increased reserves and energy resources), Affordability (Energy purchasing power to energy), Acceptability (Which is environmentally friendly and sustainable) [13].

In his book William N.Dunn mentions the policy analysis is "our problem is not doing what is right, our problem is knowing what is right". Knowledge of what is (fact), what is right (value), and what needs to be done (action), raises the need for five types of informa, ie policy, alternative policies, policy actions, policy outcomes, and achievement of wisdom They are all related. Robert C Wood in policy analysis raises five interrelated information that are issues, alternatives, actions, results and policy achievements. [14].

3. Methodology Research

3.1 System Dynamic

In this study using system dynamic method by looking at the causal relationship between the dominant factor factors in the system and considered interrelated with certain limitations or assumptions [15]. In making the model follow several stages: 1. Observation of the actual system, 2. Preparing the structure of the problem, 3. Model Making (Causal Loop Diagram and Stock Flow Diagram), 4. Validation model, 5. Simulation of BaU scenario model, 6. Simulation model of intervention scenario, 7. Interpretation and use of modal simulation results. Implementation of data modeling can be seen in the following stock flow diagram:

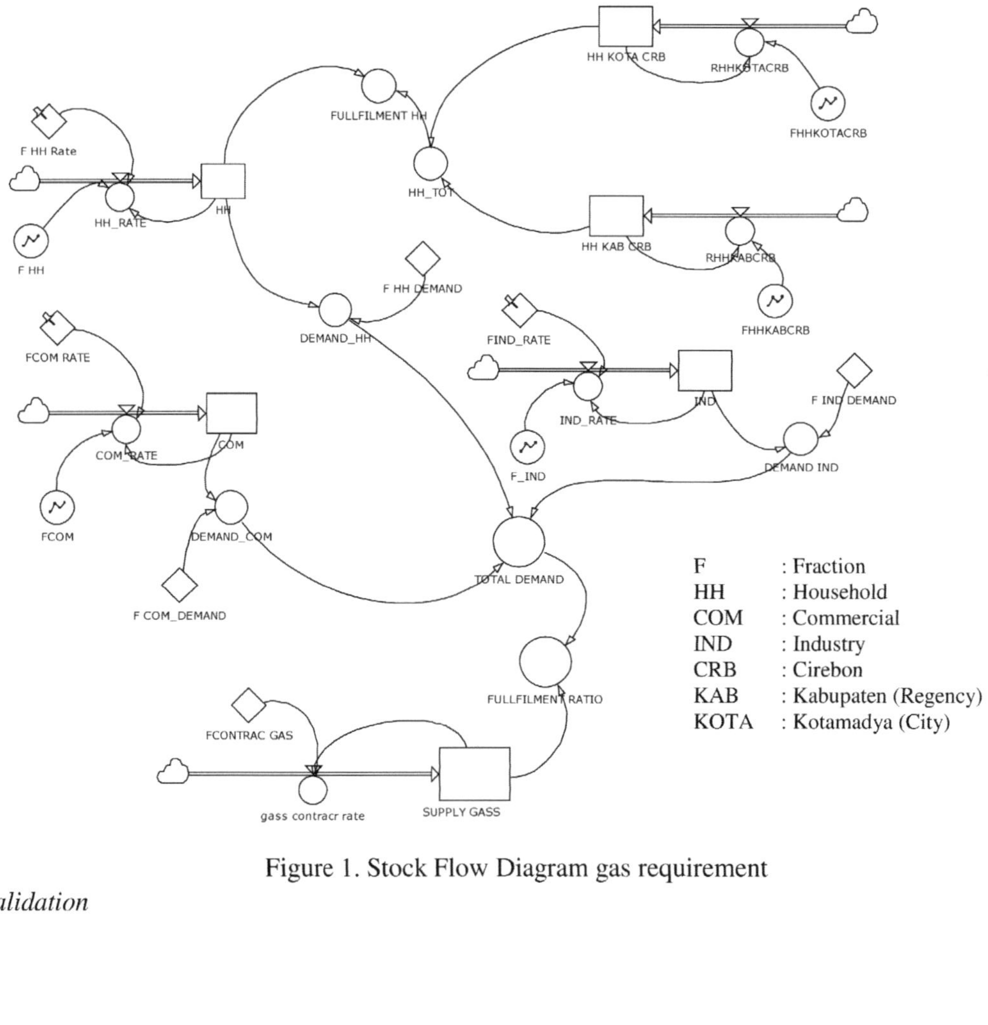

Figure 1. Stock Flow Diagram gas requirement

3.2 Validation

The purpose of the validation is the extent to which the model we created its performance equal to the performance of the real system. AME (absolute mean error) is one way of calculating the validation of the model we created by looking for deviations between the average simulation to the actual. The acceptable deviation limit is < 5 %. [16]

AME = ([S-A]) / A with S = Average value of simulation result and A = Average rating of data.

3.3 Scenario Development

In this research, two policy scenarios are made:

1. Scenario BaU (Business as Usual) National Energy Council is for the year 2015-2050 industrial sector grows 5.3% per year, the commercial sector grows 7.7% per year and the household sector grows 2.1% per year. [17].
2. National Energy Policy Scenario on national energy mix that is 2025 use of NRE 23%, 2030 use of NRE 25% and year 2050 31% use of NRE. [13].

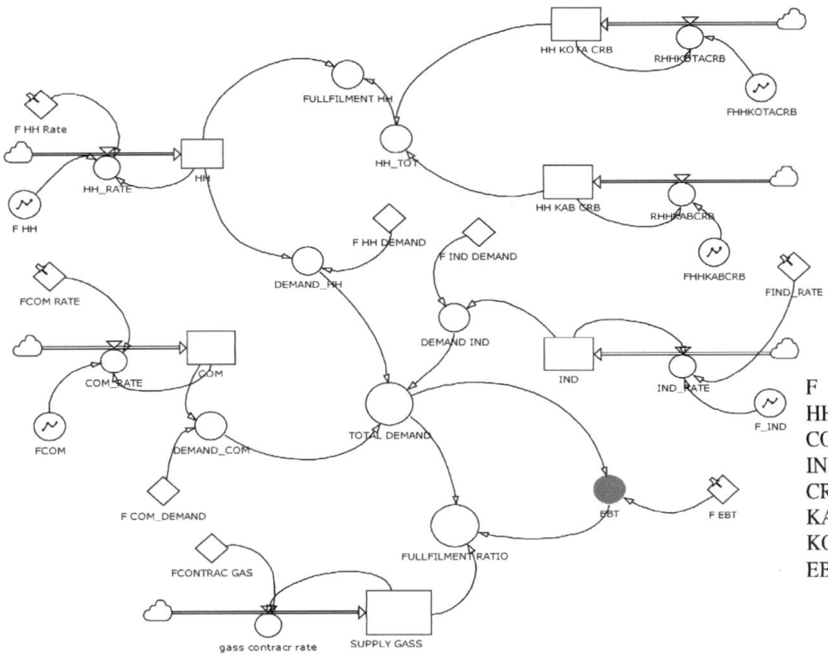

F : Fraction
HH : Household
COM : Commercial
IND : Industry
CRB : Cirebon
KAB : Kabupaten (Regency)
KOTA : Kotamadya (City)
EBT : New and Renewable Energy

Figure 2. Stock Flow Diagram with mix national energy in years 2025 used of NRE 23%, years 2030 used of NRE 25 % and years 2050 used of NRE 31%.

4. Results and Discussion

After done if the data and validation test result to the model made shows high validity number that is AME < 5%, so that model can be accepted and continued with scenario and intervention to model.

Table 1. The results of the AME validation test in percent

No	Customers	AME by customer	AME by gas used
1	House Hold	0,00097	1,15

2	Commercial	0,06906	4,13
3	Industry	0,18382	3,75

The need for the gas sector as a source of energy is getting higher every year, in the model simulation as shown in Figure 3 obtained in 2030, the PT.X Cirebon area is no longer able to supply the gas needs for its customers which is described with a fulfillment ratio (comparison between supply and demand) of less than 1 on Year 2031. From the results of the scenario BaU resulting shortage of fulfillment ratio faster, PT. X Cirebon area in 2029 was not able to meet the gas needs of its customers. While the scenario of the intervention of the National Energy Policy with the use of NRE, it is estimated that PT.X Cirebon area in 2037 is not able to meet the needs of its customers' gas as shown in Figure 4.

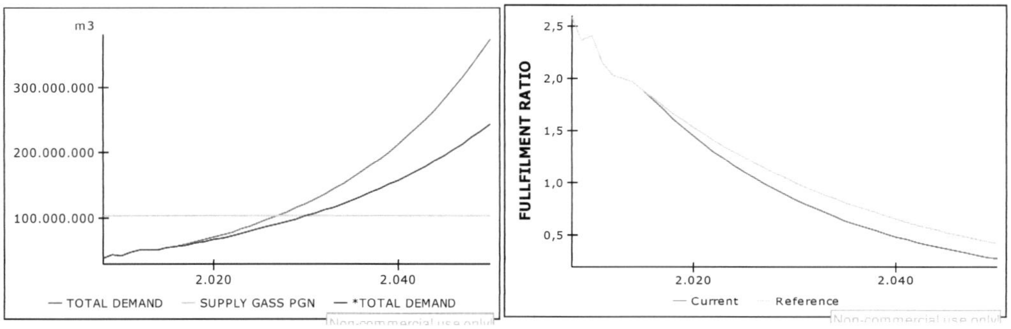

Figure. 3 Comparison ratio of total needs and ratio of gas fulfillment before and after BaU scenario

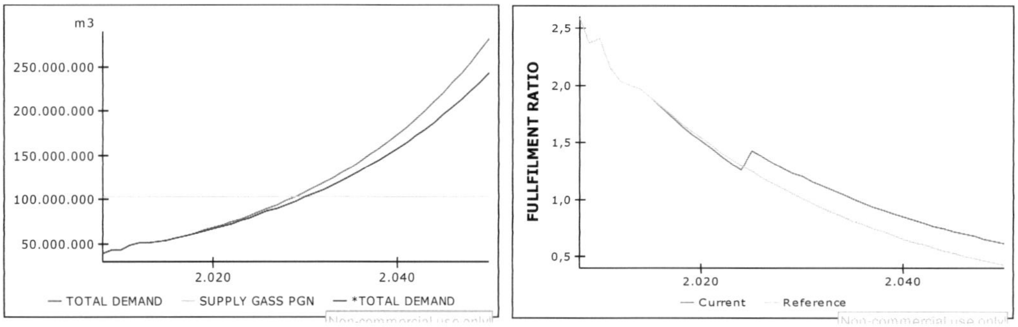

Figure. 4 Graph of the ratio of total needs and ratio of gas fulfillment before and after NRE scenario

The development of transportation infrastructure such as the construction of the Palikanci and Cipali toll roads is one of the causes of the increasing industrial development in the Cirebon region. Increased industrial sector will cause the need for human resources is increasing. The increase of industrial sector is expected to increase the economic level of the people around Cirebon area. Increased industrial sector, residence and supporting facilities such as hotels and restaurants will lead to an increase in energy demand, one of which is the need for gas. The need will continue to increase every year.

Energy scarcity is one of the inhibitors of the development wheel, although people's purchasing power of energy exists but if the energy source does not exist then the energy resistance will not be achieved. The concept of 4A in energy security says they are related to each other. Determination of policies in the face of gas scarcity needs to be done. Government intervention in national program with NRE mix target reaches 23% in 2025 and 2030 NRE blend becomes 25%, resulting in increasing energy shortage that is 2035.

According to Robert C wood in policy analysis raises five interrelated information that is problem, alternative, action, result and policy achievement. Gas is one of the fossil energy types that cannot be renewable, with continuous exploration it is not possible to make a problem in the future when the energy source runs out. Resilience Energy will be disturbed and even impossible Indonesia's integrity as a country will be affected so that the use of energy must be done effectively and efficiently, one of them is by increasing the use and production of alternative energy and one of them is new and renewable energy (NRE)

5. Conclusion

From the research conducted it can be concluded, that is as follows:

1. The gas fulfillment ratio of the model made up to 2031 is still feasible, while the BaU scenario of gas fulfillment ratio is only up to 2027. So that gas supply planning needs to be prepared for energy resilience of the region could be the use of other gas replacement energy sources
2. With intervention in the form of government policy intervention in the form of national energy mix as intervention in model that is utilization of NRE as alternative energy, it is estimated that gas fulfillment ratio until 2035. So, it is necessary to plan the source of NRE to increase the energy security of the region.

Reference

[1] Direktorat Jenderal Minyak dan Gas Bumi Kementerian Energi dan Sumber Daya Mineral. 2007. *Konversi Mitan Ke Gas,.(*Jakarta: Kementerian ESDM)
[2] Kementerian ESDM. 2015. *Rencana Strategis Kementerian Energi dan Sumber Daya Mineral 2015-2019. (*Jakarta: Kementerian ESDM)
[3] http://www.migas.esdm.go.id/post/read/subsidi-lpg-3-kg-tahun-2017-ditetapkan-rp-20-triliun.
[4] Kementerian ESDM. 2016. *Handbook of Energy & Economic Statistics of Indonesia 2016. (*Jakarta: Kementerian ESDM)
[5] Kementerian ESDM. 2016. *Statistik Minyak dan Gas Bumi 2015. (*Jakarta: Kementerian ESDM)
[6] PT.PGN(Persero).2016. *PT Perusahaan Gas Negara (Persero) Dan Entitas Anaknya/And Its Subsidiaries Laporan Keunagan Konsolidasian Interim.* (Jakarta : PT.PGN (Persero))
[7] International Energy Agency. 2014. *Energy Supply Security.* (Paris : IEA)
[8] Asia Pasific Energy Research Centre.2003 *.Energy Security Initiative : Some Aspect of Oil Security.* (Japan. APERC)
[9] Purbo, S Suwondo, 2007. *Pertimbangan-Pertimbangan Teoritik Dan Konseptual Yang Penting Untuk Perencanaan Strategik.* (Jakarta)
[10] UN. Economic Commission for Europe,2003. *Energy Security Forum.* (UN. Economic and Sosial Council : Committee on Sustainable Energy)
[11] Daniel Yergin,2006 "*Ensuring Energy Security*", in journal *foreign affairs* . (Volume 85 No. 2 March/April 2006)
[12] Florian Baumann,2008. *Energy Security As Multidimensional Concept*, (dalam jurnal CAP policy analysis, no. 1 March 2008)
[13] Dewan Energi Nasioal 2016. *Ketahanan Energi Indonesia 2015 (*.Jakarta: DEN)
[14] William N.Dunn, 1988,*Analisa Kebijakan Publik (Terjemahan)*, (Hanindita Ofset,Yogyakarta.)
[15] Forrester, J. W. 1961. *Industrial Dynamics*. (Waltham, Pegasus Communications).
[16] Muhammadi, Erman Aminullah, Budhi Soesilo, 2001. *Analisis Sistem* Dinamik *: Lingkungan Hidup, Sosial, Ekonomi, Manajemen.* (UMJ Press. Jakarta).
[17] Dewan Energi Nasinoal 2016. *Outlook Energi Indonesia 2015 (*.Jakarta: DEN)

Analysis of Environmental Friendly Library Based on the Satisfaction and Service Quality: study at Library "X"

Herdis Herdiansyah[1], Andre Satriya Utama[2], Safruddin[2], Heri Hidayat[2], Angga Gema Zuliana Irawan[2], R Immanuel Tjandra Muliawan[2], Diana Mutia Pratiwi[1]

[1]School of Environmental Science, Universitas Indonesia, Salemba, Indonesia
[2] National Resilience Studies, School of Strategic and Global Studies, Universitas Indonesia, Salemba, Indonesia.

Email: herdis@ui.ac.id

Abstract: One of the factor that influenced the development of science is the existence of the library, which in this case is the college libraries. Library, which is located in the college environment, aims to supply collections of literatures to support research activities as well as educational for students of the college. Conceptually, every library now starts to practice environmental principles. For example, "X" library as a central library claims to be an environmental friendly library for practicing environmental friendly management, but the X library has not inserted the satisfaction and service aspect to the users, including whether it is true that environmental friendly process is perceived by library users. Satisfaction can be seen from the comparison between expectations and reality of library users. This paper analyzes the level of library user satisfaction with library services in the campus area and the gap between expectations and reality felt by the library users. The result of the research shows that there is a disparity between the hope of library management, which is sustainable and environmentally friendly with the reality in the management of the library, so that it has not given satisfaction to the users yet. The gap value of satisfaction that has the biggest difference is in the library collection with the value of 1.57; while for the smallest gap value is in the same service to all students with a value of 0.67.
Keywords: user satisfaction, service quality, library, environmentally friendly, library.

1. Introduction

Library has an essential role in the development of science, technology and information. One of them is the University Library that plays an important role in the development of education, especially the fulfillment of the desire for reading materials, literature and other information in high educational relationships. University libraries are libraries contained in higher education institutions, subordinate agencies and institutions as well, which are affiliated with universities, with the primary goal of helping universities achieve their goals [1]. The general purpose of library services is to get library user satisfaction. Consumer's satisfaction is the level of someone's feelings after comparing the performance or outcomes that he feels compared to his expectations [2]. The purpose of a college library is: 1) As a provider of referral library materials at all academic levels, 2) Fulfilling the information needs of community colleges, such as faculty and students, including college administration personnel, 3) Providing an active information services not only limited to college environments but also local industry institutions. 4) Supplying reading and learning space in the library, 5) Allocating the appropriate lending services for various types of users, and libraries can support the existence of universities, because the library, as a place to manage sources of scientific information, is needed by universities [3] .

To be able to give the information services in order to achieve these objectives, the library should strive to provide various sources of information or library materials, which are necessary to be able to carry out programs of higher education activities in the field of education, teaching, research and community service. Library services can be strengthened by empowering librarians in facilitating customer reference services [4].

The library, which is discussed in this study, is a centralized of X library service. The X Library has several types of physical and digital collections. The physical collection is in the form of a book, while the digital collection is owned by X library. Digital services include desktop, online database, e-publishing, e-journal, e-book, web-portal, etc. [5].

In carrying out its functions, qualified human resources are extremely needed by the library. Central X Library has done its function in the process of library services and processing, but the need for the library remains unfulfilled, such as lack of facilities and infrastructure in the library. These deficiencies include inadequate room, lack of attention from the central leadership of the institute for the development of quality human resources, as well as limited funds to make library-based information technology.

Library is considered in a good shape if it can provide services that are able to derive satisfaction to library users, either in the field of information services or other services. Service quality can be defined as the levels of discrepancy between the user's expectations, desires and their perception[7]. Quality of service is a comparison between the expected service and received service. Thus, service quality is the outcome of a comparison between expectations and reality.

2. Theoretical Review

The Library is a collection of books that are organized in such a way as to be used for readers, consultation, and study purposes[8]. The college library is a technical implementation unit (UPT) of higher education, which is together with other units participating in implementing the " Tri Dharma of College Concept ", which is education, observation, and dedication to the society, by choosing, collecting, processing, taking care and serving information resources to its main institution in particular and academic community in general [9]. The library has functions as the education, information, research, recreation, publication, deposit, and interpretation function [10]. The purpose of the library is to build and maintain collections which is appropriate to read purpose and user information [11].

The library X is a library that already implemented an environmental friendly concept of the library including a green building that significantly designed to reduce the requirement of energy and water usage, highly efficient lightning and air ventilating system, also provide an efficient resource and service to its user. And in this research, is to valuate whether the concept was implemented as what they claimed to be by the perception of the library user.

Quality typically represents the direct characteristics of a product, such as performance, reliability, ease of use, esthetics, etc. [12]. Quality of service as a measure of how good the level of service provided can be in accordance with user expectations [13], until the quality of services can be realized through the fulfillment of customer needs and expectation along with an accurate delivery to compensate for user expectations [14].

Satisfaction is a kind of stepping away from an experience and evaluating it. One could have a pleasurable experience that can be said as dissatisfaction. Even though pleasurable, it was not as pleasurable as it was supposed or expected to be. So satisfaction and dissatisfaction is not an emotion, it is the evaluation of emotion [15].

The User's perception towards service can be categorized into five (5) [16], which are:

1. The discrepancy between customer expectations and service quality specifications established by the service provider's management
2. The discrepancy between quality and service quality specifications
3. The discrepancy between service quality and service quality specifications
4. The discrepancy between the service provided and the external communication to the customer
5. The discrepancy between the qualities of service expected and perceived.
6.

3. Methodology

This research using the quantitative approach to see how big the differences between the expectations and the reality of the library users to the quality of the central X library, which claims to be environmental friendly. The method used in this research is the Gap Analysis Method that using a simple mathematical approach formulated in four Cartesian quadrants.The population of this study is all of the postgraduate students of "Y" department whom regularly used X library.

The data was collected using questionnaires with Likert scale. The instrument validity test was performed by using Pearson Product Moment and Reliability Test that refers to Cronbach Alpha formula. Data analysis is performed using a Cartesian diagram to illustrate the degree of satisfaction, by comparison between the level of expectations and reality in the four quadrants.

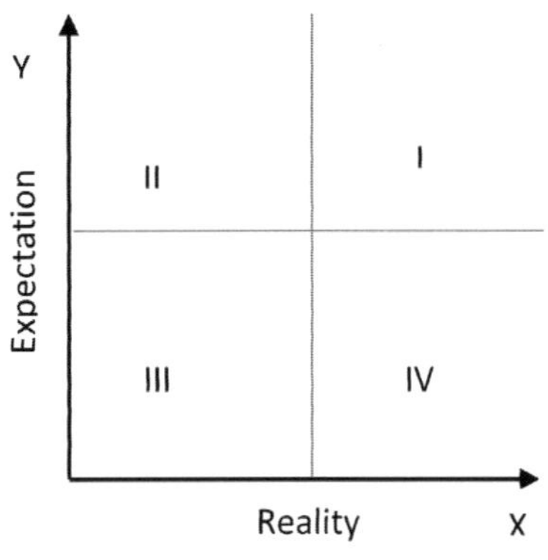

The quadrant I attribute is considered as important by the user with high expectations and so the fact is, thus the library must maintain it .

Quadrant II attribute is considered as important by user, but in reality it has not been done, so that it needs to be improved.

Quadrant III attribute is considered as unimportant and in fact the library does not implement them either.

Quadrant IV attribute is considered as unimportant, but the library executes it to the maximum.

Figure 1. Cartesian diagram

4. Results and discussions

Reliability test is a measure of a stability and consistency of respondents in answering the related matters with the question, the collision question, which is the dimension of a variable and compiled in a questionnaire form. The test of reliability can be carried out jointly against all the questions. The results show that the questionnaire answers, either the expectation or reality, is declared as reliable because the alpha value for the expectation question is 0.959 > 0.6. Likewise with the results obtained for the question of reality alpha count is 0.965 > 0.6. Validity test is used to determine the eligibility of the items in a list of questions in defining a variable. Validity test is done for all items of question either to expectation or reality. The result counts every item of all question is declared valid because it is bigger than table with sample 12 df = 10 that is 0,497.

The quality of X library services is known from the comparison between expectations and reality. The expectation average of library users 4.32; while the average reality that is in the field is 3.18.

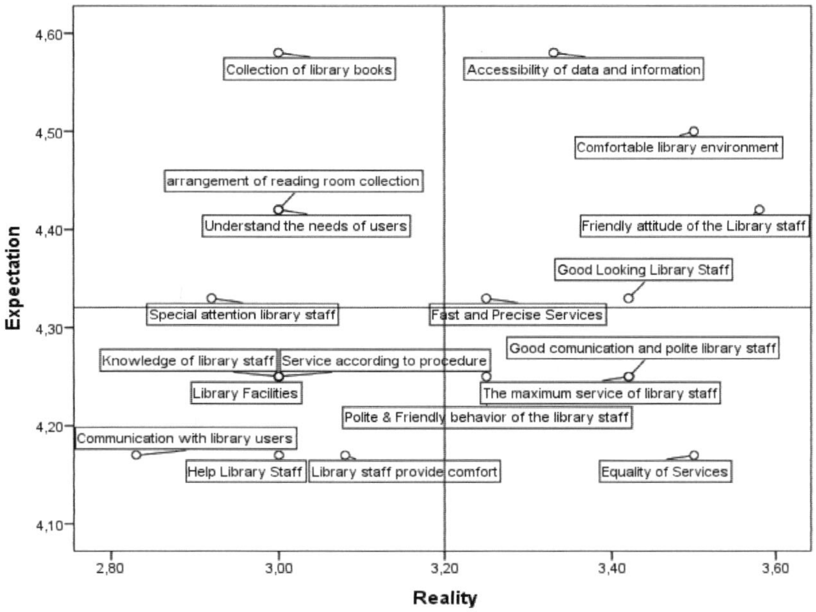

Figure 1. Expectations and realities on the quality of X center library services on Cartesian diagram.

In Figure 1, it can be seen that in quadrant 1, there is as easy access and information of library product, the comfortable environment of library, good performance of library staff, friendly attitude of library staff, and quick and proper service of library staff have high enough expectation value, moreover in reality also, it has been implemented by the library and must be maintained by the library.

Quadrant II states Collection of books, arrangement of reading room, library staff who understand the desires of the librarian and the special attention of library staff have a high expectation value above the average expectation, but in reality, it has not been implemented by the library, or the implementation is not maximal, this needs to be improved.

Quadrant III states that library facilities, service according to procedure, knowledge of library staff, communication with user, library staff's assistance and comfort given by library staff are not the main hopes of the users and in fact, readers have not been done maximally by the library yet.

Quadrant IV expresses that communication is good and polite by the library staff, polite and friendly attitude of the library staff, equality of service for library visitors is not a hope by the librarian, but in reality it has been implemented maximally by the library.

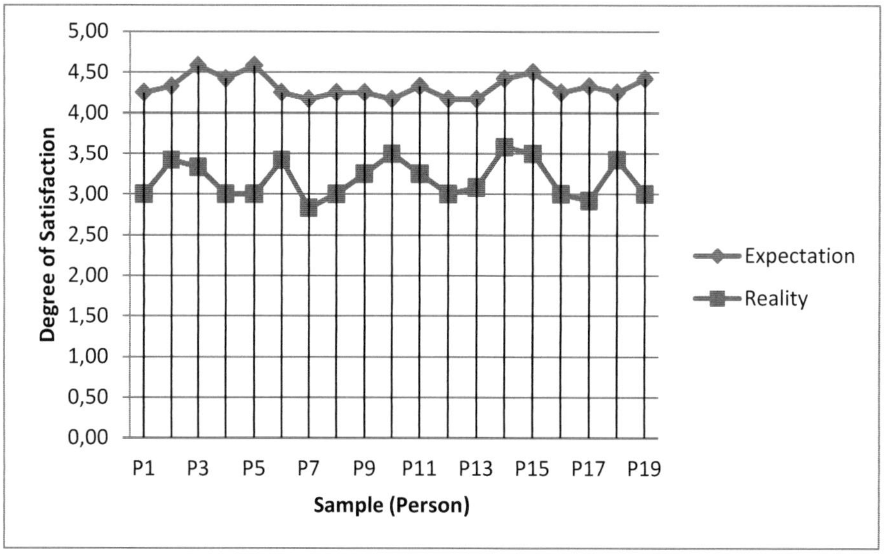

Figure 2. The discrepancy between expectations and the reality of the quality of the central X library service.

In Figure 2, it is clearly seen that there is a gap or a discrepancy among the expectations of the users of the quality of X library services center. This issue has occurred because the expectation of the user is not always directly proportional to the central X library service. In the aspect of the collection of Central X Library (on P5) it is seen that a hope towards the collection in the library Center X does not match the reality expected. The expectation, which is owned about the collection of central X libraries have averaged 4.58 and the average reality has a value of 3.00. There is a difference of 1.58.

In Figure 2, it is shown that P10 has the smallest discrepancy value. P10 is the value of equal service to all students. The average value in the expectations of the reader is 4.17 with a reality value of 3.50. The discrepancy is 0.67. The quality of X library's librarians center is good and in accordance with the readers' expectation.

5. Conclusion

The expectation of the library user for an environmental friendly library varies to high and average ones, whereas the high expectation including a quick access and information for library products, comfortable ambience, good management of the library staff, quick and precise service of the library staff and the average expectation including the collection of books, the arrangement of reading room, the library staff who understands the willingness of the users and the special attention of the library staff. In reality, most of the first point has been implemented in the Library X, but in the second point, it has not been implemented by the library or the implementation has not been maximal yet. Library facilities, service according to procedure, knowledge of library staff, communication with librarian, library staff assistance and comfort given by library staff are not the main hope of the readers and in fact, are also perceived by the librarian has not been fully implemented by the library. Good communication and polite library staff, polite and friendly attitude of the library staff, the equality of service for library visitors is not a hope by the librarian but in reality carried out maximally by the library. Then it can be inferred that the library X, that claimed had already implemented the environmental friendly system did not calculate the perspective aspect from its users.

6. References

[1] Sulistyo, B. 1994. *Pengantar ilmu perpustakaan*. Jakarta: Gramedia Pustaka Utama.
[2] Kotler, P. 1997. *Manajemen Pemasaran, Analisis Perencanaan dan Pengendalian*, Jilid 2 Edisi Kedelapan. Jakarta: Penerbit Erlangga.
[3] Sulistyo, B. 1994. *Pengantar Ilmu Perpustakaan*. Jakarta: Gramedia Pustaka Utama.
[4] Ahmata, M.A, Jaafarb, C.R.C., Azmic, N.A. 2016. The Transformation of Reference Services in Hamzah Sendut Library, Universiti Sains Malaysia. *Procedia - Social and Behavioral Sciences* 224, 6–13.
[5] Ahmad, M., Abawajy, J.H. 2014. Service Level Agreements for the Digital Library. *Procedia - Social and Behavioral Sciences*, 147, 237–243.
[7] Kotler, P. 1997. *Manajemen Pemasaran, Analisis Perencanaan dan Pengendalian*, Jilid 2 Edisi Kedelapan. Jakarta: Penerbit Erlangga.
[8] Noerhayati. 1989. *Pengelolaan Perpustakaan*. Bandung: Alumni.
[9] Qalyubi, S., et al. 2007. *Dasar-Dasar Ilmu Perpustakaan dan Informasi*. Yogyakarta: Jurusan Ilmu Perpustakaan dan Informasi Fakultas Adab.
[10] Depdikbud. 1994. *Buku pedoman perpustakaan perguruan tinggi*. Jakarta: dirjen Dikti
[11] Wortman, A.W. 1989. *Collection management*. Chicago : American Library Association
[12] Sinambela, L.P. 2008. *Reformasi Pelayanan Publik*. Jakarta: Bumi. Aksara
[13] Lewis and Booms. 1993. The marketing aspect of servie quality, emerging perspectives in services marketing. *Chicago: american marketing association*, 99-107.
[14] Tjiptono, Fandi dan Chandra, Gregorius. 2005, *Service, Quality, &Satisfaction*. Yogyakarta: Andi Offset
[15] Champion, J.C., Hunt, J.B., Hunt, T.G. 2010. The effect of retail store image on student perceptions of merchandise quality and willingness to buy. *American Journal of Bussiness Research*, 3,1.
[16] Zeithaml, V.A., Parasuraman, A., & Berry, L. 1990. *Delivering quality service: balancing customer perceptions and expectations*. New York: The Free Press.

CAPACITY ANALYSIS OF PARKING LOT AND VOLUME OF VEHICLE TOWARD SUSTAINABLE PARKING CONVENIENCE

Herdis Herdiansyah[1], Sugiyanto[2], Andrew Guntur Octavianto[2], Edison Guntur Aritonang[2], Malya Nova Imaduddin[2], Dedi[2], Magfira Rilaningrum[1]

[1]School of Environmental Science, Universitas Indonesia, Salemba, Indonesia
[2] National Resilience Studies, School of Strategic and Global Studies, Universitas Indonesia, Salemba, Indonesia.

Email: herdis@ui.ac.id

Abstract:

The development of human's population is having effect on the increase of facilities and transportation needs. One of the primary problems is the availability of parking area. This has occurred in Universitas Indonesia (UI), mainly in Salemba Campus. The availability of land is not as equal as the number of vehicles, which are to be parked, that is why the convenience of students, lecturers and employees at UI is unsatisfactory. The purpose of this paper is to know the level of parking convenience that is affected by the capacity of parking lots and the volume of vehicles in UI Salemba Campus. The results of this research indicate Salemba campus's parking index. The motor index is still in the category of medium (index 0.945) and the car parking index has less category with a parking index 0.485. While with the location of research object being behind the UI Salemba campus, the results obtained were both the motor and the car are still in the category of "enough" with the parking index of, that is 0.657 for the motor and 0.777 for the car. So theoretically, the parking management at Salemba Campus is in an unsustainable parking degree because, if there is no long-term solution, it will increase congestion in the surrounding area and intensify the dissatisfaction of existing parking users.

Keywords: Parking index, parking lot, vehicle volume, land capacity, sustainable parking.

1. Introduction

DKI Jakarta is the most densely populated province in Indonesia with a density of 15,015 inhabitants / km^2 in 2013 [1]. The growing number of working population with various activities in Jakarta has an impact on the increasing need of transportation facilities and infrastructure. That mobilization increases the complexity of managing transportation problems in Jakarta, especially during working days at the hours of entry and return. The availability of parking area becomes one of the indicators of facilities and infrastructure that has a significant influence on traffic density [2], one of which is DKI Jakarta.

Transportation facilities in the Jakarta area, especially Central Jakarta, experienced significant growth for private vehicles, such as cars and motorcycles. The minimum availability of parking spaces will influence the driver's time for searching a place to park their vehicles, and this issue can lead to congestion, increase in emissions, and traffic safety [3, 4]. The availability of this parking lot is not only stipulated in a policy within the government of DKI Jakarta, but also on all levels, including educational institutions such as the Universities in DKI Jakarta as one of the goals to mobilize the citizens. The condition of the campus's parking management in DKI Jakarta has become the main focus because it is one of the alternative solutions to the complexity of traffic congestion problems, in addition to the focus of public transportation and public road infrastructure. Some of the university have limited land that is not comparable with the number of existing vehicle, as an example we took the parking areas at Universitas Indonesia (UI) Salemba.

Universitas Indonesia (UI) Salemba is located at Central Jakarta, consist of six department, more than 3000 people including the lecturers, students, and the employees. Inadequate parking management and poor availability of parking space are major obstacles in the coverage of activity at UI Salemba. Despite of the university's area that approximately 17,252 ha, the university only provides a merely quarter or less for the parking area, meanwhile most of the students in this university bring their own vehicles, whether it is a car or motorcycle. And not only the students, but also the Lecturers and the employees. The logical consequence of increased private transportation ownership on UI Salemba campus is its relevance to the aspect of parking management and parking availability for Lecturers, Students, Employees and Guests. The purpose of this paper is to know the level of parking convenience that is affected by the capacity of parking lots and the volume of vehicles in UI Salemba Campus.

Management of parking undergone manually or with a system of cooperation with one company that has parking technology, did not give a positive contribution in solving the problem. In the period before the beginning of 2016, the management of parking on the campus UI Salemba is done manually by involving employees and Security of UI Salemba. Many complaints were submitted from parking users toward the manager of the parking at UI Salemba. Then, in the next period of early 2016 up to now, Campus UI Salemba parking management is managed in collaboration with professional parking management companies.

Parking management that includes the availability and determination of parking rates is sometimes not too popular to be accepted and known to the public, especially in many developing countries. In several cities in developing countries which have implemented parking management, they are limited to a certain area within the city. Whereas in the centers of developed countries such as Munich (Germany), and London (England), whose inhabitants have higher levels of welfare, it scales very high in terms of availability and cost on parking spaces to decide driving to a location even a decision for car ownership.

2. Theoretical Review

The Parking Space Unit (SRP) is an effective broad measurement for vehicles, including the free space and door-wide openings [5]. School or college car parks are grouped into 2 groups: worker / lecturer / teacher working in the college, and parking for students [6]. Determination of parking in general should consider [9]:

1. The width of road
2. Volume of traffic in road
3. The characteristic of velocity

4. Dimensional vehicle
5. The nature of the land allotment and the role of the road concerned

Off-street parking can be divided into two types: parking lot and parking garage. Determining the parking space requirements associated with the building function is using the L / SRP approach. Theoretically, the requirement of parking space amount can calculated using the ratio method R = L / SRP. The ratio method is based on the ratio of the floor area of the building (L) to the units of parking space (SRP), which is divided into road sections or parking blocks. The units of parking spaces are of several kinds: parallel spatial systems, cross systems, and parking space systems for two road spaces [7].

3. Methodology

The research method in this paper is done with mathematical calculations that test the relationship of land parking capacity with the understanding that is the total number of vehicles, both motor and car, parking at certain intervals. The study was conducted in October 2016 because it was assumed that there was a new, professionally managed parking management. Recording time is done in 2 shifts, i.e. morning and afternoon, but time calculation is done at break time. The observation time is 4 consecutive days and done when the lecture is happening normally. This condition was chosen because the mobility of vehicles to campus is relatively high. To determine the value of Parking space capacity in UI Salemba in one hour, can be obtained with dividing the total stall available in the

In this study the time interval is one hour. Parking space capacity is calculated to analyze how the vehicle expert fluctuates once every hour. The amount of parking space capacity here is obtained from the number of vehicles that have been using parking spaces on a certain parking lot in a certain time unit. So according to Directorate General of Land and Transportation, the equation to find the parking capacity consist of:

$$CP = (S / D)$$

Where:
- CP = Parking capacity Vehicles parked per specific time period, I.e. from Afternoon Break)
- S = Total stall
- D = Average parking (hour / vehicle)

While the volume of parking can be explained as the number of vehicles parked at the study site for a certain period of time, in this case calculations are grouped for 30 minutes. Furthermore, the survey data were analyzed to obtain parking volume in each study location for 1 hour. The formulation to calculate the volume of parking space as follows:

(Accumulation = $Q_{in} - Q_{out} + Q_s$)

Explanation:
- Q_{in} = Number of Vehicles entering the parking location
- Q_{out} = Number of Vehicles out of the parking lot
- Q_s = Number of Vehicles that have been in the parking lot before the observation is done.

4. Results and Discussions

The parking capacity at UI Salemba campus is calculated by dividing the total stall by the average length of parking. Parking capacity is calculated at 10:00 am to 14:00 pm, the peak busy hour in UI Salemba where the activity of the campus mainly started and finished.

Table 1 Parking Volume at Campus UI Salemba

Parking Location	**Total Stall**	**Parking Capacity**	**Parking Volume (An hour)**
Parking in front of Campus environment	Motor = 125 Stall Car = 55 Stall	52 (Motor) 17 (Car)	50 till 60 (Motor)/Per hour 30 till 40 (Car)/Per hour
Parking behind the Campus environment	Motor = 75 Stall Car = 78 Stall	23 (Motor) 35 (Car)	30 till 40 (Motor)/Per hour 30 till 40 (Car)/Per hour

In the previous calculation of parking land and parking volume, the data obtained from the calculation results will then be carried out into advanced analysis phase by determining the parking index of the calculation of parking capacity and parking volume. Parking Index is the comparison between parking capacity and parking volume. Parking index can be used as an assessment measure of parking space requirements, whether the parking space located within UI Salemba campus environment is still in accordance with the standards to accommodate parking requests. The parking index is divided into three categories namely Good (Index 1), Enough (Index <1 and> 0.5), Less (Index <0.5). Based on the previous explanation, the parking index of each location in the parking area in UI Salemba is shown in the table below:

Table 2 Parking Index at Campus University of Indonesia Salemba

Parking Location	**Parking Capacity**	**Volume (average)**	**Parking Index**	**Standard Index**
Parking in front of Campus UI Salemba Environment	52 (Motor) 17 (Car)	55 (Motor) 35 (Car)	0.945 0.485	Enough Less
Parking in front of Campus UI Salemba environment	23 (Motor) 35 (Car)	35 (Motor) 45 (Car)	0.657 0.777	Enough Enough

Based on the calculation of the parking index above, the location in front of UI Salemba campus environment, especially the motor parking area is still in the Enough category of parking index, in the number of 0.945 almost close to the number 1 (normal standard parking index), but different with car parking area with category of Less with parking index of only 0.485. While with the location of behind the campus UI Salemba shown that both the motor and the car are still in the category of Enough with the parking index which is, in sequence, 0.657 for the motor and 0.777 for the car.

The less ideal condition of parking especially for cars according to Rye's statement (the transportation and mobility management expert from Napier University Edinburgh) stated that "The car takes place on the road, but on average of 23 hours per day, a car occupies the parking space, And when the car is used, it takes 2 units of parking space at the origin and in the destination, so a lot of space needed for the car. Cars, which are parking, require at least 8 square meters and often more than that for maneuvering space. It is a huge need, especially in densely populated urban areas where land prices are very high. Often, cars get space for parking larger than space for humans to settle."[8].

So based on the calculation of parking load towards parking area, capacity of parking space must pay attention to the existence of a system that take to consideration the function of area, service function of road, traffic volume and direction of movement.

5. Conclusion

Based on the calculation of the parking index above, the location in front of UI Salemba campus environment, especially the motor parking area is still in the Enough category of parking index, in the number of 0.945 almost close to the number 1 (normal standard parking index), but different with car parking area with category of Less with parking index of only 0.485. While with the location of behind the campus UI Salemba shown that both the motor and the car are still in the category of Enough with the parking index which is, in sequence, 0.657 for the motor and 0.777 for the car.

With this outcome, theoretically, parking management at Salemba University is in an unsustainable parking degree because if there is no long-term solution, it will increase congestion in the surrounding area and the dissatisfaction of existing parking users.

According to the previous description and observation in the research location, then in this research, we try to give suggestions. The suggestions are:

1. It should be provided / placed an active parking attendant in each parking location to assist in the smoothing of vehicles parked so that the provision of parking lot becomes more effective.
2. It needs to be given clear parking restrictions such as giving garrison to parking lots that have not been equipped with markers, so there is no vehicle parking in the wrong place.
3. Creating vertical parking lot (Building Parking) is the best way to solve limited horizontal parking area. With this vertical parking lot, all vehicles and cars can be well managed and do not cause new problems.

6. References

[1] BPS. 2014. Population Percentage Distribution and Population Density by Province, 2000-2013. https://www.bps.go.id/linkTabelStatis/view/id/1277. Downloaded Dated 8 February 2017 8.56 o'clock.

[2] Niaraki, M.J., Malczewski, J. 2015. A group multicriteria spatial decision support system for parking site selection problem: A case study. *Land Use Policy*, 42, 492–508.

[3] McCoy, P., Ramanujam, M., Moussavi, M., Ballard, J., 1990. Safety comparison of types of parking on urban streets in Nebraska. Transport. Res. Board

[4] Shoup, D.C., 2006. Cruising for parking. *Transp. Policy* 13, 479–486. http://dx.doi.org/10.1016/j.tranpol.2006.05.005, Parking.

[5] Department of Transportation. 1998. Guidelines for Planning and Operation of Parking Facility, Directorate of Urban Transport Traffic System, Directorate General of Land Transportation, Jakarta.

[6] Department of Transportation. 1998. Guidelines for Planning and Operation of Parking Facility, Directorate of Urban Transport Traffic System, Directorate General of Land Transportation, Jakarta.

[7] Abubakar, I. et al. 1998. *City Transportation System.* Jakarta: Directorate of Urban Traffic and Transportation System.
[8] Parking Management Module: A Contribution to a Livable City. Federal Ministry for Economic Cooperation and Development (BMZ) Division 33. German.2010.
[9] Pande, Anurag, Brian Wolshon. 2016. Traffic Engineering Handbook Seventh Edition. New Jersey: Wiley.